CHEMICAL SPECIALTIES

A Symposium

Compiled by

H. Bennett

Editor-in-Chief, *The Chemical Formulary*
Director, B. R. Laboratory
Miami Beach, Florida 33140

1978

CHEMICAL PUBLISHING CO., INC.

NEW YORK, N.Y.

Chemical Specialties

ISBN: 978-0-8206-0210-3

Chemical Publishing Company:
www.chemical-publishing.com
www.chemicalpublishing.net

First Edition:

Chemical Publishing Company, Inc. - New York 1978

Second Impression:

Chemical Publishing Company, Inc. - 2011

Printed in the United States of America

Preface

This book is intended to be a guide to all those who would like to build up a chemical specialty business. Its purpose is to teach the elements of chemistry to the businessman and give the necessary business information to the chemist.

In the first place, information is given for those who are interested in chemistry but have not had the opportunity to study chemistry. There are many who have not had the advantages of a college training and technical education but who, nevertheless, started to manufacture chemical specialties and succeeded in developing a good livelihood and even a large business in this field. In most cases, they obtained the knowledge necessary for their business in a very hard way. They learned chemical compounding by reading, consulting chemists and making countless experiments.

In this book, all that many successful businessmen have learned from various sources over a great period of time has been put together and arranged for easy use. All the information that a beginner may need will be found in concentrated form here. No theory, chemical equations, or complicated mathematical formulae are used, so that everyone who can read will be able to learn how to make chemical preparations.

Not only the businessman will find this book a useful guide to build up a chemical specialty business, the chemist, also, will find it of great value. The chemist, of course, will know most of the simple chemical information given here, but the formulas listed may inspire him to prepare others of his own. He will find many valuable suggestions on the business part of specialty manufacturing, e.g., figuring costs, marketing, information on where to buy, credits, etc.

It is hoped that this book will help all those who have longed to enter this field but who have been held back by lack of sufficient information.

Preface II

All the formulas are new and up-to-date and suppliers of trademark chemicals are given. Regulations of Food & Drug Regulations covering chemical specialties have been added. Futher information is included on starting a new business.

Contents

CHAPTER PAGE

Introduction

To start a chemical specialty business, e.g. to manufacture disinfectants, stain removers, cosmetics, etc., little capital is required. No factory is needed. Manufacturing may be started in a kitchen or basement. You can start alone without other employees. The initial equipment need only be a simple balance, pots, pans, jars and wooden mixing paddles along with a source of water and heat. A batch of a particular product can be started and finished, in most cases, within a very short time. The product can then be packaged and be ready for sale. It can be seen that a chemical specialty business is theoretically, fairly simple to start. Anyone who has some business ability may enter into it. Many a business, which was started on a small scale, is of outstanding importance today. This does not mean, of course, that every small business will prosper. However, it proves that the opportunity is there for all those who are properly equipped, ambitious, and really willing to work hard and continuously, not only when they feel like doing so.

A chemical specialty business is usually based on one or more products, each of which is made in accordance with a certain formula. If the formula is good and the proper manufacturing method is used, then the finished product should be satisfactory. In choosing a formula, it must be considered that it may be good in one climate and country and may be unsatisfactory in another. A lipstick, for instance, may be excellent for the northern part of the United States or Canada but may soften when used in the southern part of the United States or Central and South America. The formula, therefore, would have to be changed to yield a product that does not soften at higher temperatures.

Once a formula is selected, it has to be tried out to see how it works. The first trial may not be quite satisfactory, yet the formula is good. The reason for the poor results may be that the raw material used is not sufficiently pure, e.g., using crude instead of a refined beeswax may lead to a discolored cream. A variation in temperature, such as overheating paraffin wax in a wax-paper-coating composition, will cause

I Compounding Chemistry

DEFINITIONS

Mixtures

A mixture is the uniform distribution of two or more substances in one another, such as a mixture of fine sand and powdered soap, or salt and sugar, or water and glycerin. The ingredients of a mixture are not chemically altered. They are the same at the end of mixing as in the beginning. Each ingredient can be removed in its original form. A good mixture, however, should have a uniform appearance. Though no chemical reaction takes place in making a mixture, very good specialties can be obtained by simple mixing.

Solution

A solution is a clear liquid obtained by dissolving a solid or semi-solid substance or a gas in a solvent. The solvent may be water, alcohol, oil, etc. A solution results, for example, when sugar is dissolved in water, or camphor in alcohol.

Suspension

A suspension is formed when a powdered substance is mixed with a liquid in such a way that the particles are suspended in the liquid without settling. A suspension can be made, for instance, by putting some bentonite (a type of clay) in water, letting it soak, and then shaking or stirring vigorously.

Emulsion

An emulsion is an intimate uniform mixture of two ordinarily immiscible liquids, e.g. oil and water. By using a special agent called an emulsifying agent, the liquids are dispersed in each other so finely that the mixture has a uniform milky appearance. So long as an emulsion is stable, the two liquids do not separate in layers. An emulsion, for example, results when water, mineral oil, and a strong soap solution are shaken vigorously. The strong soap solution acts as an emulsifying agent.

Chemical Compounds

In all previously discussed products, the ingredients did not react chemically with one another. In a real chemical compound, the ingredients are chemically combined with one another. The new compound may be entirely different in appearance and action from the original substances used. The original ingredients are changed in the chemical compound and can only be recovered by another chemical reaction. The following is an example of the formation of a chemical compound: Caustic soda and cottonseed oil are mixed together and heated. A chemical reaction takes place with the formation of soap and glycerin. Both soap and glycerin are products entirely different from cottonseed oil and caustic soda.

Every chemical is unique in its chemical and physical properties. Therefore no chemical can replace another in all its characteristics. In the following, general information is given on the properties of chemicals.

PHYSICAL PROPERTIES OF CHEMICALS

Form

Substances occur or are produced as gases, liquids or solids. These are always the same under the same conditions of temperature and pressure. Gases and liquids usually do not exhibit any variation in appearance, handling or use under similar conditions. Solids, however, do differ and may cause trouble. If they are crystals, the crystals may be large or small. The same substance may have different crystalline forms (needle-like, cubical, etc.), appearance, or bulking properties. The crystals may lose water and change into a dry powder. Powders consist of particles which may vary in size. Such variations not only affect appearance but also density, flow, agglomerating or caking tendencies, suspension, deposition, friction and other properties.

Optical Properties

Color
Clarity
Fluorescence
Phosphorescence
Iridescence

Refractive Index
Reflectance
  dull
  shiny

Color is of importance not only for appearance, but also where staining, dyeing or pigmentation will occur. The color of a substance may vary with the size of the particles. Crystalline copper sulfate is deep blue while the finely powdered material is a very light blue. Certain substances lose or change their color on being dissolved, dehydrated or on interaction with another ingredient.

Clarity refers to clearness and freedom from haze or turbidity. Sometimes commercial products which should be clear develop a haze, turbidity, deposit or a sediment, especially in metal containers. They may lose clarity even in glass containers, e.g., formaldehyde.

Odor

Pleasant
Unpleasant
Strong
Faint
Temporary
Permanent

Odor is the effect on the sense of smell produced by particles emanating from a substance. In many products such as foods, cosmetics and household articles, odor is an important factor. Where an undesirable odor cannot be eliminated, it may often be covered up by a stronger, more desirable odor.

No odor is equally pleasing to all. Certain types of pleasant odors are bland, refreshing or stimulating and are not objectionable in certain products. Unpleasant odors may be sickening, irritating or depressing. An odor may be strong or faint. Faint unpleasant odors are more tolerable than strong unpleasant odors and may be masked more easily.

Very volatile odors may only be temporary and may disappear quickly on aging, storage or use. Permanent odors must be considered as an ever present factor.

In blending various materials there may be a diminution of odor, caused by the dilution or change in character or strength of the substance. These changes may result from decomposition or interaction with another ingredient.

Taste

Sweet
Sour
Bitter
Salty
Spicy
Oily
Fruity
Neutral or tasteless
Pleasant
Unpleasant
Strong
Permanent

Taste is a factor of importance in those products that enter the mouth. Such products are foods, beverages, medicines, dentifrices and certain cosmetics for the lips.

Pleasant tastes may be sweet (as in sweet chocolate); sour (as in lemon drops); bitter (as in hops, used in beer making); spicy (as in ginger); salty (as in brine); oily (as in olive oil); neutral or tasteless (as in water); fruity (as in berries).

Just as with odors, strength and permanence are of importance, and must be given due regard. An undesirable taste may often be covered up by a stronger or more desirable taste. Certain tastes that are unpleasant when too strong are more pleasant when diluted, e.g. saccharin.

Density and Specific Gravity

Density is the weight per unit volume, e.g. pounds per cubic foot. Specific gravity is the relation between the weight of a given substance with the weight of an equal volume of water at the same temperature.

The density or specific gravity of a product will vary with its purity, porosity, size of its particles, and the process by which it was made.

Density or specific gravity are critical factors where bulking value, suspension, low cost, etc., are important. Thus calcium carbonate will vary in density or specific gravity, depending on whether it is in the form of natural limestone, marble, chalk, or a chemically precipitated product.

Viscosity

Viscosity is the resistance of a fluid to shear, agitation or flow,

More commonly, it is a measure of the rate of flow of a specific liquid as compared to that of water or any other commonly used liquid.

In some cases, viscosity is of importance because the greater the viscosity of a liquid, the lower the rate of flow, spreading, penetration, wetting, etc. and the better its suspending power. A lower viscosity, of course, reverses these properties. A viscous liquid is harder to mix, fill, pour and apply than a less viscous liquid.

Viscosity may be increased or lowered by suitable additions and treatments. Thus the viscosity of mineral oil can be increased by heating it with some aluminum stearate; the viscosity of an alkaline casein dispersion can be reduced by means of urea. Other specific methods for altering viscosity are known and these should be used when a substitute is suitable in all other respects.

Gelling or Thixotropic Tendencies

Gelling is the formation of a gel or jelly-like substance, e.g. glue or agar with water.

The thixotropic state refers to a gel that liquefies on shaking or stirring and regels on standing, e.g. iron hydroxide or certain clay suspensions in water.

Gelling may be desired in certain cases as in hectograph (duplicating) compositions, whereas in the case of a paint, gelling, which would prevent brushing or spraying, is undesirable.

Gelling may be due to the colloidal properties of a single substance in a liquid (as with gelatin and water) or may result from the interaction of one or more substances (as with sodium silicate and dilute hydrochloric acid).

Gelling may be temporary, as in the case of a cold gelatin and water jelly which becomes liquid on warming, or it may be more or less permanent, as in the case of rubber cement (rubber swollen in a hydrocarbon solvent).

Gels may be thinned or prevented from forming by the addition of suitable agents. Thus fish glue, in water, is prevented from jelling by the addition of acetic acid.

Freezing Point and Melting Point

The freezing point is the temperature at which a liquid solidifies or begins to form crystals under normal conditions. Liquids containing impurities or added substances have different freezing points than the pure liquids. Therefore, the freezing point of a liquid is a measure of its purity. If the freezing point of a substance is too high or too low, it may be altered by suitable additions.

The melting point is that temperature at which a solid changes to a liquid, under normal conditions.

Some mixtures, e.g. hydrogenated coconut oil, do not have a definite melting point but melt over a specific temperature range. Other substances soften or become plastic at certain temperatures, e.g. pitch, cellulose acetate, etc. Still others do not melt but sublime when heated sufficiently.

Vapor Pressure

Vapor pressure is the pressure of any vapor above its liquid or solid form. The greater the vapor pressure of a substance, the greater is its tendency to evaporate when exposed. High vapor pressure is desired in products which are expected to evaporate or dry quickly, as in cleaning fluids and lacquer thinners. Low vapor pressures are desired in product that should not change in bulk or dry out, as in flexibilizers for glue, casein, etc. or plasticizers for lacquers or plastics.

Sublimation

Sublimation is the direct vaporization of a solid that does not first liquefy, e.g. camphor or naphthalene.

Substances that sublime are useful when volatilization at certain temperatures is desired. Certain substances (camphor and naphthalene) sublime at ordinary temperatures. Of course, this means that they will gradually disappear when exposed. Where such volatilization is undesirable, subliming substances should not be used.

Boiling Point

The boiling point is the temperature at which the vapor pressure of a liquid equals the atmospheric pressure. Pure liquids have a definite boiling point. Commercial products, which contain impurities, boil over a range of temperatures, which is known as the boiling range. Thus pure water boils at 100°C at 760 mm pressure. Commercial methyl oleate boils at 200 to 215° C at 15 mm pressure.

Low-boiling liquids volatilize readily and disappear. This, of course, is advantageous where quick drying is necessary, as in the case of rubber cement or hair lacquers. High-boiling liquids are specified where volatility is to be kept at a minimum to prevent drying out, brittleness, shrinkage, etc. as in the use of glycerin in Cellophane or castor oil in ethyl cellulose.

The boiling point can be varied by dissolving soluble materials in the liquid or by mixing it with another liquid of a different boiling point. In the former instance, the boiling point is raised whereas in the latter it is either lowered or raised, depending on the boiling point and solubility of the added liquid.

Certain mixtures form a constant boiling mixture (azeotropic mixture) at a temperature different from that of any of their components. Here the distillate has the same composition as the substance being distilled.

Solubility

Solubility is the weight of a substance that can be dissolved in a definite weight of solvent at a given temperature. Thus 100 g. water will dissolve 35.8g salt at 10 °C.

Most substances are more soluble in hot water or other solvent than in cold. Therefore, it is important to determine the solubility of a substance in the solvent in which it will be used, in the temperature ranges to which it will be exposed. Otherwise, precipitation may result on later cooling or heating.

Many substances form a true solution when added to a solvent, e.g. salt in water or menthol in alcohol. Other substances disperse colloidally, as does soap in water or nitrocellulose in acetone. Colloidal substances usually swell slowly; therefore, a new colloidal material should not be hastily discarded on account of poor solubility. Let it stand overnight with the solvent, before stirring.

Certain substances may not dissolve in either of two solvents but will dissolve in a suitable mixture of the two, e.g. nitrocellulose in ether and alcohol.

Hardness

Hardness can be measured by the ability of a substance to abrade or scratch other materials or conversely by the ability to be abraded or scratched by other materials; or by the depth of penetration of a sharp edge or point under a definite weight or pressure.

Thus, if a coating is to be applied to a floor, it should be sufficiently hard to withstand a certain amount of wear. For this reason, in a wax polish, carnauba wax, which is harder than paraffin wax, is preferable.

An abrasive for rough polishing of quartz must be harder than quartz; corundum would be satisfactory.

A typical scale of abrasion hardness is that of Mohs given below, with 1 as the softest and 10 as the hardest.

1. Talc
2. Gypsum
3. Calcite
4. Fluorite
5. Apatite
6. Orthoclase
7. Quartz
8. Topaz
9. Corundum
10. Diamond

Penetration tests of hardness made with a penetrometer or durometer are expressed, in the case of the former, in units referring to a standard material under certain conditions and, in the case of the latter, in units on a calibrated dial. Rockwell and Brinell hardness values are used for referring to the hardness of metals, alloys, and similar materials.

Tenacity or Cohesion

Tenacity or cohesion is the property which prevents a substance from breaking into pieces when struck, pressed, or pulled strongly.

Certain materials possess tenacity or cohesion inherently, as rubber or fresh sisal fiber. Others require the addition of a plasticizer, e.g. as in cellulose acetate, Cellophane, etc. Tenacity varies in degree in different materials and also varies with the ingredients with which it is mixed and the process of manufacture. For example, unvulcanized or soft vulcanized rubber is tenacious and coherent, whereas highly vulcanized hard rubber is brittle.

Plasticity and Ductility

Plasticity is that property of a material that permits the altering of its shape or size by the use of pressure, tension, heat, or a combination of these forces, and does not permit a return to the original shape with the removal of these forces. Examples of plasticity are seen in butter and putty. When heat is one of the elements producing plasticity in a substance, the substance is said to be thermoplastic. When the shape or size becomes fixed and cannot again be altered by the original or any other forces, the substance is said to be thermosetting, e.g. Bakelite. When maintenance of shape and size is important, the use of plastic deformable materials should be avoided, whereas if rigidity is not desired and alterations in shape or size are immaterial or quite desirable, a plastic deformable body should be used.

Ductility is that property that permits the drawing out of a body in the direction of its length, into a continuous thread, wire, tube or rod, e.g. rayon thread, steel wire, lead tubing or brass rod. When continuous threads, wires, rods or tubes are to be made by the drawing out of a material, it is important to use only those materials which exhibit a high degree of ductility.

Flexibility

Flexibility refers to the ability to bend repeatedly within limits, under certain conditions, without cracking or breaking. Soft rubber qualifies as a flexible material but hard rubber is inflexible and brittle. Most materials are more flexible at higher than at lower temperatures. The addition of suitable plasticizers or flexibilizers increases flexibility in many commercial products, e.g. the addition of dibutyl phthalate or Glaurin in lacquers and certain plastics.

Elasticity

Elasticity is the property of recovering original shape and dimensions after stretching. Squeezing or twisting. Soft rubber is the classic example of an elastic material. Elasticity of many substances will vary with age, use and admixture with other ingredients.

Length

Length refers to the stringiness of a fluid when it is poured or when a rod is dipped into it and then pulled out. For example, mineral oil possesses length after it has been heated with a certain amount of aluminum stearate; also a strong "solution" of gum karaya in water.

Length is required in certain products to decrease their rate of flow and penetration so that they will remain in place for a longer time.

Stickiness

Stickiness is the ability of a substance to adhere to anything that it touches. It is advantageous in adhesives, glues, cements, etc., but is detrimental in many other uses. Certain products require adhesiveness on one side and non-adhesiveness on the other. Thus, a paint should stick to surface to which it is applied and dry, free from adhesiveness, on its exposed surface.

In some cases stickiness must be permanent, as in fly paper. In other cases it must be temporary, as in rubber cement for paper. In the latter case, it eventually dries without losing its adhesiveness and can be peeled off without damaging the paper.

Adhesiveness may be increased by suitable additions or processing, e.g. by the addition of an alkaline casein solution to rubber latex. Conversely, stickiness may be decreased, as in the case of the addition of Acrawax C to the polyvinylbutyral resin that is used to replace rubber for coating raincoats.

Adhesion

Adhesion is the attraction that causes two surfaces to stick to each other, so that they cannot be separated easily. This is usually brought about by means of an agent which solidifies by drying, cooling, chemical or physical change.

Adhesion varies with the composition of the surfaces to be united, their smoothness or roughness, and other factors. The choice of an adhesive material is dependent on whether it is to be applied hot or cold, wet or dry, in aqueous or other solution; whether it is to set slowly or quickly; whether it is to be temporary or permanent. The choice is also dependent on conditions of temperature, stress, and contact with water, solvents, or chemicals, and effect on the materials with which it is used.

Slipperiness

Slipperiness is the ability of a substance to slide or move with very little friction.

Certain materials, such as oils, greases, waxes, talc and graphite are inherently slippery. Of course, they vary in degree and with the nature of the surfaces on which they are used, and with temperature and load.

Slipperiness can be increased by the addition of the substances mentioned above. Conversely, slipperiness can be decreased by the addition of adhesives or abrasives, e.g. rubber or sand respectively. Slipperiness may be temporary, as in the case of linseed oil, which dries to a non-slippery solid, or permanent, as in the case of talc, which is not readily affected by age or oxidation. Suitable additives are available for increasing the useful life of a slippery material.

Drying Qualities

Most coating materials, such as paints, lacquers, varnishes, etc., must dry fairly rapidly so as not to mar, stick or hold dust, insects, etc. Rapid air-drying at normal temperatures is a pre-requisite for them. Other products, such as lubricants or fly-paper, must be non-drying.

Drying may be due to evaporation, absorption, oxidation, polymerization, or other factors. Time of drying may be modified by admixture with other substances, change of temperature or pressure, chemical treatment, etc.

Feel or "Hand"

Feel or "hand" refers to the sensation felt by the fingers or other part of the body in contact with a material. Thus, jute is rough and rayon is smooth. Wood, a poor conductor of heat, is warm to the touch, whereas aluminum, a good conductor of heat, is cool to the touch.

Rough materials can be made smooth by mechanical processes such as grinding, polishing, weaving, compressing or coating with wax, resins or starch. Smooth materials may be made rough by grinding, sandblasting, garnetting and other mechanical processes, or by treatment with certain chemicals. Materials that feel warm can be modified by combining with good conductors of heat, as by weaving a fabric from textile and metal threads. Conversely, a metal can be made to feel warmer by mixing the powdered metal with a plastic or fiber powder and compressing until a homogeneous body is formed.

Cryoscopic Properties

The cryoscopic property of a substance is its ability to lower the freezing point of a liquid. Water which ordinarily freezes at 32°F is protected against freezing to about −7.4°F by the addition of 50% glycerin. Most chemicals that dissolve in water or other solvent affect its freezing point.

In many cases a frozen product will revert to its original state of uniformity when it thaws out. In other cases, as in certain emulsions, particularly those made with vegetable gums, freezing breaks the emulsion and the latter dces not re-emulsify on thawing out.

Hygroscopicity and Efflorescence

A material is said to be hygroscopic when it can absorb moisture from its surrounding medium (usually air), efflorescent when it loses moisture when exposed to dry air.

While hygroscopicity is often useful in preventing drying out or embrittlement, it may be undesirable where caking, lumping of dry materials, or liquefaction results.

Efflorescent materials tend to give up moisture, dry out and lose their original bulk or crystal form.

Both of these properties can often be neutralized by coating dry materials with waxes, fats, oils or other filming materials which prevent the ingress or egress of moisture.

Typical of hygroscopic products is calcium chloride. A typical efflorescent product is sal soda, or sodium carbonate crystals.

Inflammability

The flash point of a substance is a measure of its inflammability.

Specifically, it is the lowest temperature at which a substance in an open vessel gives off enough combustible vapors to produce a momentary flash of fire when a small flame is passed near its surface.

The flash point must be considered not only from the standpoint of a fire-hazard, but also from the standpoint of local, state and national regulations for storing, shipping and using products that are inflammable. Thus a hi-flash naphtha is less hazardous than a lower-boiling naphtha or gasoline.

Burning point is the lowest temperature at which a substance will burn when a source of heat is applied under specific conditions. Different woods will burn at different temperatures, depending on their nature, dryness, etc.

Explosiveness

Explosiveness refers to the tendency of a material or its combustion or decomposition products to burst or expand violently. An explosion is usually initiated by impact, heat, pressure, an electric spark, contact with another substance or a catalyst. Certain explosive materials are quite safe if carefully handled, whereas others are very sensitive and extremely dangerous. An example of the former is dynamite and of the latter is mercury fulminate.

Toxicity

Toxicity refers to harmful physiological effects of a substance on living organisms. Its effect not only on human beings but also on animals, fish and plants with which it may come in contact must be considered. Nearly every substance is toxic, if taken in or applied beyond a certain limit. For example, salt is not harmful in small quantities, but larger amounts will produce vomiting and other harmful effects. Similarly lemon oil, which is used in baking, candy and other internal preparations, becomes very toxic and corrosive when used in larger doses.

Some substances exert a toxic effect through their vapors, particularly in confined spaces e.g. mercury; others when taken through the mouth, e.g. sodium fluoride; and others by absorption through the skin, e.g. para-toluidine.

By providing proper conditions and utilizing all safety measures, many toxic substances may and are being used with a minimum of hazard.

Effect on Skin, Hair, or Finger-Nails

Some materials produce undesirable effects on the skin, hair or finger-nails. These effects, especially in small amounts, may not be dangerous but they are nevertheless undesirable.

As examples, cresols irritate and cause the skin to crack; sulfuric acid corrodes the skin; picric acid stains the skin; acetone and other solvents cause brittleness of finger-nails; alkali, e.g. lye, causes roughening of the skin and hair; alum or formaldehyde causes toughening of the skin; vegetable oils cause softening of skin, nails and hair.

All these effects are dependent on concentration of the substance and time of contact. If a substance is present in only small amounts and in contact but momentarily, it is not necessarily objectionable. But substances of this nature should be carefully evaluated before use.

Edibility

A substance is considered edible if it possesses food value and is not harmful when eaten in certain amounts. The use of glycerin in cakes is permitted, because it has food value and is not harmful in the amounts used. Mineral oil, while not harmful, is not permitted in mayonnaise because it has no food value.

Physiological data should be obtained from the supplier or from authoritative tests before using a new material for edible purposes.

Thermal Changes

Temperature can produce many effects on different materials. Heat may change a liquid to a gas, e.g. ether; a solid to a vapor, e.g. naphthalene; a solid to a liquid, e.g. paraffin wax. Cold will, of course, reverse the above changes. Crystalline materials may melt and fuse to a solid mass and not revert to crystals when cooled, e.g. abietic acid crystals. Solid objects may flow out of shape on heating, with a resultant change in appearance, e.g. thermoplastics, like cellulose acetate.

Most materials shrink on cooling, but water expands on being frozen. Most materials expand on heating but ice shrinks on being melted. Each material shows a definite amount of expansion or shrinkage on heating or cooling. This information may be of importance in storage and use of certain products.

Heat will decompose or alter some substances, e.g. egg white. The effect of heat on color, odor, taste, viscosity, and all other properties mentioned in this chapter should be noted.

Cooling will often cause precipitation or thickening. Freezing will often break emulsions, which in many cases, will not revert to their original form on thawing.

Heating may also produce changes in stickiness, adhesion, drying, optical properties, odor, taste, density, viscosity, length, vapor pressure, solubility, hardness, flexibility, tenacity or cohesion, feel, slipperiness, elasticity, hygroscopicity, inflammability, explosiveness, drying qualities, plasticizing properties, chemical reactivity, stability, toxicity, effect on skin, hair or finger-nails, edibility, pressure, solvency, bacterial content, emulsifiability, surface tension, wetting properties, dispersing properties, adsorption and absorption, electrical properties, radio activity, etc. It would be prudent, therefore, to refer to each of the above named headings, and try to judge how temperature changes in any substance may affect its final use.

Effects of Pressure

Increased pressure usually increases the solubility of gases in liquids, e.g. carbon dioxide in beverages. It may also increase the speed of chemical reaction. It usually increases density, viscosity, hardness, and rigidity. Conversely, it decreases vapor-pressure, flexibility, elasticity, absorption, and size.

Solvency

Solvency refers to the ability of one material to dissolve another at a given temperature. For example, the solvency of water for borax is 14.2 g per 100 cc at 55°C.

An ingredient may be used primarily to thicken or to prevent drying, but its solvency must also be considered. If its solvent properties are ignored, precipitation or crystallization may result. For example, corn syrup may replace glycerin as a thickener but its solvent powers for vanillin, borax and many other ingredients are much lower and may cause possible difficulties.

Plasticizing or Flexibilizing Properties

A plasticizer of flexibilizer is used because of its ability to prevent brittleness or cracking of a solid material, e.g. glycerin with glue or dibutyl phthalate with pyroxylin.

A good plasticizer should be compatible with the material in which it is used and be sufficiently non-volatile to remain in it for its useful life. It should not change appreciably the properties desired in the finished product. The amount of plasticizer used is often critical, too much will give a soft or sticky product, while too little will not be sufficient to plasticize properly.

Interaction with Other Materials

Chemical reaction between two ingredients may produce changes which rule out the finished product. It is best to avoid the use of a substance which may react with any of the substances already present.

Acids may produce esters or salts or induce hydrolysis. Adding acetic acid to alcohol will produce some ethyl acetate on standing. The addition of phosphoric acid to a caustic soda solution will form sodium phosphate. Ethyl lactate will be hydrolyzed by the addition of an aqueous acid, such as hydrochloric acid.

Alkalies may form soaps, induce hydrolysis or form hydroxide gels or precipitates. The addition of caustic potash to a solution of castor oil in alcohol will produce a castor oil soap. An ester such as methyl oleate will be hydrolyzed by an alkali such as sodium carbonate. The addition of ammonium hydroxide to a solution of ferric chloride will produce a colloidal precipitate or gel of iron hydroxide.

Salts may produce electrolytic effects, such as increased conductivity of an electric current or generation of an e.m.f. which may induce corrosion. Distilled water is a poor conductor of electricity but the introduction of an electrolytic salt, such as sodium chloride, increases its conductivity greatly.

Oxidation or reduction of other ingredients or of the newly added material by the other ingredients may produce many changes in properties.

Hydrogen peroxide will oxidize many organic materials. Reduction with hydrogen (produced from metal and acid) will also often bring about undesirable changes.

Double decomposition of two substances with resultant precipitation or formation of new gaseous or liquid substances may occur in certain cases. If barium chloride is added to a solution containing a sulfate, such as sodium sulfate, a heavy, white precipitate of barium sulfate forms. If sodium acid sulfate is added to sodium bicarbonate, a gas, (carbon dioxide) is formed.

Stability

Stability refers to the ability of a substance to remain substantially unaltered over a certain period of time. A composition is subjected to so many detrimental conditions that it is really surprising that commercial products which remain in a dealer's stock for extended periods are still suitable for use when purchased. This is not due to luck but to judicious testing and aging.

Water, when present, is useful in many products but detrimental to others. Thus, a dry starch preparation will keep indefinitely. The introduction of even a small amount of water may cause bacterial decomposition which will ruin the product.

Air, which is so necessary to living things, is detrimental to many substances. Thus, cottonseed oil tends to absorb and react with the oxygen in the air and becomes rancid.

The effects of heat and cold have been indicated under, Thermal Changes, 9.00.

Sunlight may improve some products, e.g. irradiated foods, and injure others. An example of injury by sunlight is its effect on rubber, which becomes brittle and cracks.

Motion often produces detrimental effects. A powdered mixture consisting of materials of different densities, e.g. bentonite, soda ash and soap, may separate into different layers because of the motion produced during transportation.

Age may also produce various changes. Certain crystalline rearrangements or other changes may take place.

The electric discharge or corona effect, producing ozone, is detrimental to many organic products. Rubber-insulated electrical ignition wires deteriorate rapidly when exposed to electrical discharges.

Abrasion is a factor that must be considered when selecting a material that is subject to friction or pounding. While marble makes beautiful steps, it wears down much more rapidly than a harder material like granite.

The material in which a composition is to be packaged is of utmost importance. It would be foolish to package a caustic paste in aluminum tubes as the latter would be readily attacked and soon destroyed. Even ordinary water solutions corrode, through pin-holes, beneath the tinplate in cans. To avoid these corrosion effects, either other packaging materials are selected or, when possible, suitable inhibitors are added to stop or delay such action.

Bacterial Content

While a material may not be susceptible to bacterial decomposition, it may act as a carrier for bacteria which may affect other ingredients.

With compositions containing organic matter, on which bacteria may thrive, it is necessary to avoid the introduction of bacteria or to destroy them by sterilization or antiseptic action.

Emulsifiability

Emulsifiability is the ability to form an emulsion. Cottonseed oil is emulsifiable with water and gum tragacanth. The latter is the emulsifier. Different substances have different degrees of emulsifiability and therefore different types and amounts of emulsifiers as well as different processing methods.

Vegetable oils, for example, can usually be emulsified with ammonium linoleate. Paraffin wax, in the presence of salts, is usually emulsified with gelatin or gum arabic. Toluol with acetic or other acids requires a special emulsifier that will act in the presence of acid, such as Emulgor A.

For a certain purpose, a water-in-oil or oil-in-water type of emulsion may be desired. The introduction of a new ingredient may produce a type of emulsion opposite to the one desired. Sometimes a change in type can be made by changing the emulsifier, the proportions of one or more ingredients, or by reversing their order of introduction and mixing.

Variations in quantity and identity of ingredients and methods of emulsification will also produce variations in particle size, color, viscosity and other properties.

Surface Tension

Surface tension refers to the tendency of the surface of a liquid to contract to the smallest area possible. The higher the surface tension, the greater the contraction. A low surface tension produces the opposite effect.

Low surface tension permits rapid and thorough wetting of an insoluble material. Thus, the addition of a small percentage of a wetting agent like Sulfatate (a sodium salt of a sulfonated hydrocarbon) to a water solution of a flame-proofing agent like sodium borophosphate permits a fabric to be impregnated quickly and thoroughly.

The addition of a wetting agent to various solutions makes them spread more quickly and evenly on smooth surfaces such as glass, steel, etc.

Certain oils are made to penetrate more quickly and deeply into crevices when a suitable surface-tension-reducing agent is added to them.

Dispersing Properties

The dispersing property of a substance is its ability to suspend insoluble particles in a fluid. For example, colloidal carbon is suspended in water with gum arabic to form India ink. Pigments are suspended in drying oils by suitable grinding with "Glaurin" or glycol naphthenate.

Proper dispersion of insoluble particles in a liquid is important to avoid deflocculation or settling, and to form uniform films and coatings as in paints and pigmented lacquers.

Adsorption and Absorption

Adsorption is the taking up (concentration) of a substance on the surface of another substance. This is illustrated by the taking up of odors in a refrigerator by charcoal. Absorption is the taking up of a gas or vapor by a fluid in which it dissolves, e.g. the dissolving of acetylene in acetone.

Adsorptive materials may take up certain materials and thus throw a formulation out of balance. They may also remove or lighten the color or odor of a product.

Lighting Power (Candle Power)

Lighting power is the ability of a substance to produce illumination. Materials such as illuminating gas, kerosene, etc. arc used for producing illumination.

Metals such as tungsten or zirconium alloys produce light when heated sufficiently, e.g. by an electric current. Carbon and other conductive refractories produce light when an electric arc is formed between two electrodes in an electrical circuit.

Gases subjected to an electric discharge are also used to produce light, e.g. as in neon tubes.

II Classes of Chemicals, Their Chemical Properties, Reactions and Uses

When two or more substances are mixed, the following may occur:

1. The substances do not react with one another; each substance will exist in its unchanged original form.

2. The substances react with one another, forming new products or compounds. In this case, three different results may be obtained:

(a) If both the combining substances and the reaction products are in the liquid state, the chemical change is visible only if there is a change in color, clarity, viscosity, etc.

(b) Substances combine, forming new products and releasing a gas. The bubbles formed in the liquid by the escaping gas make the reaction noticeable.

(c) Substances in solutions can combine, forming a precipitate, a solid substance which is insoluble in the reaction mixture.

It is very important to know what reaction may be expected when mixing chemical substances or their solutions. In some cases, substances are brought together to obtain a new product. In other cases, identification of a substance is the purpose. There are cases, however, when a change in the compounded substances is undesirable and all additions which would alter the original product should be avoided. Any substance whose addition would result in an undesirable change of the original product is said to be incompatible.

Almost all chemical substances have typical group reactions and also specific reactions.

In the following are given reactions characteristic of certain chemical groups; a few specific reactions are also listed.

Most of the specific reactions given here may be considered as incompatibilities.

ACIDS AND ACID-REACTING MATERIALS

Acids are chemical compounds which in general have a sour taste. They can be recognized chemically by using an indicator, e.g. litmus paper or solution. The color of litmus turns from blue to red when it comes in contact with an acid. Acids corrode metals. They combine with bases or metals to form salts. For example, hydrochloric acid combines with caustic soda to form sodium chloride (common salt).

Acids are weak or strong. Their strength depends on the chemical nature of the acid and on its concentration. Strong acids are more active and generally more corrosive than weak acids.

There are inorganic and organic acids. The organic acids contain carbon and many of them are found in living cells (plants, animals). Nearly all organic acids may be considered as weak acids.

An acid is concentrated when its water solution contains the highest possible amount of the acid. Special care should be exercised in working with concentrated inorganic acids, such as hydrochloric acid, nitric acid, phosphoric acid, and sulfuric acid. All these concentrated, strong inorganic acids burn holes in any material with which they come in contact. Concentrated hydrochloric and nitric acid release vapors easily. The vapors are very corrosive and the bottles should always be well closed to prevent the vapors from escaping. When nitric acid comes in contact with the skin, the skin becomes brown. Sulfuric acid develops great heat when mixed with water. The mixture may boil and spatter around. This is very dangerous. Sulfuric acid should always be diluted with water in such a way that the acid is poured into the water, *never* the water into the acid.

Examples of commercially used acids are:

Name	*Properties*	*Use*
Acetic Acid	Clear, colorless liquid; pungent odor; soluble in water, alcohol, ether.	Solvent; in making vinegar; in medicine.
Acetylsalicylic Acid	White crystalline powder; soluble in alcohol, ether, chloroform; slightly soluble in water.	In medicine.
Boric Acid	White crystals; soluble in water and alcohol.	In medicine.
Citric Acid	Colorless, odorless crystals; soluble in water, alcohol, ether.	In medicine; for flavoring extracts, confectionery, soft drinks, effervescent salts.
Hydrochloric Acid (Muriatic Acid)	Colorless or yellowish fuming liquid, pungent odor, poisonous; soluble	In organic synthesis, dye-manufacture, bleaching, metallurgy, sugar manufacture,

Name	*Properties*	*Use*
	in water, alcohol, and ether.	etc.
Lactic Acid	Yellow or colorless thick liquid; soluble in water, alcohol, ether.	In the leather industry; textile industry; in medicine; organic synthesis; manufacture of cheese, sauerkraut, syrups; as solvent.
Nitric Acid	Colorless or yellowish, fuming caustic liquid; corrosive and suffocating; soluble in water and alcohol.	In the manufacture of dyes, explosives, drugs; in photography, etching; in medicine.
Oxalic Acid	Colorless clear crystals, poisonous; soluble in water, alcohol, ether.	In ink and rust removers; in polishes; as cleansing agent; in engraving and lithography.
Phosphoric Acid	Transparent, colorless liquid or crystalline solid; soluble in water.	In synthesis of dyes; in manufacture of fertilizers, jellies, essences, soft drinks; in pharmaceuticals; in engraving and lithorgaphy; for water softening; in rubber industry; in textile industry.
Picric Acid	Yellow crystals; very poisonous; explosive; soluble in water, alcohol, chloroform, benzene, ether.	In explosives; in the manufacture of dyes; in medicine.
Salicylic Acid	White needles; soluble in acetone, oil of turpentine, alcohol, ether.	In dyes; in perfumes; as antiseptics; in medicine; in rubber industry.
Sufuric Acid	Strongly corrosive thick oily liquid, colorless to dark brown; soluble in water in all proportions with release of heat.	In the manufacture of organic and inorganic compounds; in metallurgy for electric batteries; in rubber and textile industry; in the manufacture of glue and gelatin; in sugar industry; in engraving and lithography.
Tannic Acid	Yellowish amorphous powder; soluble in water, alcohol, acetone.	In tanning; in textile industry; in electroplating; in medicine; in wine manufacture; in photography.
Tartaric Acid	Clear, colorless crystals; soluble in water, alcohol, ether.	In manufacture of soft drinks; in baking powder; in medicine; in textile industry; in silvering glass mirrors.

Group Reactions of Acids

Acids react with

1. Alcohols	Forming ester and water.
2. Alkaloids	Forming salt.
3. Amines	Forming salt.
4. Bases (Alkalis, Metal Hydroxides, Metal Oxides)	Forming salt and water.
5. Carbohydrates (Of High Molecular Weight)	Splitting them into smaller molecules.
6. Carbonates	Forming new salt and carbon dioxide (a gas).
7. Metals	Forming salt and hydrogen (a gas).
8. Proteins	Causing coagulation.
9. Salts, Including Carbonates Soaps Sulfides	Setting free the acid of the salt if this is weaker than the added acid.
10. Soaps	Forming new salt and setting free the fatty acid.
11. Sulfides	Forming new salt and hydrogen sulfide (a gas).

Specific Reactions of Acids

Acid	*Reacts or is incompatible with*	*Results*
Acetic Acid	Ferric salt.	Red colored product.
Acetyl Salicylic Acid	Water, acids, alkalis, carbonates;	Decomposition.
	hexamethylenetetramine, antipyrine, phenol; ferric salt.	Violet-red product.
Boric Acid	Salicylic acid together with alkaloids.	Precipitate.
Citric Acid	Calcium, strontium, lead salts.	Precipitate.
Hydrochloric Acid (Muriatic Acid)	Lead, silver, mercury salts; antimony and potassium tartrate solution.	Precipitate.
Lactic Acid	Oxidizing agents.	Aldehyde.
Oxalic Acid	Calcium salt.	Precipitate.
Phosphoric Acid	Silver, lead, mercury salts; many iron, alumi-	Precipitate.

Acid	*with*	*Results*
	num and calcium salts.	
Salicylic Acid	Lead, mercury, silver, muth, zinc salts; together with boric acid and alkaloids.	Precipitate.
	Ferric salt;	Violet color
	alkalies in air.	Reddish-brown discoloration.
Sulfuric Acid	Organic matter;	Decomposition of.
	strontium, calcium lead, barium salts.	Precipitate.
Tannic Acid	Salt of heavy metals; alkaloids, glucosides;	Precipitate.
	oxidizing agents.	Decomposition.
Tartaric Acid	Potassium, calcium, lead salts.	Precipitate.

ALCOHOLS

Alcohols are organic componds. Some of them are water-soluble. Examples of commercially used alcohols are:

Name	*Properties*	*Use*
Butyl Alcohol	Colorless liquid; soluble in water; miscible with most of the common organic solvents; inflammable.	Solvent for camphor, resins, shellac, varnish, gums, waxes; used for manufacture of rayon; flavoring extracts; paint and varnish removers; perfumes; in photography; in medicines.
Diethylene Glycol	Colorless, almost odorless, clear syrupy liquid; very hygroscopic noncorrosive; miscible with water, acetone, alcohol; immiscible with carbon tetrachloride and benzene.	Solvent for resins, oils, gums, nitrocellulose, dyes; in printing of textiles; in manufacture of explosives; to remove moisture; as lubricant and softener: in cosmetics.
Ethyl Alcohol (Usually Called Alcohol)	Volatile, colorless liquid; soluble in water and ether.	Solvent for essential oils, and many organic and inorganic compounds; in medicine; in the manufacture of liqueurs and perfumes.
Glycerol	Colorless, or pale yellow, clear, syrupv, odorless liquid; soluble in water and alcohol; insoluble in ether.	In medicine; in cosmetics; in flavoring tobacco; in liqueurs; in ink; in binder for cement; in photography; in liquid soaps; in

Name	*Properties*	*Use*
		leather industry; as lubricating agent; antifreeze.
Isopropyl Alcohol	Clear, mobile, colorless liquid; inflammable; soluble in water, alcohol and ether.	Solvent for organic and inorganic compounds, essential oils, oils, alkaloids, resins, gums, cellulose derivatives; in cosmetics; in medicine; for organic synthesis.
Methyl Alcohol	Volatile, colorless, mobile, clear liquid; soluble in alcohol, water, ether; inflammable; very poisonous.	Solvent for many organic compounds; in organic synthesis; as paint, varnish remover; in polishing and cleaning preparations; as anti-freeze.
Terpineol	Colorless liquid with lilac odor; soluble in ether and alcohol; insoluble in water.	Solvent for waxes, oils, resins, gums and cellulose products; in perfumes and soap.

Group Reactions of Alcohols

Alcohols react with

1. Acids	Forming esters and water.
2. Emulsions	Breaking them (in many cases).
3. Oxidizing Agents	Being oxidized to aldehyde, ketone or acid.
4. Proteins	Coagulating and rendering them water-insoluble.

ALDEHYDES

Aldehydes are organic chemical compounds. They occur as gases, liquids or solids.

Examples of commercially used aldehydes are

Name	*Properties*	*Use*
Acetaldehyde	Light, colorless, liquid with a pungent fruity odor; inflammable; miscible with water, benzene, with water, benzene, gasoline, alcohol, ether, turpentine, acetone.	As disinfectant; in photography; in medicine; in organic synthesis; in the manufacture of dyes.
Benzaldehyde	Volatile, colorless or yellowish oil with odor of bitter almonds; miscible with alcohol and ether.	Solvent for resins, oils, and some cellulose produets; in organic synthesis; in the manufacture of perfumes

Name	*Properties*	*Use*
		and dyes.
Butyraldehyde (Also Called Butyl Aldehyde)	Colorless liquid with pungent odor; inflammable.	In rubber industry; in tanning extracts; in perfumes; in organic synthesis.
Formaldehyde	Colorless clear liquid with pungent odor; aqueous solution of formaldehyde gas; poisonous; miscible with water and alcohol.	In medicine; as disinfectant; in explosives; as insecticide; in organic synthesis; for silvering mirrors; in metallurgy; in rubber manufacturing.
Furfuraldehyde (Also Called Furfural)	Mobile, odorless liquid with penetrating odor; soluble in alcohol, ether and benzene; lsightly soluble in water.	Solvent for lubricating oils, rosin, and other organic material, and cellulose products; as fungicide, wetting agent; synthesis of resins.
Paraformaldehyde	White powder with silght odor of formaldehyde; almost insoluble in alcohol, ether and water.	As preservative; for the manufacture of artificial horn and ivory; as fumigant.

Group Reactions of Aldehydes

Aldehydes react with

1. Acids (In Small Amounts)	Forming polymerization product.
2. Alcohols	Forming acetal.
3. Alkalis And Alkali-Carbonates (In Small Amounts)	Forming polymerization product.
4. Fehling's Solution	Forming cuprous oxide (red precipitate).
5. Heavy Metal Salts (Gold, Silver, Etc.)	Reducing them.
6. Hydrocyanic Acid	Forming addition product.
7. Hydrogen	Forming alcohol.
8. Oxidizing Agents	Being oxidized to acid.
9. Sodium Bisulfate	Forming addition product.
10. Ultra-Violet Light	Being split, giving off carbon monoxide.
11. Water	Forming addition product.

Specific Reactions of Aldehydes

Aldehyde	*Reacts or is incompatible with*	*Results*
Formaldehyde	Silver ammonium nitrate;	Precipitate of metallic silver.
	gelatin and glue;	Becomes insoluble in water.
	oxidizing agents;	Destruction.
	phenol, urea, thiourea.	Formation of resins.

ALKALOIDS

Alkaloids are organic compounds containing nitrogen. Most are extracted from plants. Some alkaloids can be synthesized. Alkaloids are toxic and, with a few exceptions, used only in medicine. They are soluble in alcohol. They form salts with acids and become water-soluble; most of them are used in the form of their salts.

Examples of commercially used alkaloids are

Name	*Properties*	*Use*
Caffeine	White mass or crystals.	As stimulant; in heart diseases.
Cocaine	White crystals.	As local anesthetic.
Nicotine	Colorless oil, turning brown on exposure to the air.	As insecticides; for tanning.
Pilocarpine	Hygroscopic, colorless or yellow crystals.	In hair tonics.
Quinine	White crystals or powder, odorless; with bitter taste.	In malaria; as antipyretic; in colds.
Strychnine	White crystals with very bitter taste.	In medicine as stimulant; as poison for animals and rodents.
Theobromine	White crystalline powder.	As diuretic; in diseases of blood circulation.

Group Reactions of Alkaloids

Alkaloids react with

1.	Acids	Forming salt.
2.	Alkalis	Being precipitated from salt-solutions.
3.	Fröhde's Reagent (Sod. Molybdate in Conc. Sulfuric Acid)	Giving color reactions.
4.	Gold Chloride	Forming precipitate.
5.	Mercury Potassium Iodide	Forming precipitate.
6.	Iodine in Potassium Iodide	Forming precipitate.
7.	Mercury Chloride	Forming precipitate.
8.	Phospho-Molybdic Acid	Forming precipitate.
9.	Picric Acid	Forming precipitate.
10.	Tannic Acid	Forming precipitate.

AMINES

Amines are organic compounds containing nitrogen, bound to hydrogen in most cases. Most of the amines give an alkaline reaction.

Examples of commercially used amines are

Name	*Properties*	*Use*
Amylamine	Colorless liquid; inflammable.	In manufacture of pharmaceuticals, dye-stuffs; as emulsifying agent; de-sizing and anti-corrosion agent.
Aniline	Oily colorless liquid; turns brown on exposure to air and light; slightly soluble in water; soluble in alcohol and ether.	In manufacture of many organic compounds; in printing.
Ethylene Diamine	Colorless, volatile liquid; soluble in water.	As textile lubricant; as anti-corrosive; solvent for albumin.
Methenamine (Hexamethylene Tetramine)	White crystalline powder; soluble in water and alcohol.	In medicine; in manufacture of synthetic resins and other organic chemicals.
Methylamine	Gas; soluble in water, alcohol and ether; inflammable.	In manufacture of organic compounds; in treatment of rayon; in tanning.
Morpholine	Colorless liquid; soluble in water and most organic solvents.	Emulsifying agent; as solvent for dyes, resins, waxes.
Triethanolamine	Yellow, viscous liquid; soluble in water and alcohol	Emulsifying agent; plasticizer; as penetrating agent; in cosmetics, adhesives, dry cleaning.

Group Reactions of Amines

Amines react with

1.	Acids	Forming salts.
2.	Fat, Oil, Wax	Forming soap (only in some cases).
3.	Aldehyde	Forming resin (in some cases).

Specific Reactions of Amines

Amine	*Reacts with or is incompatible with*	*Results*
Aniline	Chlorinated lime.	Blue or purple coloration.
Methenamine (Hexamethylene Tetramine)	Lithium benzoate;	Liquefaction.
	Acetylsalicylic acid, antipyrin, gelation.	Decomposition.

BASES

Bases and basic compounds are chemical substances which give an alkaline reaction, i.e. turn red litmus blue (see acids). They form salts with acids. Alkalies are strong bases that attack the human skin; they give a soapy feel and taste to the aqueous solutions. They form soaps with fatty acids, oils or waxes.

Examples of commercially used bases are

Name	*Properties*	*Use*
Ammonia	Gas with strong odor, mostly used in aqueous solution; soluble in water.	In refrigeration; in manufacture of chemicals; in rubber industry; water treatment; oil refining.
Caustic Potash (Potassium Hydroxide)	Hygroscopic white pieces, lumps, sticks or pellets; soluble in water and alcohol; slightly soluble in ether; produces heat when dissolved in water.	In manufacture of soap and matches; in engraving and lithography.
Caustic Soda (Sodium Hydroxide)	Hygroscopic white pieces, lumps, sticks or pellets; produces heat when dissolved in water.	In manufacture of soaps, rayon; in rubber, paper, textile industries; in petroleum refining; in organic synthesis.
Potassium Carbonate	White, hygroscopic powder; soluble in water; insoluble in alcohol.	In glass, textile and soap industry; in engraving and lithography; in electroplating.
Soda Ash (Sodium Carbonate)	Grayish or white powder or lumps: soluble in water,insoluble in alcohol.	In glass, soap, paper and textile industry; as water softener; in metallurgy.

See also **Amines**, which are organic bases.

Group Reactions of Bases

Bases (alkalies, metal hydroxides and metal oxides) react with

1.	Acids	Forming salt and water.
2.	Esters	Forming salt and alcohol (saponification).
3.	Fatty Acids, Fats, Oils, Waxes	Forrning soap.
4.	Proteins	Altering their structure and solubility.
5.	Resins	Forming soap.
6.	Salts	Replacing and releasing the base of the salt in case this is weaker than the added base.

Specific Reactions of Bases

Base	*Reacts with or is incompatible with*	*Results*
Sodium Hydroxide	Alkaloid salts;	Precipitate.
	mercury chloride (calomel);	Black coloration.
	atropin, cocaine, aconitine.	Decomposition.

Carbohydrates (Sugars)

Carbohydrates are compounds of carbon, oxygen and hydrogen. Most carbohydrates have a sweet taste, or develop one, when their molecules are partly decomposed.

Examples of commercially used carbohydrates are

Name	*Properties*	*Use*
Cellulose	Essential constituent of all vegetabre tissues and fibers, e.g., cotton; purified cellulose is a white substance.	In manufacture of paper, textiles, powder, plastics.
Dextrin	White or yellowish powder, soluble in water; insoluble in alxohol	In adhesives, sizings, and ink.
Glucose	Yellow to brown, syrupy liquid; soluble in water.	In confectionery; medicine; tobacco treatment.
Gum Tragacanth	Yellowish or white powder; swells up with water; insoluble in alcohol.	In the manufacture cf emulsions; in printing.
Lactose (Milk Sugar)	White crystalline powder with sweet taste.	In medicine; in infant foods; in margarine manufacture.
Starch	White tasteless powder or lumps; insoluble in cold water, alcohol and ether; when dispersed in hot water, it forms a jelly.	In adhesives, sizings, food; in medicine; in cosmetics.
Sugar (Saccharose)	White crystals or lumps with sweet taste; soluble in water; very slightly soluble in alcohol.	In food.

Group Reactions of Carbohydrates

Carbohydrates react with

1.	Ammoniacal Silver Nitrate solution	Reducing it to silver.
2.	Fehling's Solution	Reducing it.
3.	Iodine	Giving color reactions.

4. Oxidizing Agents	Being oxidized to acids.
5. Phenyl Hydrazine	Forming osazones.
6. Bacterial (Yeast)	Fermenting the carbohydrate.

ENZYMES

Enzymes are substances produced by living organisms. They are of great importance in biological processes. Most of the reactions taking place in the living organism are influenced or governed by the activity of enzymes. The action of every enzyme is specific. Most of the enzymes are destroyed by heat, acids, alkalies, alcohols, or heavy metals.

Examples of enzymes are

Enzyme	*Substance upon which enzyme acts*
Diastase	Starch
Invertase	Sucrose
Lipase	Fat
Pectinase	Pectin
Pepsin	Protein
Protease	Protein
Rennin	Casein

ESSENTIAL OILS

Essential oils are volatile oils of pleasant odor usually composed of esters, terpenes, alcohols and other organic compounds.

Examples of essntial oils are

Citronella Oil	Lavender Oil	Peppermint Oil
Eucalyptus Oil	Lemon Oil	Rose Oil

ESTERS

An ester is a compound of an alcohol and an acid. Some are water-soluble but most are wate-insoluble and oil-and organic-solvent-soluble. Esters are solvents for many organic materials. They are split (broken down) by alkalies, acids, and often by water alone.

Examples of commercially used esters are

Name	*Properties*	*Use*
Amyl Acetate	Colorless liquid with banana-like odor; very slightly soluble in water;	For flavoring; as solvent in lacquers, paints, perfumes; in the manufacture of cel-

Name	*Properties*	*Use*
	soluble in alcohol and ether; inflammable.	luloid.
Dibutly Phthalate	Oily colorless liquid; slightly soluble in water; miscible with organic solvents.	In lacquers, perfumes; in manufacture of safety glass; in inks as solvent.
Ethyl Acetate	Colorless liquid with fragrant odor, soluble in organic solvents; inflammable.	As solvent; in confectionery; in manufacture of plastics, rayon, pharmaceuticals.
Glyceryl Monostearate	Wax-like solid; dispersible in water; soluble in oils and hydrocarbons.	As emulsifying agent; for protective coating of hygroscopic powders.
Methyl Lactate	Liquid, soluble in water and organic solvents.	As solvent for cellulose; in lacquers.
Tributyl Phosphate	Colorless liquid.	As solvent for cellulose; in plastics and lacquers.
Tricresyl Phosphate	Colorless odorless liquid; miscible with most organic solvents; soluble in oils; insoluble in water.	As solvent in lacquers; in waterproofing composition; as plasticizer.

Group Reactions of Esters

Esters react with

Alkalis	Forming salt and alcohol (saponification).

Specific Reactions of Esters

Ester	*Reacts with or is incompatible with*	*Results.*
Methyl Salicylate	Ferric chloride.	Purple color.

ETHERS

Ethers are organic compounds prepared, e.g. by the dehydration of alcohols. Most of them are liquid, insoluble in water but soluble in organic solvents.

Examples of commercially used ethers are

Name	*Properties*	*Use*
Butyl Ether	Colorless liquid with ethereal odor; soluble in organic solvents; insoluble in water.	As solvent for hydrocarbons, fats; in synthesis.
Ethyl Ether	Colorless, volatile very inflammable liquid with aromatic odor; very	In medicine; as solvent for fats, oils, resins, alkaloids, etc.; in synthesis.

Name	*Properties*	*Use*
	slightly soluble in water, soluble in organic solvents; vapors are very narcotic.	

FATS

Fats, in the chemical sense, are esters of higher fatty acids and glycerol. They are insoluble in water and sparingly soluble in cold alcohol, but soluble in ether and other organic solvents.

They can be split into glycerol and fatty acid by boiling with inorganic acids and alkalies, or by treating with superheated steam. Fats heated above 250°C. decompose, releasing acrolein. which has an intense, sharp odor.

Examples of fats, in the chemical sense, are

Olein	Glycerol ester of oleic acid
Palmitin	Glycerol ester of palmitic acid

Fats, in the ordinary sense, are mixtures of the above-mentioned fats, but contain, in addition, salts, flavors, etc.

Examples of fats, in the ordinary sense, are.

Butter
Tallow
Cocoa Butter
Lard
Wool Fat (Lanolin)

GUMS

Gums consist of several carbohydrates. Most of them are water-soluble. They are on the market in white to browncolored lumps or powders. They are used in adhesives, inks, confectionery, medicine and textile industry.

Examples of commercially used gums are

Acacia (Arabic) Gum	Locust (Carob) Bean Gum
Cherry Gum	Shiraz Gum
Karaya (India) Gum	Tragacanth Gum
Guar Gum	Sodium Carboxymethyl Cellulose
Methyl Cellulose	Sodium Alginate

KETONES

Ketones are organic compounds containing a special group consisting of carbon and oxygen and called a carbonyl group.

Examples of commercially used ketones are

Name	*Properties*	*Use*
Acetone	Colorless fragrant liquid; soluble in water, alcohol, ether, chloroform; inflammable.	As solvent in lacquers, dyes; in synthesis.
Acetophenone	Colorless pungent liquid; has narcotic action.	As solvent for cellulose-derivatives, resins; in medicine.
Camphor	White translucent mass; soluble in alcohol, ether, chloroform, and volatile oils.	In medicine; in manufacture of cellulose; in photography; in explosives.
Phorone	Yellow liquid.	Solvent for nitrocellulose; in lacquers; in organic synthesis.

Group Reactions of Ketones

Ketones react with

1.	Oxidizing Agents	Forming acids.
2.	Reducing Agents (Strong)	Forming alcohols.
3.	Sodium Bisulfate	Crystalline addition product.

OILS

Oils can be divided into 3 groups. The first group includes the fatty oils. These are derived from animal and vegetable sources and consist mainly of esters (especially glycerol esters) of higher fatty acids.

The second group embraces mineral oils obtained from petroleum, coal, shale. The third group contains the essential oils.

Oils are water-insoluble, soluble in ether, benzene, etc.

Examples of Oils of Group I	*Examples of Oils of Group II*
Castor Oil	V. M. & P. Naphtha
Coconut Oil	Mineral Seal Oil
Cod Liver Oil	Mineral Oil, Heavy
Linseed Oil	
Neatsfoot Oil	
Red Oil	
Sperm Oil	

OXIDIZING AGENTS

Oxidizing agents are agents which release oxygen easily. This oxygen can combine with another compound or can take hydrogen away from another substance. Oxidizing agents have a bleaching and disinfecting action.

Examples of oxidizing agents are:

Name	*Property*	*Use*
Benzoyl Peroxide	White solid; insoluble in water, slightly soluble in alcohol, soluble in other organic solvents.	As bleaching agent, as catalyst, in rubber compounding.
Chlorine	Greenish-yellow gas or liquid with irritating odor; very poisonous.	As bleaching agent; in synthesis, water purification, metallurgy.
Hydrogen Peroxide	Colorless liquid; available as an aqueous solution; has to be kept cool and dark.	As bleaching agent, antiseptic; in cosmetics.
Nitric Acid	See under **Acids.**	
Potassium Dichromate (Potassium Bichromate)	Yellowish-red crystals; poisonous; soluble in water; insoluble in alcohol.	As bleaching agent, preservative; in pyrotechnics, electroplating.
Sodium Chlorate	White crystals; soluble in water and alcohol.	As explosive; in medicine, leather industry.
Potassium Permanganate	Dark red crystals; soluble in water.	As disinfectant, deodorant, bleaching agent; in textile industry.

PHENOLS

Phenols are aromatic alcohols (see **Alcohols** and **Aliphatic Compounds).**

Examples of commercially used phenols are:

Name	*Properties*	*Use*
Cresol	Colorless or yellowish liquid; slightly soluble in water, soluble in alcohol and ether.	As disinfectant; in manufacture of synthetic resins; in explosives.
Phenol (carbolic Acid)	White crystals which turn pink or red when exposed to light; absorbs water from the air; soluble in water, alcohol and ether.	As disinfectant; in manufacture of many organic products.
Pyrogallol	White crystals; soluble in water, alcohol and ether.	In photography; in synthesis of dyes; in medicine.
Resorcinol	White crystals becoming pink on exposure to light; soluble in water, alcohol, ether.	In manufacture of dyes, celluloid; in medicine.
Thymol	White crystals of aroma-	In medicine; as antioxidant.

Name	*Properties*	*Use*
	tic odor; soluble in organic solvents.	

Group Reactions of Phenols

Phenols react with

1. Acids	Forming esters.
2. Alcohols	Forming ethers.
3. Aldehyde	Forming insoluble resins.
4. Ferric Chloride	Giving red, blue, violet, green or brown coloration (in most cases).
5. Reducing Agents (Strong)	Being reduced to hydrocarbons.
6. Sodium Hydroxide	Forming phenates.

PROTEINS

Proteins are complex organic, nitrogen-containing compounds formed in living organisms.

Examples of commercially used proteins are

Name	*Property*	*Use*
Albumen (Albumin)	Yellowish-white; soluble in water; coagulated by heat.	In adhesives; textile and leather industry; as clarifying agent.
Casein	Yellowish-white powder; soluble in aqueous akali solutions.	In paper coating, sizings; in medicine; in plastics.
Gelatin	Purified glue (see glue).	In food products; in medicine.
Glue	White to brown sheets, flakes, etc.; absorbs cold water with swelling; dissolves in hot water forming a jelly on cooling.	As adhesive, sizing and stiffening.
Zein	Protein derived from corn; resinous substance.	In plastics, coatings, inks, photographic films.

Group Reactions of Proteins

Proteins react with

1. Acetone	Forming precipitate.
2. Acid	Forming precipitate.
3. Alcohol	Forming precipitate.
4. Aldehyde	Forming precipitate.
5. Mercury In Nitric	Giving reddish-brown precipitate.

Acid (Millon's Reagent)	
6. Salt	Forming precipitate.
7. Sodium Hydroxide And Copper Sulfate	Violet coloration (with some) (Biuret reaction).
8. Ferments (Pepsin, Trypsin)	Decomposing proteins.
9. Heat	Coagulates Proteins.

REDUCING AGENTS

Reducing agents are agents that release hydrogen or take up oxygen easily.

Examples of commercially used reducing agents are

Name	*Properties*
Carbon	Black powder.
Formaldehyde	Colorless liquid with pungent odor.
Hydrogen	Colorless gas; very inflammable.

RESINS

These can be divided into natural and synthetic classes.

Natural resins are obtained from certain plants and trees. Some are hard and more or less brittle, insoluble in water soluble in certain organic solvents. Others, the so-called oleoresins or balsams, are viscous fluids.

Synthetic resins are complicated organic compounds, and replace natural resins. Resins are used in the manufacture of plastics, varnishes, and other coatings.

Examples of resins are

Natural Resins	*Oleoresins*	*Synthetic Resins*
Copal	Copaiba Balsam	Phenolics
Dammar	Peru Balsam	Ureas
Kauri	Tolu Balsam	Melamines
Mastic		Polyestess
Rosin		Polyamides
Sandarac		Epoxies
Shellac		Urethanes
		Styrene Butadienes
		Alkyds

SALTS

Salts are compunds formed by the reaction of an acid with a base (see under acids). Salts usually have a salty or saltybitter taste. Some are corrosive. They can be white or colored. The salts of heavy metals, e.g. iron, copper, etc., are usually colored.

Examples of commercially used salts are

Ammonium Chloride
Calcium Acetate
Copper Chloride
Iron Sulfate
Magnesium Sulfate (Epsom Salt)
Potassium Citrate
Sodium Chloride (Table Salt)
Sodium Phosphate

Group Reactions of Salts

Salts react with

1. Acid — In some cases forming new salt.
2. Alkali — In some cases forming new salt.
3. Protein — In some cases forming precipitate.

SOAPS

Soaps are the metallic salts of palmitic, stearic, oleic acids, etc. Soaps in which the metal is sodium are hard, while those containing potassium are soft. The alkali salts of resins are called resin soaps. Soaps can be split by heating with a stronger (especially inorganic) acid than the acid they contain.

Example of commercially used chemical soaps are

Ammonium Linoleate
Calcium Stearate
Lead Oleate
Lithium Stearate
Sodium Ricinoleate
Sodium Stearate
Zinc Stearate

Soaps, in the common sense, are alkali salts of a mixture of higher fatty acids. In addition, they contain small amounts of glycerin and other ingredients.

III Chemicals Classified by Use

ABRASIVES AND POLISHING AGENTS

Abrasives are usually hard mineral particles for roughly grinding away uneven surfaces. Polishing agents are fine mineral powders for producing a fine smooth finish or polish, or removing discoloration.

Typical Commercial Abrasives	*Typical Commercial Polishing Agents*
Carborundum	Bentonite
Corundum	Cuttlefish Bone
Diamond Dust	Rouge (Iron Oxide)
Emery	Silica (Fine Powder)
Garnet	Talc
Sand	Tripoli
	Cerium Oxide

ADHESIVES AND BINDERS

Adhesives and binders are materials that cause one or more bodies to adhere or stick together.

Typical Commercial Adhesives and Binders

Casein	Rubber Latex
Dextrin	Synthetic Resin Emulsions
Glue	Resins
Gum Arabic	Rubber
Starch	Sodium Silicate

ADSORBENTS

Adsorbents are substances that attract and hold gases or liquids on their surfaces.

Typical Commercial Adsorbents

- Alumina, Activated
- Bauxite
- Carbon, Activated
- Charcoal
- Silica Gel

ANTIFREEZES

Antifreezes are materials which lower the freezing point of water.

Typical Commercial Antifreezes

- Alcohol, Ethyl
- Alcohol, Methyl
- Calcium Chloride
- Ethylene Glycol
- Glycerin

BLEACHES AND DECOLORIZERS

Bleaches and decolorizers are substances which decolorize or lighten a color by physical absorption or chemical change.

Typical Commercial Bleaches and Decolorizers

- Bauxite
- Benzoyl Peroxide
- Carbon, Activated
- Chlorine
- Hydrogen
- Hydrogen Peroxide
- Sodium Chlorite
- Sulfur Dioxide
- Chlorcyanuric Acid Salts

CLEANERS AND DETERGENTS

Cleaners and detergents are materials used in solution to remove dirt, discoloration and other foreign matter.

Typical Commercial Cleaners and Detergents

- Ammonia
- Caustic Potash
- Diglycol Laurate
- Soapless Soaps, e.g., "Aerosol," "Gardinol," "Nacconol"
- Soap
- Soda Ash
- Sulfonated Oils
- Trisodium Phosphate
- Sulfated Fatty Alcohols, Esters, Amides, Ethoxylated Esters, Phenols, Alcohols

DEODORANTS

Deodorants are substances that remove odors. This effect can be obtained by adsorption of the odor, as is the case when activated carbon is used, or by changing the odorous substance chemically in such a way that the odor disappears.

Typical Commercial Deodorants

Aluminum Chloride	Formaldehyde
Carbon, Activated	Oxidizing Agents (see)
Potassium Permanganate	Reducing Agents (see)

DYES

Dyes are organic coloring matters that are soluble in water or other solvents. Many are sensitive to heat, alkali or acid.

Typical Commercial Dyes

Auramine	Methylene Blue
Fluorescein	Naphthol Yellow
Malachite Green	Nigrosine
Tartrazine	

EMULSIFIERS

Emulsifiers or emulsifying agents are substances that cause two non-miscible (or insoluble) fluids to mix to give a uniform milky liquid. Thus mineral oil and water do not mix. If, however, some diglycol oleate is added to the oil and then the water is added slowly, while mixing, a uniform milky emulsion results.

Typical Commercial Emulsifiers

"Aerosol" O.T.	Lanolin
Ammonium Linoleate	Propylene Glycol Laurate
Bentonite	Saponin
Calcium Stearate	Soap
Diglycol Stearate	Sulfated Fatty Alcohols
Glyceryl Monolaurate	Sulfonated Castor Oil
Gelatin	Triethanolamine Oleate

Also those listed under Commercial Cleaners

ETCHES

Etches are substances that discolor, corrode, or eat away that part of a surface on which they are placed.

Typical Commercial Etching Agents

Acids, e.g. Sulfuric, Hydrochloric or Hydrofluoric Acid
Alkalies, e.g. Caustic Soda or Potash
Oxidizing Agents
Solvents, e.g. Acetone on Cellulose Acetate or Celluloid
Zinc Chloride

FILLERS

Fillers are powdered inactive materials included in a product to give hardness, weight, or bulk, to reduce activity, close pores, or reduce cost.

Typical Commercial Fillers

Barytes	Infusorial Earth	Silica
Calcium Sulfate	Lithopone	Talc
Clay	Pumice	Whiting

FLAMEPROOFINGS

Flameproofings are substances that reduce flammability or make a substance non-flammable.

Typical Commercial Flameproofing Bases

Ammonium Sulfate	Chlorinated Naphthalene
Ammonium Sulfamate	Tin Oxide, Precipitated
Borax and Boric Acid	Zinc Borate
Organic Phosphorus Compounds	Antimony Oxide

FLAVORS

Flavors are substances that impart a distinctive taste or odor.

Typical Commercial Flavors

Clove Oil	Lemon Oil	Peppermint Oil
Coumarin	Methyl Salicylate	Vanillin

GREASEPROOFING SUBSTANCES

Greaseproofing substances prevent passage of oil, fat or grease into or out of a material.

Typical Commercial Greaseproofing Bases

Casein	Glue	Sodium Silicate
Dextrin	Glycerin	Starch

INSECTICIDES

Insecticides are substances that kill insects.

Typical Commercial Insecticides

Arsenic Trioxide	Derris	Paradichlorbenzene
Calcium Arsenate	Lethane	Pyrethrum
DDT	Nicotine Sulfate	Sodium Fluoride
"Lindane"	"Endrin"	"Malathion"

LUBRICANTS

Lubricants are substances that diminish friction and resultant heat.

Typical Commercial Lubricants

Fat	Oil, Fish	Sugar Syrup
Glycerin	Oil, Mineral	Talc
Graphite	Oil, Vegetable	Tricresyl Phosphate
Soap	Wax (Soft)	Silicones
Fluorocarbons	Molybdenum Disulfide	

MOISTURE CONDITIONERS

Moisture conditioners are substances that absorb moisture or give it up under certain conditions. They are important in many products to avoid drying-out or absorption of too much moisture. Different materials can absorb varying amounts of moisture, and suitable ones must be selected for different products.

Typical Commercial Moisture Conditioners

"Sorbo" (Sorbitol Syrup)	Glycerin
Calcium Chloride	Lithium Chloride
Diethylene Glycol	Propylene Glycol

Triethanolamine

ODORANTS AND PERFUMES

Odorants and perfumes are materials used to impart a characteristic odor or to cover or mask any unwanted odor.

Typical Commercial Odorants and Perfumes

Amyl Acetate	Lavender Oil	Nitrobenzol
Clove Oil	Lemon Oil	Peppermint Oil
Coumarin	Methyl Salicylate	Rose Oil
	Vanillin	

PIGMENTS OR COLORS

Pigments are finely powdered white, black or colored substances used to impart color.

Typical Commercial Pigments

Carbon Black	Lead Chromate	Prussian Blue
Chrome Green	Lithopone	Red Lead
Iron Oxide	Ochre	Titanium Dioxide

PRESERVATIVES

Preservatives are materials that prevent decomposition or undesirable change in a product. The following are important classes of preservatives.

1. Antiseptics or germicides which prevent bacterial growth.
2. Anti-oxidants which prevent changes due to the action of oxygen in the air.
3. Insecticides that prevent attack by insects.

Typical Commercial Preservatives

Class No. 1	*Class No. 2*	*Class No. 3*
Alcohol (Ethyl Alcohol)	Amines	Copper Napthenate
Benzoic Acid	Beta-Napthol	Paradichlorbenzol
"Dowicide"	Guaiac Resin	Pyrethrum
Glycerin	Hydroquinone	Sodium Fluoride
Hydroxybenzoic Ether ("Paraben")	Lecithin	Sulfur
Quaternary Ammonium Compounds		

Salicylic Acid
Sodium Benzoate

The following table shows the concentration of various preservatives necessary to prevent bacterial growth:

Ammnium Fluoride	0.04 %
Benzoic Acid	0.18 %
Borax	0.90 %
"Chloramine" T	0.05 %
Chlorbutanol	2.00 %
Chlorthymol	0.02 %
Cresol	0.10 %
Mercuric Chloride	0.02 %
Ortho-chlormetacresol	0.1 to 0.2 %
Ortho-phenylphenol	0.10 %
Oxyquinoline Benzoate	0.025 %
Para-hydroxybenzoic Esters	0.10 %
Phenol	2.00 %
Phenyl Mercuric Nitrate	0.01 %
Salicylic Acid	0.21 %
Sodium Benzoate	0.1 to 0.2 %
Sodium Bisulfite	1.00 %
Sodium Fluoride	0.80 %
Sodium Salicylate	0.10 %
Sodium Silicofluoride	0.15 %
Thymol	0.04 %

Germicidal Values of Miscellaneous Antiseptics

	Germicidal Value (Phenol = 1)
Acetic Acid	0.60
Acetylcoumaric Acid	4.50
Alcohol	less than 0.10
Amyl-meta-cresol	280.00
Aniline	0.57
Benzoic Acid	5.00
Boric Acid	less than 0.10
Catechol	0.48
Chinosol	0.15 to 0.30
Chlorbutanol	1.20
Chlorine Water	28.00
Copper Sulfate (anhydrous)	2.00
Cresylic Acid	3.70
Eucalyptol	1.20
Formalin	0.55 to 0.75
Formic Acid	5.70
Guaiacol	0.90
Hydrochloric Acid	1.58
Hypochlorites (on basis of available Cl)	146.00 to 220.00
Iodine Trichloride	94.00
Lactic Acid	1.80

	Germicidal Value (Phenol = 1)
Mercuric Chloride	400.00 to 3540.00
Metacresol	2.00
Monochlor-phenol	2.00 to 3.00
Ortho-toluidine	1.00
Ortho-cresol	2.10
Para-cresol	2.40
Pentabrom-phenol	500.00
Phenol	1.00
Picric Acid	6.00
Potassium Permanganate	42.00
Pyrogallic Acid	0.22
Quinol	1.10
Resorcinol	0.30
Silver Nitrate	15.80
Sodium Bisulfate	4.10
Tribrom-phenol	46.00
Trichlor-phenol	25.00
Zinc Chloride	0.15

RUST PREVENTIVES

Rust preventives are substances that protect a metal against oxidation or rusting.

Typical Rust Preventives

Amine Soaps
Lanolin
Mineral Oil
Petrolatum
Reducing Agents
Silica Gel

SOFTENERS AND PLASTICIZERS

Softeners and plasticizers are materials that decrease hardness and brittleness in films, fibers, and coatings.

Typical Commercial Softeners and Plasticizers

Castor Oil
Dibutyl Phthalate
Diethylene Glycol
Glucose (Corn Syrup)
Glycerin
Mineral Oil
Tricresyl Phosphate

SOLVENTS

Solvents are liquids that can dissolve other substances. Alcohol dissolves camphor, water dissolves sugar or salt, etc. A solvent may consist of one or more liquids. Sometimes a mixture of several liquids is a stronger solvent than a single liquid would be. A mixture of solvents may be chosen also to reduce the costs, e.g., the dilution of an expensive solvent with an inexpensive one. Ethyl acetate, for example, can be diluted with toluol and used as a solvent for pyroxylin.

Typical Commercial Solvents

Acetone	Carbon Tetrachloride
Alcohol (Ethyl)	Ethyl Acetate
Benzol	Gasoline
Butanol	Methanol
Butyl Acetate	Naphtha
Trichlorethylene	Methylene Dichloride
Xylol	

STIFFENERS AND SIZES

Stiffeners and sizes are materials that increase stiffness, strength, weight or smoothness of yarns, fabrics, paper, leather or other products.

Typical Commercial Stiffeners and Sizes

Dextrin	Lacquer	Sodium Silicate
Gum	Resin	Starch
Synthetic Resins	Sugar	

THICKENERS

Thickeners are substances which make a liquid more viscous or form a jelly.

Typical Commercial Thickeners

Water Thickeners	Solvent and Oil Thickeners
Diglycol Stearate	Metal Soaps, e.g., Aluminum
Guar Gum	Stearate
Gum Karaya	Resin, Synthetic
Gum Tragacanth	Rubber
Methyl Cellulose	Sulfur
Pectin	"Thixin"
Quince Seed	"Cab-O-Sil"
Sodium Carboxy ethyl Cellulose	
Cellulose Gum	
Guar Gum	

WATERPROOFING AGENTS

Waterproofing agents are substances that when added to a material prevent the penetration of water.

Typical Commercial Waterproofing Agents

Asphalt
Metal Soap, eg. Aluminum Stearate
Paraffin Wax
Paraffin Wax Emulsion and Aluminum Acetate
Resin, Synthetic
Rubber, Synthetic
Silicones

WETTING AGENTS

Wetting agents are substances which, when added to a liquid in small amounts, cause more rapid or better wetting, spreading, or penetration of the liquid on a dry material or surface.

Typical Commercial Wetting Agents

"Aerosol"
"Gardinol"
"Nacconol"
"Triton"
"Tergitol"

IV Raw Materials

The most important factor concerning raw materials is their availability. Whenever there is a choice of raw materials, select the one which is produced in large quantities and by more than one manufacturer.

No matter how good the material is, it is useless if not available in the desired quantities at the desired time. Therefore, the first step is to make sure that the materials to be used are in plentiful supply, that they are preferably of domestic and not foreign origin, and that suppliers will be able to take care of quantity requirements. What is available today may become unavailable tomorrow. Since all consumers are subject to this uncertainty, everyone is on an equal footing.

UNIFORM REPLACEMENT

Uniform replacement refers to the ability to obtain the same grade of product each time it is ordered. Slight variations when unavoidable may be compensated for. Great variations may make a product unusable. Traces of copper are undesirable in materials used in rubber compounding; consequently a material which is the same in all respects as in previous deliveries would be ruled out if contaminated with copper or its compounds. A material containing 0.1% iron may be suitable for certain purposes, but if the iron content increased to 0.5%, it may no longer be usable.

Raw materials should not be bought in too great quantities and should be so stored that they will not deteriorate. Before use, they should be inspected and if they do not appear uniform or in perfect condition, they should not be used. When a change in the source of supply is made, the material should be tested for purity and workability to make sure that it is suitable. The final and conclusive test is how it will affect the finished product. This can only be seen by making up a small experimental batch. Of course, time may bring out some defects that are not immediately visible. Various aging tests will be given in the section on *Testing Finished Products.*

Raw materials vary in quality. Always buy similar raw materials, ordering on the same specifications as given in previous satisfactory purchases.

Do not try to save money by buying odd lots of chemicals or opened containers. No initial saving can compensate for spoiled products and loss of customers' good will. It is best to buy raw materials from a well-known supply house or manufacturer who can be depended on to furnish uniform materials at all times.

GRADE OR PURITY

The following grades of chemicals may be available:

C.P.	Natural	Unofficial
Commercial	Synthetic	N.F.
Technical	U.S.P.	N.N.R.
Special	B.P.	

C.P. stands for chemically pure. Each container usually bears a label of analysis indicating the amounts of impurities present. This grade is usually the purest grade of the chemical available. It is generally more expensive than the other grades but it is not specified except when high purity is required.

Commercial is the most common grade of chemicals sold. Any chemical which does not bear a grade designation can be considered of commercial grade.

Technical is the ordinary commercial grade. This grade should not be used for food, drug or cosmetic purposes without investigation.

Special refers to a particular grade made for a particular consumer or industry. It differs from all other grades; it may be more or less pure. Its form and packaging may also be different.

Natural refers to a crude or refined product of vegetable, mineral or animal origin, e.g., crude or resublimed iodine or camphor.

Synthetic refers to a chemical which is built up from a number of different chemicals by a chemical process, e.g., synthetic menthol. A synthetic chemical usually contains fewer impurities than the corresponding natural product.

HANDLING

The selection of a new material may introduce new problems in handling. Often this is of minor importance. Sometimes it may require new or changed equipment and special precautions.

LEGAL RESTRICTIONS

The purchase, ownership, storage, use or sale of many materials and products is covered and bound by national, state, city or local regulations.

Foods, drugs and cosmetics must conform to the regulations of the U.S. Food and Drug Administration and State and City Departments of Health.

Habit-forming drugs such as cocaine and morphine are strictly regulated by the federal government under the Narcotics Act.

Certain states and cities require licenses for the manufacture and sale of drugs, cosmetics and other products.

Explosive regulations of the U.S. Bureau of Mines require a license for owning any type of explosive material, even though it is to be used in a non-explosive product.

The use of many materials is not only subject to the above regulations (which are far from complete in scope) but to state and local laws on production of disagreeable odors, toxic materials and other hazardous products, that may increase the fire or explosion hazard, or injure anyone in or near the factory.

Since many substances and their uses are patented, it is also well to ascertain whether the use of a material for commercial purposes will mean infringement of a patent.

FORMULATION

The use of a new material may involve a slight or a radical change in formulation. If only a change in proportions (see Proportions) is necessary, then reformulation is relatively simple.

A more complicated case is the use of polyvinyl alcohol as an emulsifying agent for a carnauba wax-water emulsion. Ammonium linoleate or other fatty acid derivatives can function equally as well as polyvinyl alcohol as an emulsifying agent, but they are not suitable because they are not film-forming adhesives. The latter quality is necessary in a particular coating problem. The nearest logical substitute would be a filmforming adhesive type of emulsifier such as methyl cellulose. Different proportions of emulsifier, wax and water mixed under different conditions would have to be tried to get optimum results. It might be necessary to add a secondary emulsifier or stabilizer and perhaps even a plasticizer or solvent coupling agent, such as "Carbitol."

The replacement of dibutyl phthalate as a plasticizer in lacquers is also an example of a more complicated case. When "Glaurin" (diethylene glycol monolaurate) is substituted for the former, it is necessary to use a smaller amount of the latter and also to vary the resin and solvent proportions in order to get a good lacquer.

Sometimes, the use of a new material may entirely change the character and use of a product. For example, trichlorethylene, a low boiling liquid used in degreasing of metals, may be replaced by sodium abietate, a dry crystalline material which is used with water. Here is an example of a new material totally different in its properties and method of use but serving the same function.

PROPORTIONS

It is seldom that a new ingredient in a composition can replace the original material pound for pound. It may be necessary to use a greater or a smaller amount of a new material to produce the desired result.

Thus, where a soap was formerly made from coconut oil and alkali and the former is to be replaced by castor oil, a larger amount of castor oil would be needed for complete saponification. This is due to the difference in combining weights of these two oils. Combining weights are of importance in determining amounts to be used when a chemical reaction takes place.

Where another material is to be used in the same volume as the material which it is to replace, then the finished product will usually have a different unit weight. Thus in a composition containing 2 quarts of glycerin per gallon of solution, wherein the glycerin is to be replaced by 2 quarts of propylene glycol, the difference in weight per gallon of solution produced by this change will be appreciable.

Certain materials that have great strength or activity are used in smaller amounts when substituting for weaker materials. If hydrochloric acid is used in place of acetic acid, a much smaller amount may be needed. Similarly caustic soda, which is a much stronger base than triethanolamine, would be used in a smaller amount. Conversely, when a weak material replaces a strong material a larger amount may be needed, e.g., in replacing phosphoric by lactic acid.

METHODS OF MANUFACTURE

The different properties of a new or alternative material may present many problems in the manufacture of a finished product. In some cases this problem may be minor and in others more serious.

For example, if sugar is to be replaced by molasses, the latter may require a tank for storage, pumps and valves for delivering it to the mixing tank, exhaust fans for drawing off the odor, special materials of construction to prevent corrosion, cleaning and sterilization of all tanks, pipe-lines, valves, pumps, etc., to prevent fermentation. A sludge (due to impurities present in molasses, not present in sugar) may also require special filtering apparatus or settling tanks. If color is a factor, it may be necessary to heat the molasses with a decolorizing carbon or other materials and then filter it.

In cold weather the storage tank may have to be heated to reduce viscosity and permit ready flow of the molasses. None of these factors is present when sugar is used, because it is a dry powder or crystalline material, that can be dumped into the kettle from sacks or barrels, and presents none of these problems.

When a corrosive material replaces a non-corrosive material, as in the case of calcium chloride in place of activated alumina for moisture absorption, the calcium chloride may corrode the metals formerly used and require the installation of materials at points of contact which are resistant to it.

COSTS

The relative price per pound of a new material to that of the material which it replaces is not the sole economic criterion in making a choice. While propylene glycol is more expensive than glycerin, it may be more economical to use in making an imitation vanilla flavor because of its far greater solvent power for vanillin. Accordingly, a smaller amount of it is needed.

Sometimes a higher cost for a new material may be justified if the finished product is superior in use or salability. For example, a stamp pad ink made with glycerin varied too much in consistency during dry and damp weather. Furthermore, the rubber stamps had a limited life period. Replacing the glycerin with glyceryl monoricinoleate produced an ink which varies very little with atmospheric moisture changes and prolongs the life of the rubber stamps considerably.

Replacements which are usually introduced by necessity are sometimes continued in use after the necessity has ceased to exist. The reasons for this may be ease in handling, uniformity, more sources of supply, shorter manufacturing time, lower maintenance and labor costs, reduction in insurance, avoidance of license fees or patent suits, governmental regulations and record-keeping and other possible advantages.

USE OF THE FINISHED PRODUCT

The introduction of alternate materials may produce such changes in a product as to necessitate changes in the method of use by the consumer. This, while highly undesirable, is sometimes unavoidable.

For example lemon extract as used by bakers consists of a solution of lemon oil in alcohol. When the alcohol is replaced by emulsifying the lemon oil in water with a vegetable gum, a thick messy emulsion results which is less easy to handle than the limpid alcoholic lemon extact. Once the user learns how to handle and mix the lemon emulsion into his baking batter, he gets equally good results. Educating the user, however, is a slow and expensive process.

STORING ESSENTIAL OILS

With few exceptions, essential oils keep quite well if they are in well-purified condition and are stored under proper conditions. Conditions of storage and use are extremely important, however, and complaints concerning instability of oils have often been traced back to bad stockroom conditions.

Since most essential oils consist of mixtures of more or less volatile components, more or less sensitive to various types of chemical change, precautions must be taken to avoid unfavorable changes in the oils.

Evaporation is an occasional source of loss in storing oils. Leaky or insecurely stoppered containers will result in loss. This loss does not consist, however, merely of evaporation of the whole oil; since the oil is a complex mixture, some components will naturally evaporate more quickly than others. As a result, the residual oil will undergo definite changes in composition and may change its odor characteristics quite radically. Incidentally, loss by evaporation should be guarded against when oils are being used, because of this possible change in odor value. The perfume material should not be allowed to stand in open vessels for any length of time before being added to the batch, and the cream temperature should be low before perfumes are added.

Oxidation is probably the most common type of chemical change taking place in essential oils. The aldhydes, which are important constituents of many perfume and flavoring oils, are very sensitive to air, being easily oxidized. Oxidation usually produces the corresponding acid which are either odorless or have an odor quite different from that of the original aldehyde. Terpenes, especially those characterized as unsaturated, are also quite sensitive to the action fo air, and may produce discoloration of the oil and even cause the odor to deteriorate. Unsaturated acids present in some oils may act in the same way.

Resinification is another type of chemical reaction that tends to spoil oils. This phenomenon does not depend solely on the presence of air, since it may be caused by traces of impurities present in badly purified oils from doubtful sources. Light may also be active in promoting resinification.

The presence of minute traces of impurities is a very important factor in the acceleration of chemical reactions. Traces of metals are especially active in promoting oxidations; many reactions will not start except in the presence of traces of moisture; ordinary dust may start a chain of reactions. As an example affecting essential oils, the citral content of lemongrass oil is reduced much more rapidly in a wet than in a dry oil.

A small rise in temperature will usually cause a substantial increase in the speed of the reaction, and is to be considered as a factor in storing sensitive chemicals.

From these general observations, we may draw certain conclusions regarding the best method of storing essential oils.

The oil should be as pure and dry as possible, with suspended particles of any kind removed by careful filtration. Not only must the oil be clean when first stored, it must also be kept clean during use. Dust, cork particles, and all other foreign substances must be kept out of the bottles.

To avoid evaporation, bottles must be tightly stoppered with a clean, new cork or other type of closure. If a cork must first be removed with a corkscrew, it should not be used again, but should be replaced by a new one.

Oxidation is retarded by the use of a tight stopper and by keeping the bottle as full as possible to minimize the volume of air over the liquid. If oil is to be taken from bottles at intervals, it may be advisable to add clean glass marbles to the bottle to raise the liquid level and cut down the air space. If oil is received in drums and then taken as needed from the drums, the last half of the oil may be subject to considerable action from air over the liquid. With oils known to be sensitive it will be advisable to transfer the essential oil from the drum to several bottles, completely filled, using only one at the time.

Oils may often be kept in perfect condition in clean, inert metallic drums, transfer to bottles being recommended only to avoid oxidation caused by the large air space present in halffilled drums.

To be on the safe side, oils should be protected from the action of light by being kept in amber bottles. Direct sunlight is especially active, and bottles should not be exposed to its action for even short periods of time.

The temperature of storage should be uniform, neither too high nor too low. High temperatures will initiate or accelerate deterioration of the oil, while low temperatures may result in precipitation of some of the important constituents. Temperatures between 60 and 65°F and generally most satisfactory.

The use of antioxidants and retarders in essential oils has apparently not been extensively investigated, but a few observations may be noted. The use of from five to ten per cent of anhydrous alcohol added to some essential oils or aromatics has been recommended for preservation. A small amount of phenolphthalein added to citronellal acts as an effective retarder, reducing oxidation to a minimum. In the same way, traces of hydrocyanic acid present in natural oil of bitter almonds definitely retard the oxidation of the benzaldehyde. We have seen no references to the use of hydroquinone, pyrogallol, and similar organic antioxidants in essential oils, but we believe that research might prove them to be very useful in some of the more easily oxidized products.

The following tabulation and grouping was kindly supplied by Fritzsche Brothers, Inc.

The storage of essential oils is a very important factor in the preservation of their quality. The oils fall into various groups and their stability may be measured by the ease with which they undrego oxidation, polymerization, decomposition and interaction. These detrimental factors affecting the oils can be greatly minimized by correct storage.

Group I

The citrus oils (Orange, Lemon, Lime, etc.) are very prone to oxidation because of their high terpene content. The terpenes in general are unsaturated hydrocarbons and readily absorb oxygen from the air. This oxidation si catalyzed by light, metals such as iron or tin, and heat. It is best that this type of oil be kept in well-filled containers, preferably amber glass, securely closed and away from light and excessive heat.

Certain other oils fall in this group because of their high terpene content. They are Abies Alba Oils, Pinus Sylvestris Oil, Pinus Pumilio Oil, Juniper Berry Oil, Turpentine Oil, etc.

Group II

The oils that contain large amounts of aldehydes and in some cases ketones are readily polymerized. This, of course, is a normal chemical reaction undergone by these two functional groups. Some of the oils which fall in this group are: Almond Bitter, Cassia, Cumin, Cherry Laurel, Spearmint, etc.

These oils are also subject to oxidation with the corresponding acid being formed. The method of storage is the same as for the citrus oil but with special emphasis on the glass container.

Group III

This group will be constituted by those oils that contain large amounts of alcohols. Included in the group may be Bois de Rose Oil, Cedarwood Oil, Citronella Oil, Coriander Oil, Geranium Oil, Peppermint Oil, Vetiver Oil and many others.

These oils are fairly stable since the terpene alcohols are not easily oxidized by exposure to air. Therefore the oils of this group may be safely stored in metal containers but the same precautions should be taken concerning light and heat.

It might be well to mention here that some oils, although not seriously affected chemically, may be detrimentally altered in color due to exposure to light, air, or heat. It is therefore wise to be cautious even with the more stable groups.

Group IV

This group is composed of the fixed oils such as Sweet Almond, Peach Kernel, Croton, etc. The abundant fatty acids present in these oils make them very readily oxidizable. The resulting damage is commonly known as rancidity. Storage in metal containers is permissible but absolute exclusion of air is very important.

Group V

This group will include those oils that contain large amounts of phenols, phenol ethers or oxides. Some of the oils included might be Bay, Cajeput, Clove, Eucalyptus, Sassafras, Origanum, Marjoram, Thyme, etc.

These oils, like those in the alcohol group, are fairly stable and may be stored in a similar manner. This group, however, seems more readily affected by color change, especially the last three oils mentioned. The light factor must therefore be carefully considered.

Group VI

This group includes those oils that contain large amounts of esters. The only real danger here is the decomposition of the ester into acid and alcohol, creating an undesirable pungency due to the odor of the acid. This group includes the oils of Lavender, Linaloe, Bergamot, Wintergreen, etc. Ordinary storage methods will prevent this slow process of breakdown.

Group VII

This is a miscellaneous group and under it several oils may be considered.

Mustard Oil, upon exposure to light or air, readily decomposes and deposits sulfur.

Certain oils improve with age probably because of interreaction between the various functional groupings contained in the oil. Classed in this category might be Rose Oil, Patchouly Oil, Vetiver Oil, Ylang Ylang Oil, etc.

In general, oils containing large amounts of terpenes, aldehydes, fatty acids, and certain ketones are very sensitive. Those containing large amounts of alcohols, phenols, phenol ethers, oxides and certain ketones are fairly stable. Those containing large amounts of esters are very stable and often improve with age.

WATER

Water is used in varying quantities in different preparations. It may be a source of annoyance and difficulty in many cases. Water is not a pure or uniform material. It has in solution different and varying amounts of salts, bacteria, gases, etc. It varies from place to place and from time to time in the same place.

Certain natural so-called hard waters are troublesome when they are used with soaps, fatty acids and other ingredients. Such hard water must be softened by suitable treatment or replaced by rain, distilled or deionized water.

When materials are used that ferment or are decomposed by bacteria that may be present in the water, the water should first be boiled to kill such bacteria. Boiling will also tend to remove many odors and will often precipitate or coagulate other foreign matter present, which should be filtered out.

V Processing Procedure and Equipment

THE LABORATORY

The laboratory may consist of a large building containing much apparatus and materials costing thousands of dollars or it may be a nook in a kitchen, basement or garage. Just as a good workman may accomplish much with a few tools in a small space, so a good compounder may achieve much with little apparatus and materials in any small space.

While the description given later of a laboratory and its contents will tend toward the ideal, it will not be necessary in all cases to fit it up as completely as is indicated.

The bare necessities of a laboratory consist of a work-bench or table, water, a source of heat (gas, oil, coal or electricity) and certain pots, pans, bottles, cups, glasses and cutlery, most of which you already have.

Let us assume that you intend fitting up a laboratory in the basement of your house. First select a corner where you have the best light and are nearest to a water and drain-pipe. Here should be placed the laboratory table. The latter should be 36 to 40 in. high and about 8 ft long. It should have a number of drawers and cupboards for keeping apparatus, materials, etc. On the wall near the table shelves should be built for holding containers of chemicals and other materials. At one end of the table attach any second-hand sink (enamel or galvanized preferred) and connect to the water inlet and to the drain. A gas line should be drawn to the table or, if inconvenient, a rubber gas tube can be used instead. This gas tube is then attached to a small gas stove or Bunsen burner. If convenient, a small electric stove or hot-plate may be connected to a nearby electrical outlet.

The other necessary equipment includes a scale or balance, thermometer, measuring glasses or graduates, stirring or mixing rods of glass and wood and an assortment of pots, pans, cans, knives, forks, spoons, glasses, bottles, jars, corks and caps.

A full-length rubberized or oil-silk apron and a pair of rubber and cloth gloves will be useful in preventing staining or injury to the clothing and hands.

Additional miscellaneous supplies include a note-book, a black and colored pencil, pen and ink, scratch-pad, small labels, a towel and dish rag, soap, scouring powder, steel-wool and small brushes.

With chemicals, this equipment will furnish a working laboratory that can be used for making small amounts of many useful commercial products and for testing. Additional equipment can be added as needed. Much of it can be obtained at any hardware or specialty store at very low prices.

Cleanliness and Neatness

Good habits of cleanliness and neatness formed at the beginning will pay big dividends in good products and avoidance of waste in materials and apparantus. Keep your table, its contents, and all apparatus and containers as clean as possible.

First find a definite place for everything and try to keep an orderly arrangement all the time. When you stop work, put everything back where it belongs and clean up anything that may have spilled. Wash out all containers and apparatus, allow them to drain and then wipe them dry whenever possible. Periodically dust and wipe off everything.

To avoid spillage, pour materials slowly and carefully and do not overfill vessels in which ingredients are mixed or heated. To avoid spoilage, do not leave containers open in the sunlight or in an overheated or very cold (below freezing) room. When taking an ingredient out of a container always brush or wipe off any particles or drops which adhere on the outside.

In cleaning away spilled mateials, try water if it is known to be water-soluble (like sugar or salt) or alcohol or naphtha (for rosin or oils). Where a material is not dissolved readily, try rubbing it with fine sand or steel-wool and water or a solvent.

Particular care should be taken with the balance or scale. The knife-edges or suspension points should be wiped with a dry soft rag or brushed with a soft brush periodically to remove dust. It goes without saying that anything spilled on the former should be removed completely as quickly as possible.

THE COMPOUNDING ROOM

The compounding or manufacturing room should, wherever possible, not be in a dwelling. Preferably it should be at street level, with windows for light and ventilation and at least two entrances, one at each end. It should have a source of gas or electricity, water and a waste-drain. A garage, store or separate small farm building can be readily fitted up to serve. The floor should preferably be of cement or stone and so pitched that any liquids spilled or employed in washing will drain off.

The size of this room will depend on the scale of manufacturing that will be done. Wherever possible, extra space should be present to be used for storage of raw materials, containers and finished products as well as for possible moderate expansion.

Since the compounding room is really only an enlarged laboratory it would be well to review the section on the laboratory. The equipment, of course, will depend on the type of products to be manufactured.

OPERATIONS

Measuring and Weighing

Most of the materials used in compounding are weighed, especially solids. Liquids may either be weighed or measured out in a liquid measure. Since volume (space occupied) is not usually the same as weight (except in the case of water where it is approximately the same), care must be taken to determine whether the quantity expressed in a formula is by volume or weight.

Small amounts (less than 10 g) are usually weighed on a balance and larger amounts on a scale. Small volumes of a liquid are usually measured with a pipette or small graduate, and larger volumes with a graduated cylinder or measuring glass.

Since small differences of weight or volume may often produce unwanted variations in a product it is important that all ingredients be measured or weighed as accurately as possible.

In making weighings, a tin can, glass or other container is put on the pan of the balacne or scale and balanced exactly with the weights on the beam. The weight of this empty container is called the tare.

Let us assume that the tare (weight) of the empty container is 50 g and that you wish to weigh out 20 g of sulfur. Push the sliding weight on the scale beam to 70g. Then pour some of the sulfur from its container on a sheet of paper. Next hold the latter over the container on the scale pan and tilt it so that the sulfur slides down slowly. Stop the addition of sulfur when the scale is nearly balanced. Then add or take away sufficient sulfur to balance the scale.

Weights of chemicals are usually in the metric (grams or kilograms) or avoirdupois (ounces or pounds) systems. A table for converting from one system to the other is given near the end of this book.

Measurements by volume (drams, fluid ounces, pints, quarts, gallons, cubic centimeters or liters) are made with various measuring devices, such as pipettes, glasses, graduates or cans. These are marked with lines showing their capacity or volume at a certain level. A pipette is used for measuring and delivering small quantities of a liquid. For larger quantities the other measuring containers of a suitable size are used.

Temperature is measured with different types of thermometers, using the Fahrenheit or Centigrade systems (see tables for changing from one system to the other).

Since thermometers are usually made of glass, they are easily broken by shock or too sudden changes in temperature. Accordingly, thermometers should not be struck against the sides of a vessel or other solid and should be laid down gently after use. They should not be plunged suddenly into very hot liquids nor from a hot liquid to a very cold liquid lest they break. Before inserting a thermometer into a hot liquid hold it above the liquid to absorb a little of the heat above the liquid and so warm up gradually. Conversely, when taking a thermometer out of a very hot liquid, swing it through the air so that it may cool down before placing it in a relatively colder liquid.

Metal sheaths for glass thermometers protect them against breakage from shock but not against wide temperature changes. Of course, such metal sheaths (or all-metal thermometers) must not be used where they may be corroded by the medium in which they are placed and so cause contamination because of the corrosion products.

Concentration or strength of a solution is readily determined by measuring its specific gravity or density as compared to water. This is usually done with a hydrometer, a glass device which sinks into a liquid to a certain level. By reading the number at the point where the surface of the liquid meets the hydrometer and referring to a table or chart of specific gravity the concentration or strength of the solution is obtained.

Heating

Gas, steam, electricity, coal, oil, etc., may be used as sources of heat.

The most common source of heat in the laboratory is the Bunsen (gas) burner. The size and intensity of the flame is readily regulated and it can be moved about easily.

Heat should always be applied gradually and carefully to glass or enamelled vessels to avoid cracking and chipping of the vessel, as well as over-heating or boiling-over of its contents.

In the case of materials which tend to discolor, decompose or char if overheated, it is advisable to use a double boiler or water bath, if the latter is available. For similar gradual heating at temperatures higher than that of boiling water, a bath of molten salts or a readily fusible metal or alloy is used.

Inflammable materials should never be heated with an open flame, or fire and explosion may result. In such cases, heat the water in the double boiler to a boil, melting or warming any of the non-inflammable ingredients first; then turn out the flame and stir in the inflammable ingredient.

Most ingredients are poor conductors of heat and unless you stir while heating, local overheating results. Certain mixtures give off gases or vapors on heating and tend to froth and foam over the sides of the vessel. Therefore, do not overfill vessels and heat gradually, with stirring, when working with a new combination of materials.

Filtration and Clarification

Filtration is the process of removing undissolved particles of ingredients and foreign matter from a liquid. For very coarse particles, cheese-cloth or muslin may be used. For finer particles any finely woven fabric may be used. All fabrics should first be boiled with soap and then thoroughly washed to remove starch or other sizing and non-fast dyes. For very fine particles filter paper of suitable porosity is used.

The cloths may be sewed into bags and suspended through a small hoop. They, as well as the filter paper, may be folded and supported on a funnel. Filter paper is usually folded twice, into four parts, and put into the funnel. Wet it with a little water or other liquid so that it will stick to the funnel. Pour enough liquid into the filter so as to wet all of the filter paper and press the paper so that no air pockets form between the paper and the funnel. Then add more of the liquid to be filtered, slowly, without splashing. Stop adding the liquid when the level is about ¼ in. below the top of the filter paper. If the liquid gets too near the top, some will creep over and run down into the clear, filtered liquid. More liquid is added when the level in the filter drops.

Sometimes, when speed is no factor and insoluble matter can settle down in a liquid, the latter is allowed to stand overnight or longer. Dirt or other insoluble matter will settle out and the clear liquid on top may be poured off or drawn off by a rubber-tube siphon.

Cooling

Cooling is produced by air, water, ice, "dry-ice," or special freezing mixtures, depending on how rapid and how low a temperature drop is desired. Since metals are the best conductors of heat, metallic vessels are used wherever rapid cooling is desired. Stirring a compound greatly increases the rate of cooling. Spreading in thin layers also increases the rate of cooling.

Dissolving and Mixing

Most materials dissolve more readily when warmed with water or other appropriate solvent. Stirring also hastens dissolving. Small particles dissolve more rapidly; accordingly, all large pieces should be broken or ground as finely as possible. Certain materials which form thick solutions are best dissolved by soaking them in just sufficient water to wet all particles and then gradually adding more water to effect more complete solution.

Mixing may be done by shaking (in the case of small, closed containers), or stirring with a wooden stick, spoon, fork, eggbeater or electric-mixer. In mixing, care must be taken not to splatter material out of the container. This not only causes loss of material but may also change the proportions of ingredients and spoil the finished product.

Mixing of heavy pastes in small amounts can be done with a flat knife or wooden or bone stick on a non-porous slab of marble or metal. Larger amounts require a geared-motor mixing unit, fitted with a suitable paddle or mixing arms.

Decolorizing and Bleaching

The most common method of decolorizing is to add 1 to 5 parts by weight of decolorizing carbon to 100 parts of liquid. Stir while warming. Allow to stand for a while and then filter. The amount of decolorizing carbon, the temperature and time of heating depend on the nature and quantity of coloring material present and can only be determined by trial.

When this method does not give the result desired or is not feasible, it is necessary to resort to bleaching. While bleaching may be faster and, in certain cases, gives a lighter color, it may produce unforeseen changes in the product through chemical reaction.

The most common bleaches are: hydrogen peroxide and chlorine and its compounds.

Hydrogen peroxide is used by adding, with stirring, 1 to 2 % at room temperature. Chlorine or its compounds, such as bleaching powder (chloride of lime or calcium hypochlorite), or Javelle water, are added in small amounts with stirrings. Bleaches work very rapidly and their effect can be seen quickly. Certain colors bleach very easily, others are more difficult, and some are not amenable to ordinary bleaches.

Grinding and Pulverizing

Large pieces are first roughly crushed and the smaller pieces then divided into smaller units. This process is continued until pieces of the size desired are obtained. Thus, very large lumps can be put in a bag and broken into small lumps with the flat side of an axe or other broad, heavy object. If the material is friable, putting a board over the bag and putting one's weight on it will crush the material. By means of a sieve or piece of screen, the small particles may be sifted out and the large pieces crushed again as above.

Of course, where a mortar and pestle or a mechanical grinder is available it should be used. When using a mortar and pestle only, put a small quantity of the material to be ground in it. Pound with an up and down motion to break up the large pieces. Fit a piece of paper or cloth with a hole in its center over the pestle and mortar to prevent pieces from flying out of the mortar. Then grind with a circular motion to get finer pieces and particles.

TESTING

Raw Materials

This book is not intended to give explicit methods of chemical and physical testing. Such methods can be found in the standard books on testing.

The following general information may prove useful, as a starting point in eliminating unsuitable substitute or alternate materials.

Data from manufacturers' literature, technical handbooks or dictionaries will usually give some information and should be consulted.

Color, clarity, odor, form, homogeneity, grade, etc., can often be checked, quickly, with a small sample. Heating a sample in a test tube will give some indication of changes that may be expected in color, odor, form, taste, density, melting-point, boiling-point, viscosity, vapor pressure, flexibility, cohesion, stickiness, adhesion, drying, slipperiness, elasticity, composition, solvency, stability, bacterial activity, surface tension, explosiveness, firehazard, etc.

Strong cooling with dry-ice or other freezing mixtures will show changes in many other properties. Manipulation with the fingers will give a quick rough estimate of hardness, flexilbiity, tenacity or coherence, stickiness, adhesion, feel, slipperiness, elasticity, etc.

A rough estimate of solubility is made by dropping a little of the material and the other ingredients in a test tube with water or with one of the solvents to be used, shaking to see if it dissolves. Solvency is determined in the same way, only in this instance, the solvent is the substitute.

Interaction with other ingredients is tested by using the susbtance in the finished product letting it stand (or even warming it) and determining any apparent change in the properties mentioned in this chapter.

Acidity or alkalinity (pH) is tested with pH papers or a pH meter.

Whirling a sample in a centrifuge will cause the separation of finely divided particles. This separation might not otherwise appear for some time.

The effect of heat is determined by placing a sample in an open or closed tube on a steam or electric plate for a working day or overnight. Heat speeds up most reactions and a test of this nature is often indicative as to how a material will change on aging.

The effect of sunlight can be checked fairly rapidly by exposure to the actinic rays in such devices asf the Fade-o-Meter or the Laundero-meter.

The effect of moisture or dryness is seen by exposure in a closed vessel, containing either a dish of water or a desiccant (Drierite or sulfuric acid).

The effect of oxygen is rapidly determined by the oxygen bomb test or treatment with an active oxidizing agent such as hydrogen peroxide.

Finished Products

Age affects chemical mixtures in various ways. These effects may appear as discoloration, change in viscosity or consistency, gas or odor formation, separation into layers, hardening, settling, precipitation, drying-out, alteration in taste, effect on use, corrosion, etc.

Changes due to chemical interaction can often be speeded up by heating the mixture in a large closed container (to prevent evaporation) at about 100 °C. Exposure to direct sunlight (or better yet to an ultra-violet lamp) will accelerate certain changes. Tumbling the filled container, from time to time, or for an extended period in a mechanical device, will show up any changes that will later be brought about in products that are shipped and handled roughly.

No product should be marketed before it has been tested as above or passed upon by someone who is competent to do so. Failure to do this is certain to cause disaster to one without experience in making and marketing a specific product. This is a typical example when haste can only be made slowly.

EQUIPMENT

The most frequently used equipment consists of mixing vessels, agitators (mixers) and filters. These can be simple or complex. Because a new business is often limited as to capital and because the products are more or less uncertain it is not wise to invest in high-priced equipment. It is usual to improvise or use devices that can be bought cheaply and adapted to your needs or to buy used equipment.

Instead of buying a tank or mixing unit, an empty drum or barrel often serves well. As an agitator, a shaft with a homemade paddle or mixing blades can be attached to a motor. To the latter a clamp is welded so that it can be fastened to the drum or barrel. This makes a good portable agitator.

Where it is not feasible to build a piece of equipment it is often possible to buy a used unit from a dealer in second-hand equipment. A list of such dealers is given in the appendix immediately after the list of equipment manufacturers.

In buying equipment, it is always best to state to the manufacturers exactly how it is to be used. The manufacturers usually have had many many years of experience in designing and adapting standard equipmnet to meet special conditions. They have trained engineers who are ready to help you, at no extra cost. Some manufacturers even run tests on small units to see if their equipment will fully meet your needs. Take advantage of this service. Even if you have had some experience with the type of equipment in question, place your problem before them. As specialists they may be able to improve on your ideas or even point out some disadvantage or how money can be saved in initial or operating cost.

The manufacturing operations that have been discussed in the preceding pages are for very small quantities, using handy, available or improvised equipment. When finances permit, larger or more suitable equipment can be bought.

Scales and Measuring Devices

More than one size or type of scale is necessary, in order to weigh with accuracy and efficiency. For weighing small amounts carefully a pharmacist's torsion balance is useful. For weighing from 1 oz to 50 lb, a counter-indicating scale such as is used by butchers is needed. For weighing drums and barrels a platform scale weighing up to 600 lb is required. Low-grade or inaccurate scales, no matter how cheap, are a false economy. Too sensitive, high-priced scales are a needless luxury. A good dealer or manufacturers' representative can advise you best as to the proper unit for your needs.

Always get a guarantee from the seller that, if the device is not approved by the Weights and Measures inspector, he will refund the purchase price or make a satisfactory replacement. Any reliable seller will give such a warranty.

Tablet Machines

The machines described herein produce tablets equal, in both appearance and quality, to those compressed by larger machines. They produce tablets of all sizes up to 1/2 in. in diameter with output up to 100 tablets per minute. The machines are of simple design, easy to clean, easy to change over for different tablet sizes, with simple adjustments for weight and hardness of tablets. They are of exceptionally rugged construction, made from a special close-grained cast iron of high tensile strength and built to give years of satisfactory service at little or no maintenance cost except that necessary to correct normal wear in the punches and dies.

These tablet machines are widely used for manufacturing tablets from private formulas, for compressing pharmaceutical specialities in small scale manufacturing operations and for experimental work in research and industrial laboratories.

Two models, hand-operated and motor-driven, are available. Both are compact and require little bench space. The power-driven machine has a 1/4 H.P. motor, current connections for which are made to the nearest lighting circuit. No special wiring is necessary.

The motor is mounted on an adjustable bracket. The machine is driven by a V-belt which can be conveniently removed, if desired, and the machine operated by hand. This machine is also equipped with an over-size hopper.

Three sets of round punches and dies are furnished with each machine, those generally included being 7/32, 5/16, and 3/8 in. Punches and dies of other sizes in graduations of 1/32 in. are available.

TABLET PRODUCTION

Most materials to be compressed require some simple treatment to insure proper operation of the compressing machine and a product of good appearance. Very fine powder does not feed evenly. It also imprisons air, which tends to split or *cap* the tablets. These faults are overcome by *granulating,* which is usually accomplished by moistening the powder with a suitable liquid, then screening and drying, or by compressing the fine powder into larger tablets or slugs which are subsequently broken down into a granulation by passing through an oscillating granulator.

If a binder is required to make the tablet hold its shape after compressing, sugar, gelatin, gums, starch, or other binding material is introduced into the mixture by dissolving in the granulating fluid. In case materials have a tendency to stick to the inside of the die or the faces of the punches, a small amount of lubricant is added to the otherwise finished granulation. Powdered stearic acid and hydrogenated fat are among the most efficient lubricants. Other lubricants are powdered soap, boric acid, talc and refined mineral oil.

For the actual compressing, single-punch machines are used for limited production and rotary machines for larger production. The speed of the presses varies with the nature and compressibility of the material. Standard materials usually compress to a thickness of one-half the maximum depth of fill.

The choice of a tablet machine depends usually on the maximum diameter of the tablet and its thickness or weight, as well as the material and the production desired. Should the material require high pressure to make a tablet sufficiently hard, sometimes a model larger or heavier than indicated in the machine specifications should be selected.

Not every tablet machine will be satisfactory for all products. Manufacturers of such machines know exactly what to suggest for each particular job, provided they are given complete and accurate information. A suitable questionnaire for this purpose, as compiled by a leading manufacturer, follows.

TABLET QUESTIONNAIRE

To assist us in developing your formula, and working out a satisfactory manufacturing technique, we ask that you fill in the following questionnaire.

This information will be held strictly confidential. *We suggest, if possible, that sample tablets or materials be submitted.*

Pupose or use of tablet: ....

Desired daily or hourly production: ....

Description of tablet:

If round, diameter (inches, mm.)..................

Thickness (inches, mm.) ....

Weight (grains, grams, ounces) ....

Surface of tablet (convex, concave, flat, bevel-edge) ....

If design or lettering, is this raised or sunken? ....

If special shape, give dimensions, show sketch

Physical Properties Desired:

Density or hardness (hard, medium, soft) ....

Should tablets dissolve in water? ....(fast or slow) ....

Should solution be clear? ....

Are tablets to be coated, fired, molded or otherwise treated after being compressed?

....

....

Other properties: ....

....

List Ingredients,

giving proportions or weights, if possible:

....

....

Properties of Ingredients:

What is particle size of material? ....

Can it be supplied in a granular, free-flowing condition? ....

Are materials harmful to those working with them, poisonous, explosive irritating, etc.? ....

If so, which ones? ....

What are the usual precautions? ....

....

If materials are heat-sensitive, advise maximum permissible temperature (°F, or °C.) ....

To what extent, if any, do moisture, atmospheric oxygen, or carbon dioxide affect materials? ....

....

It may be necessary to add a small percentage of one or more of the following materials to your formula in preparing it for compressing, to secure a binding or lubricating effect. *Please cross out those that cannot be used.*

Water	Starch	Boric Acid
Alcohol	Resins	Talc
Other Solvents	Gum Acacia	Vegetable Oil
Sugar	Other Gums	Mineral Oil
Gelatine	Stearic Acid	Graphite

Have you any specific suggestions for fillers, or binding or lubricating agents?

....

....

Additional information which may prove helpful to us: ....

....

....

Submitted by:

Company ....Address..............................

Individual ....City ....State ....

(Additional copies of this questionnaire will be mailed on request)

F.J. STOKES MACHINE COMPANY,

TABOR ROAD, OLNEY P.O., PHILADELPHIA, PA.

(Do not fill in blanks below)

TEST No............ ORDER No .... *Date* ....

CONTAINERS AND PACKAGING

Finished products are put into containers such as tubes, cans, bottles, jars, bags. etc. The choice of container depends on which container will best hold the product; which will be opened, closed and used most easily; and which will have the greatest visual appeal to the buyer. For additional protection and sales appeal, the container is enclosed in a suitable carton or wrapper.

Manufacturers of containers and packages have a wealth of useful information which they are glad to make use of in helping you to make a proper selection. Give them all the information you can so they can give you intelligent aid.

Collapsible Tubes*

Collapsible tubes for the packing of products are made chiefly from either of three different materials, tin, lead alloy, or plastics.

Tin tubes are particularly effective for shaving creams, dental creams, ointments, food products, pharmaceuticals and toilet preparations.

Lead alloy tubes are practical for certain shoe polishes, glues, cements, tinting colors, etc.

A recent development is the shell type tube. There is no cap and the tube is flat-ended with the usual clipped end for closing. These shells are suitable and economical for powders and pastes where it is desired to package a measured amount to be consumed all at one time. The shell is opened by cutting or tearing off the flat end, thus destroying it.

A similar development is the single application tube. This tube

*The following information is reprinted by courtesy of National Collapsible Tube Co.

provides an attractive and economical container for liquids in the measured single-application field. The shoulder is tapered with a circular piercing area which permits of easy opening and dispensing.

Both of these styles of tube may be obtained in tin and lead.

The tube capacities that follow are for fluid measure with the tube closed. The weight of the closed tube content will vary with the specific gravity of the product contained in it.

Table of Capacities of National Tubes
(Tin and Lead Alloy)

Diameter In.	*Length, In.*	*Fluid Capacity*	*Openings*
*½	2	1½ drams	
½	2½	1¾ drams	
½	3	2 drams	Nos. 8*, 12
½	3½	2½ drams	
	2½	2¾ drams	
⅝	3	3 drams	Nos. 8, 12*,
⅝	3½	3½ drams	14
*⅝	4	4 drams	
¾	2½	3¾ drams	
¾	3	4½ drams	Nos. 12*, 14,
¾	3½	5¼ drams	16
*¾	4	6 drams	
⅞	3½	7 drams	
*⅞	4	1 ounce	Nos. 12, 14,
⅞	4½	1⅛ ounces	16*
⅞	5	1¼ ounces	
*1	4	1¼ ounces	
1	4½	1½ ounces	Nos. 12, 14,
*1	5	1¾ ounces	16*, 20, 28
1	5½	2 ounces	
1	6	2¼ ounces	
*1⅛	5	2 ounces	
1⅛	5½	2¼ ounces	Nos. 14, 16*,
1⅛	6	2½ ounces	20, 28
*1¼	5	2½ ounces	
*1¼	5½	3 ounces	Nos. 14, 16*,
1¼	6	3¼ ounces	20, 28
1¼	6½	3½ ounces	
1¼	7	3¾ ounces	

*These sizes maintained in stock having openings as denoted.

Diameter, in	*Length, in.*	*Fluid Capacity*	*Openings*
1⅜	5	3⅛ ounces	
1⅜	5½	3½ ounces	Nos. 14, 16*,
*1⅜	6	3⅞ ounces	20, 28
1⅜	6½	4¼ ounces	
1⅜	7	4⅝ ounces	
1½	5	3½ ounces	
1½	5½	4 ounces	Nos. 16, 20,
*1½	6	4½ ounces	28*
1½	6½	5 ounces	
1½	7	5½ ounces	

SPECIAL OPENINGS

Common types of special purpose openings are: ophthalmic (eye tip), for eye preparations; nasal tip of larger size, for nasal jellies; screw-eye tips for liquid cement and solders.

CAPS

Metal caps have, until quite recently, been the most common type of tube closure but with the rapid progress of molded Bakelite, the latter material is now largely supplanting tin for caps.

Bakelite caps in a variety of colors can be furnished for all openings. Shapes of Bakelite caps are round and hexagonal for all openings. Caps are made with a valve seat for round openings, eliminating the necessity of using a cork liner. Where it can be used, this style of closure is recommended.

The standard tin cap is available for all openings. Tin caps are supplied with cork liners; Bakelite caps contain valve seats unless cork liners are specified. Cork liners are supplied in plain cork, tin-faced cork, Vinylite-faced cork, or other materials for special purposes.

A popular type of cap, attached to the tube, has been developed. The advantages of this type are obvious; it does away with fishing down the drain, hunting under bath tubs and wash basins for the elusive disappearing type of cap. It makes repeat sales for a product by the gratitude it always wins.

Another style of closure is one where the orifice is exposed by rotation of the ball valve. A washer under the ball provides the seal and balls can be furnished in most colors.

When inquiring for or ordering tubes, be sure to specify the diam-

*These sizes maintained in stock having openings as denoted.

eter, length, type of metal, opening No., cap style and the quantity desired. Also, if possible, furnish a copy of your design and state the number of colors to be used in printing, if decorated tubes are wanted.

The customer who wishes to have his trade name embossed on the shoulder of the tube may obtain this at no added cost beyond that of the original embossing die.

Collapsible tubes are sold by the gross and are packed on the basis of this unit in individual compartment boxes as follows:

$\frac{1}{2}$ to 1 in. diameter inclusive .... 1 gross per box
$1\frac{1}{8}$ to $1\frac{3}{8}$ in. diameter inclusive .... $\frac{1}{2}$ gross per box
$1\frac{1}{2}$ in. diameter .... $\frac{1}{4}$ gross per box

Tube Filling Equipment

For hand-filling tubes a simple and efficient machine is available. This machine is tinned all over and is easily cleaned or refilled. Nozzles of nickel plated brass to take all sizes of tubes are available and are easily interchanged. One size of nozzle is furnished with each machine.

Filling, Closing and Sealing Equipment*

In selecting tube filling and sealing equipment or machines for filling cans and jars, various factors must be considered. Production desired, tube size or container shape, physical characteristics and consistency of the product, whether material to be filled may cause corrosion in machine parts, cost of labor, all these have direct bearing on the choice of machine, nozzle and hopper that will best meet specific requirements.

Hand-Operated Tube and Jar Filler

An available tube and jar filler is a piston-type machine, crank-operated, designed to handle a range of products from liquids to light pastes, any material that will flow by gravity, and to fill tubes, jars and cans. It is recommended for all liquids, self-levelling semi-liquids and pastes such as colors, adhesives, creams that are filled while warm, light ointments, etc.

The piston-type pump operates through a three-way valve with large opening (to facilitate filling light pastes). It has an adjustable "suck-back" action to minimize any tendency to drip when handling liquids and semi-liquids. The entire mechanism is especially designed for quick and easy cleaning. Capacity is up to 5 fl. oz. with one turn of

*The following information is reprinted by courtesy of F. J. Stokes Machine Co.

the handle. Output is as fast as the handle can be turned and containers moved under the filling nozzle. Production rates of 30 gross or more per day are readily maintained. Larger containers are filled by turning the handle several times.

This machine may be equipped with horizontal nozzles for filling containers from the bottom upward with pastes; or with vertical nozzles for filling liquids and semi-liquids. One nozzle is supplied as standard equipment. Additional nozzles may be obtained as needed. Provision is made for inserting an electric heating element or cartridge to keep the pump mechanism warm when filling materials which must be packaged in a melted, semi-liquid condition. Standard material of construction is brass, with galvanized iron hopper. Special metals may be employed.

Tube Closing Machine

The tube closing machine is used to close the open end of the tube after it has been filled, the filled tube being placed in a cup between the jaws and the handle turned to produce a neat, tight single or double fold.

This device handles any size tube up to 1½ in. by 8¾ in. Many of the parts are the same as used in larger machines. Adjustments are simple and construction is rugged. A holder for one size tube is supplied as standard equipment. Additional holders or cups may be obtained as required.

Sealing and Clipping Press

After tubes are closed they are sealed in a press. This machine makes various types of closures when suitable inserts for the jaws are used. It will also squeeze firmly into position any size clip required for tubes up to 2 in. diameter when plain jaws are used.

Operation is fast, both hands of the operator being free for handling tubes. Construction is rugged, of semi-steel. Two standard models are available, foot-operated or with motor drive.

Vacuum Bottle Filler*

The Hy-Speed bottle filler can be operated within 5 seconds, is adjusted to bottles of any size in 30 seconds. It cleans itself for use with different liquids in 60 seconds. This performance is of definite value in

*The following information is reprinted by courtesy of Alsop Engineering Corporation.

any plant, and is an absolute necessity where many different items are produced.

The approximate filling speed is 1000 quart-, 1500 pint-, 1500 half-pint-, and 3000 ounce-bottles per hour. This depends to some extent on the liquid used and the speed of the operator.

Imperfect bottles with the slightest pin-hole or crack are rejected instantly. This valuable feature prevents waste of liquid and imperfect shipments.

With little or no experience any operator can turn out thousands of absolutely uniformly filled bottles per day, regardless of size or shape. Oils, chemicals, acids, fruit juices, toilet preparations, in fact practically any and all liquid products are filled better, faster, more easily, and at less cost with this small, compact, inexpensive filling machine.

Siphon Bottle or Can Filler

The Hy-Speed filler has been built to meet the needs of the modern bottling plant for an efficient, sanitary, fool-proof unit of this tried and proven type.

It handles bottles, jugs and cans of all sizes, shapes and styles from 1/2 pt. up to gallons. Every container is filled evenly to any desired height without any foaming or slopping. The filling height in the bottles is adjusted by a simple finger and thumb wheel control which raises or lowers the rubber-encased bottle rest.

The filling tank, triple galss-coated inside and outside, with its close-fitting cover, is always spic and span and is exceptionally easy to keep clean and germ-free. In addition, this dust-proof cover prevents any foreign matter from getting into your liquid.

Connection to the supply tank is made with a 6-foot length of pure gum spirit hose supplied with each unit.

The entire machine can be varied in height to suit the convenience of the operator. This exclusive feature results in greater speed, ease, and efficiency on the part of the operator because of less fatigue.

The spouts are instantly removable for cleaning or sterilizing. There are no pins, nuts or screws to undo. Costly sloppiness and dripping are eliminated by a special self-sealing arrangement which closes each spout tightly when the full bottle is removed.

A one-piece columnar stand with a rigid cone base occupies less floor space and is very much more sanitary because it is easier to keep clean than any other type of angle iron or case frame stand used today.

The entire filling machine is of rust-proof construction throughout, insuring against any possibility of contamination of the liquid. Always clean and sanitary, this most up-to-the-minute siphon filler is a unit of which the modern bottling plant owner can be justly proud.

Screw Capping Machine

The fundamental purpose of any capping machine is to make it impossible for a loosely capped leaky container (and it only takes one) to get out with any shipment and spoil the whole case or carton before it reaches your customer.

Extreme accuracy in the degree of tightness of any particular kind or size cap is absolutely essential and we all know that the required tension may vary on every job, depending on the kind of cap or container used.

The cap-tightening action is automatically positive regardless of the operator. The very necessary adjustment of the tension can be instantly made with the touch of a finger.

To have to make a careful pre-arrangement of the containers is a waste of time, and that is where the flexibility and wide range of capping action of these machines is such a distinct advantage when it comes to speed and cutting out lost motion.

The standard-type tapered cone chucks will handle all but the most unusual shape or style of caps. The average chuck will take different size caps within range of approximately 4 mm. and a complete assortment of chucks to fit any caps up to 70 mm. is always obtainable. When necessary, the chuck can be changed for a different size in a few seconds without any special tools.

The caps, even the most delicately finished, are fully protected against damage or marring by the specially molded, toughened composition rubber liners furnished with each chuck.

Aerosols

An aerosol consists of a gaseous liquid or fine powder so packaged with a gas (propellent) that it is forced out in a fine spray when a release valve is opened. Because of ease of application, many products, such as polishes, insecticides, paints, cleaners, hair treatments, shaving cream, deodorants, etc., are now being marketed as aerosols.

Different propellents are indicated for different products and conditions. The most frequently used propellents are the chlorofluoro methanes, e.g. "Freon." Others are low boiling hydrocarbons, carbon dioxide, and nitrogen. The selection of the proper propellent, container and filling equipment should be left to an experienced contract manufacturer (loader). The chemical specialty manufacturer should confine himself to compounding his formula and sending it to the loader for packaging. This will also avoid capital investment in loading equipment and eliminate manufacturing hazards.

Bottle Labeling*

Care in the design and specification of labels, in the planning and purchase of glass packages, in the regulation and maintenance of machines, and in the selection and use of adhesives is the key to successful labeling. However, success depends, to a large degree, upon the skill, dilligence, and intelligence of the operator, as well as upon the age and condition of any labeling equipment used.

Hand Labeling

The highest speeds and greatest satisfaction in strictly hand work operations are obtained by the lay-out method. The procedure is to thin the labeling paste with the amount of water (if any) prescribed by the adhesive manufacturer, and brush out a liberal film on a glass, stone, or seasoned hardwood surface. A number of labels (depending upon the speed at which the paste sets and the operator applies the labels) may then be laid backs downward on the pasted slab to soak and temper.

A heavy, damp cloth can be placed over the layout of labels to hold them flat on the table while soaking, and to prevent the paste from setting prematurely. This cover cloth can be rolled back to expose one line of labels at a time when ready for application to the bottles, and should be rinsed in clean water before re-using. If the labels are heavy and springy, with a tendency to curl or blister, the tempering time should be prolonged until the stock reaches its maximum wet expansion and becomes limp. When the labels have been properly registered on the bottles, they should be wiped down with a soft damp rag. Wait until labels dry on the bottle before judging the job. Usually they shrink tightly into place when the moisture has evaporated.

Table Gumming Machines

For greater speed of labeling and versatility as regards the types of adhesives that can be used, hand or motor-driven table gummers are widely employed in labeling operations. The labels are fed into these gummers by hand and applied to the bottles by hand, usually followed by a wipe-down with a damp rag.

---

*The following information is reprinted by courtesy of National Adhesives Division of National Starch Products, Inc.

Generally the gummed-off labels are applied immediately to the bottles, so that relatively fast-setting glues can be used in this method of labeling. In fact, where embossed labels are to be applied, it is best to employ fast-setting adhesives that are relatively high in solids (low in water content) so that the paper will not be softened to the point where the embossing will be lost in the wipe-down action.

Semi-Automatic Labeling Machines

Whereas the bottles are fed into and removed from these machines by hand, the equipment automatically applies body and neck labels (also foil if desired), so that a greater speed of production can be obtained than in hand applications. There are two general types available, those which utilize a glued stencil to pick the labels out of the hopper and those which employ a vacuum pick-up. These labelers provide a greater glue mileage, with a resultant lower adhesive cost, and automatically ensure precise label registration. They offer great versatility in the handling of stocks and adhesives as well as bottle and label sizes. Moreover, they permit the efficient use of fast-setting glues so that properly labeled bottles may be nested immediately without the labels shifting.

Fully Automatic Labeling Machines

The highest production speed is provided by fully automatic labeling machines geared to the output of bottle filling and capping operations. These machines provide the lowest labor cost per unit of production and are usually tied in with automatic packing and case sealing equipment designed for mass production plants. They are available in several types: straight-away labelers (both single and double action), rotary labelers, and gang labelers accommodating as many as 10 bottles at a time. Depending upon the make of the machine, fully automatic labelers employ three basic gluing principles, a stencil transfer to the labels, a stencil transfer to the containers, or a direct glue roll application to the labels. Equipment of this type can be used to apply as many as four separate labels to the front and back of bottles simultaneously.

Package Selection

Glass bottles and jars should be designed so that they will be entirely suitable for labeling. A reputable glassware supplier should be consulted and a quality package specified. The walls of glass containers should be uniform in thickness, and the outer surface free from imperfections and irregularities. Shapes having a tendency to tip, turn, or wedge in the conveyor leading to automatic labeling machines should be avoided. To be properly labeled, glass containers should have no projecting fins, no pronounced bulges nor deep concave surfaces. In short, the size and shape of the package should be practical, both to produce and to handle in the labeling operation. Accordingly, if labels are to be applied automatically, the recommendations of the machine manufacturer should influence the selection of glass containers.

Label Selection

The following considerations should be borne in mind when designing a label:

1. Attractive form and design
2. Pleasing color or colors
3. Trade-mark and name
4. Description and uses
5. Directions for use
6. Cautions (if any)
7. Conformation with legal requirements

Examine as many labels as possible of the leading competitive products before deciding on actual design of a label.

The recommendations that follow offer some general guide rules, particularly to the inexperienced. It should not be inferred, however, that labels which do not conform strictly to these specifications will invariably prove troublesome. Successful labeling depends upon many factors and it would be impossible in a book of this type to cover completely all the aspects of label selection.

In the average plant, 50 lb. to 60 lb. stock (weight of a ream, 500 sheets, of paper, size 25 by 38 in., the basis commonly used by paper manufacturers and printers) is most generally and successfully used for bottle and jar labels. Heavier stocks may also be satisfactory, providing the label is not too heavily calendered, inked, or varnished. However, thinner translucent stocks allow the labeling adhesive to show through, giving the effect of discoloration or staining, so that while 45 lb. stock might be satisfactory, the results will depend upon the opacity of the sheet. A paper of moderate but uniform absorption qualities is preferred to either blottery or hard sized stocks. Occasionally, due to spotty sizing or to the offset of oily printing inks or varnish coatings, water will *crawl* on some sections of the label stock and soak readily into others. This uneven penetration of moisture may lead to blistering and destroy the adhesive film continuity essential to successful labeling.

The consensus of opinion among leading manufacturers of labeling machines, labels, and adhesives is that the grain direction of paper labels should be horizontal, running from side to side, parallel with the printing. To avoid difficulties in application, it is advisable to specify a horizontal grain direction when ordering labels.

As successful machine labeling requires that the size and shape of bottle and jar labels bear a proper relationship to the package, it is very important that label design first be discussed with the manufacturer of your labeling equipment; otherwise, proper label registration may be difficult to obtain, and the wiper or pressure action may be ineffectual at the edges of the labels.

Labels for angular bottles should not overlap a corner by a very slight margin, as that places too much tension on the glue at the very edge of the label. For this reason, it is well to avoid die-cut designs which leave points to be glued. On the other hand, tapered and other odd-shaped bottle surfaces are usually best labeled with die-cut stock designed to conform with the panel or container surface to which they are applied. Wherever possible, avoid combinations of heavy stocks, solid ink backgrounds, and heavy varnish or lacquer coatings.

Obviously, a wise precaution would be to patronize a reputable label manufacturer and to discuss your plans freely in advance with your machine and adhesive suppliers. If possible, labels comparable to those you are considering should be tested under actual operating conditions before ordering.

Anyone operating old, worn machines must be especially careful in the selection of labels, and should seriously consider

Adhesive Considerations

Labeling adhesive, to be truly efficient, must serve two important functions: first, they must apply the labels properly in the plant, keeping pace with production requirements; second, they must hold the labels securely in the field, where bottles and jars are subjected to a great variety of trying conditions, depending upon the nature of the product packed. Unless this two-fold olbigation is fulfilled by the adhesive, the packer's investment in labels, machines, labor, glue and, most important, in the brand advertising with which the labels are expected to tie up at the point of sale, may all be lost, along with the good will and good business of his customers.

Accordingly, only quality labeling adhesives should be used. Bear in mind that initial glue cost is not a true index of label gluing cost, and that a higher-priced adhesive will often prove more efficient and economical in the long run. For example, it may apply more labels per gallon than some cheaper grade, or ensure a higher labeling speed with fewer interruptions, or hold the labels more securely. Any one of these advantages would more than justify the purchase of a quality adhesive.

In the final analysis, glue, regardless of its selling price, is the smallest item of cost in the entire labeling set-up. On the other hand, it is of prime importance to the success of the whole labeling operation, and therefore deserves careful attention.

As the individual glue user cannot be expected to make a proper independent selection of labeling adhesives, it is best to rely upon the recommendation of a reputable adhesive manufacturer.

In order to recommend a satisfactory labeling glue without actually seeing the job, the adhesive supplier should be furnished with the following information:

1. Is the bottle labeled wet or dry? Hot or cold? Clean or greasy?
2. Is the bottle plain, stippled, frosted, flint glazed, or opal? Is the labeling surface flat, round, concave, convex, angullar, or tapering?
3. What labeling method or machines do you employ?
4. How many bottles per minute do you label on each machine?
5. In addition to body labels, do you apply neck labels? Foil to bottle tops?
6. What products are you packing in glass?
7. To what conditions will your products logically be subjected in the field? (Chilled in water coolers, stored in steaming bathrooms or unheated warehouses, shipped to hot, humid countries?)
8. Are the bottles to be returned for re-use? If so, what type of soaker do you operate? How many alkali compartments has it? What is the water temperature and caustic concentration of each compartment? What are your state regulations with regard to water temperature and caustic concentration?
9. What type of labels do you use and who supplies them? (Be sure to send a half dozen samples of each label to your glue supplier with the above information.)

Handling of Adhesives

Storing

All labeling adhesives should be stored at a moderate temperature (in warm locations in winter and cool locations in summer). Winter shipments should be brought to normal room temperature before diluting and using. Always use glue shipments in the order in which they are received, placing new supplies in the background so that the oldest batches will be consumed first. Keep glue containers covered when not in use. Place a piece of wet burlap or other heavy unstarched cloth under the cover of barrels and kegs after opening, to prevent the glue from drying out. It is best to tack the cloth to the inside of the cover to prevent it from coming into direct contact with the adhesvie. As metal drums permit air-tight closures, the use of wet cloths applies only to wooden or fibre containers.

Diluting

Follow the adhesive manufacturer's dilution instructions carefully, weighing both the glue and the water until the proper proportions are established and recorded in some convenient manner, such as the marking of the correct glue and water levels on mixing pails.

All glues should be agitated before they are diluted. Only tepid water should be used when water dilutions are called for, and mixtures should be made in a separate clean container before pouring the diluted glue into the glue pot of the labeling machine. Use a round mixing pail, as water collects in the corners of square cans and makes proper mixing difficult. This is one reason why diluting glue in the pot of a labeling machine is not recommended. Never pour the full complement of water into the glue before agitating. Instead, add the water slowly and mix in thoroughly. When diluting glue already in the glue pot of a labeling machine, best results are obtained by adding a thin solution of the same glue, prepared separately as previously described.

Running

When running fast-setting dextrin-type glues, it is well to keep the glue in the glue pot at a reasonably low level, so that frequent additions of fresh material at the proper consistency may be made, thus compensating regularly for evaporation, and minimizing the tendency of the glue "to skin over" in the pot. Always machine a labeling glue for five or ten minutes before starting to label, as this churning will smooth out the adhesive and permit it to follow the roller evenly.

General

Advise your glue supplier whenever a change is to be made in the type of label you use, the source of label supply, or method of labeling. Test the adhesion you are obtaining in your labeling operations frequently to make certain the glue and the application are proper. Never pour diluted or used glue back into the original glue container.

Regulation and Care of Machines

While the mechanics of this subject are fully covered by the machine manufacturers in their customer instruction books, a few salient recommendations can bear repetition, for the suggestions which follow tend to ensure proper labeling results from the standpoint of the most efficient use of adhesives. Of course, all bottle labeling machines do not operate on the same principle nor employ the same attachments, so that some of the parts discussed will not be found on certain makes and models.

Glue Roll

As accurate glue film regulation is essential to successful labeling, the glue roll must at all times be truly cylindrical. If the roll becomes unevenly worn due perhaps to carelessness in adjusting the film control mechanism, or if the label comb is permitted to rest too firmly upon the roll, thus etching its surface, the glue roll should be sent back to the mnufacturer of the equipment or to a reliable machine shop to be "turned down" on a lathe.

Scraper

The term *scraper*, as used here, refers to any mechanism by which the glue film may be regulated. To obtain the best results in labeling, the scraper should be adjusted close to the glue roll. In order to regulate a very thin, even film, the thumb screws should be adjusted to provide a uniform clearance throughout the width of the glue roll.

Too much glue is worse than not enough. A heavy application slows down the tack and setting seed of the glue, promotes the crystallization and resultant failure of ordinary types of dextrin glue, and leads to the smearing of bottles, as the excess glue will be squeezed out from under the label by the wipers or pressure pads.

When running jelly-type labeling gums at a heavy consistency in semi- or fully-automatic machines, it is necessary to use wing or "U" type scrapers designed to prevent cohesive materials of this type from climbing up on the axle of the glue roll and overflowing the glue pot. Such scrapers are standard equipment on most new machines, or can be obtained from the manufacturer of your machine.

TRANSFER ROLLER

If the surface of a rubber transfer roller becomes worn or uneven, prompt replacement is necessary; otherwise, an improper film will be transferred, with resultant labeling difficulty. Transfer rollers should be washed clean whenever the machine is shut down, even for a relatively short time. During idle periods, the transfer roller should not be left resting either on the glue roll or on the pickers. Failure to observe this rule may reuslt in tearing or loosening the rubber face of the transfer roller.

PICKERS

Label pickers should at all times be smooth, flat, and on the same plane so as to receive identical glue films from the transfer roller. The adjustment of pickers to accommodate labels of various sizes should be made only within the confines of their original design. Whenever widely different label sizes and shapes are used, special sets of suitable pickers should be ordered from the machine manufac turer.

WIPERS, PRESSURE PADS, ETC

Wiper pressure should be regulated to apply the maximum force that will not tear or smudge the label or destroy its embossing. Pressure wipers should be adjusted to be effective over the entire surface of all labels. Brushes, rubber wipers, and pressure pads should be cleaned frequently with warm water. If labeling is to be resumed promptly, dusting the wiper surfaces with talcum powder will prevent their smearing or engaging the labels. Flexible-type wipers characteristic of some semi-automatic labelers should be replaced when they have lost their ability to snap back into proper wiping position. Brushes, too, should be replaced when the bristles become worn, spread, or thinned out.

LABEL HOPPER

The prongs of the label hopper should be adjusted to allow the label stack to ride freely and to lie flat against the prong hooks. The hooks should be kept clean and smooth at all times. The quantity of labels in the label hopper should be sufficient to provide proper spring tension on the follower plate, which, in turn, should be of a size that will insure an even, over-all pressure of the label stack against the prong hooks.

CLEANING

All gluing parts of bottle labeling machines, including the glue pot and the label box prongs, should be cleaned frequently with hot water. The glue pot should be emptied and thoroughly washed before any prolonged shutdowns, such as at the close of the working day, and especially at the end of the working week. However, where this is impractical during the week, a heavy piece of wet cloth should be placed over the glue pot, after all rollers, scrapers, pickers, etc., have been washed clean and dried. Always empty the glue pot and clean the machine thoroughly before changing from one type or brand of glue to another. Never use a knife or sharp instrument to clean the glue from the rollers.

OILING

Labeling equipment should be oiled and greased in the manner and frequency prescribed by the manufacturers of the machine. Again, we recommend that the machinery instruction book be read frequently so that no important detail that may lead to inefficient operation will be overlooked. Care should be taken to keep lubricants away from rubber parts of the machine (rollers, wipers, etc.), and from contaminating the glue.

Labeling Difficulties

In order to fully understand the suggestions which follow, be sure to read the preceding pages first. Not all the listed *causes* of difficulty will apply in every case, so that the reader must decide, after a careful investigation of the possibilities, which of the *recommendations* should be tested. In all events, before changing the adjustment of any labeling machine mechanism, be sure the original position is marked, so that, if necessary, the unit may be easily returned to its proper place. While the treatment of each subject in this section is not complete, an effort has been made to cover all the more common conditions that arise in the various industries labeling glass containers.

CURLIING OF LABELS AWAY FROM BOTTLES

Causes	*Recommendations*
1. Application of too much glue.	1. Apply the thinnest possible even film by adjusting the scraper close to the glue roll.
2. Inefficient wiper action.	2. Adjust wipers to apply proper pressure, slightly beyond all

Causes	*Recommendations*
	edges of the labels.
3. Use of stiff, springy, labels that are vertically grained.	3. (a) "Roll" or flex the labels before running them, by bending label stacks sharply in the direction of application. make sure, however, that they lie flat in the babel hopper of the machine. (b) Humidify or stipple the labels before gluing. (c) If hand labeling, prolong the tempering time until the stock is limp. (d) Specify labels with a horizontal grain direction for machine application.
4. Use of glue with insufficient tack.	4. (a) Apply the glue concentrated, adding only the amount of water prescribed for proper machining. (b) Inquire of glue supplier for a suitable grade of a tackier glule.
5. Labeling wet bottles.	5. (a) Humidify or stipple the labels. (b) Change to a jelly-type labeling gum. (c) Install an air blower on the conveyor line to the labeling machine.

WRINKLING OR BLISTERING OF LABELS

Causes	*Recommendations*
1. Use of a glass package having a two directional curve of the labeling surface.	1. (a) Design lables to conform as snugly as possible with the contour of the labeling surface. (b) Humidify or stipple labels applied automatically or prolong the tempering time of

Causes	*Recommendations*
	labels applied by hand. (c) Wherever possible, design the package so that the labeling surface curves in but one direction.
2. Application of too much glue.	2. Apply the thinnest possible even glue film by adjusting the scraper close to the glue roll.
3. Use of vertically grained labels.	3. (a) Humidify or stipple the labels before gluing. (b) Specify horizontal graining on future label orders.
4. Uneven absorbency of label paper, due to spotty sizing or offset of inks and varnish coatings.	4. (a) Humidify the labels before gluing. (b) Use a slow-drying labeling glue to provide time for uniform penetration of the stock.
5. Presence of too much moisture in machine labeling glue.	5. Use a labeling adhesive with high solids (low water content), and run as concentrated as possible.
6. Use of fast-drying glue on labels of high moisture sensitivity.	6. (a) Use a tacky but slowdrying glue to permit shrinkage of the labels as moisture evaporates. (b) If labels have been stored in a warm, dry location and have consequently dried out, they should be humidified before running.
7. Applying glue to all but a small portion of the label.	7. Use picker plates or center gummer pads of the proper size for the labels being applied by machines operating on the stencil principle.
8. Inefficient wiper action.	8. Adjust wipers to apply proper and even pressure throughout the label area.

Causes	*Recommedations*
9. Labeling wet bottles.	9. (a) Humidify the labels. (b) Change to a jelly-type labeling gum. (c) Install an air-blower on the conveyor line to the labeling machine.
10. Glue on grip-finger of the machine.	10. Wash grip-finger thoroughly and dry.

FAILURE TO PICK LABELS OUT OF HOPPER

Causes	*Recommendations*
1. Lack of proper contact between pickers, pads, or bottles and labels.	1. (a) Check that label follower is riding freely and the follower plate is covering the entire label area. (b) Adjust pickers to the same plane or level.
2. Application of inadequate or excessive glue to the pickers, pads, labels, or bottles.	2. Adjust scraper to apply a thin, even glue film.
3. Use of glue lacking sufficient initial tack to pick labels out of hopper.	3. (a) Run glue concentrated, applying a minimum even film. (b) Consult adhesive supplier for tackier label glue.
4. Too tight gripping of labels by prongs or hooks of label hopper.	4. (a) Wash dried glue from prongs and hooks. (b) If prong hooks are rough, slant and smooth with fine emery cloth. (c) Check adjustment of prongs so that the stack rides freely.
5. Insufficient supply of labels in the label hopper.	5. Maintain a cushion of labels in the hopper to insure a proper spring tension on the follower plate.

Spotty Adhesion

Causes	Recommendations
1. Use of pickers which are not on the same plane or are bent.	1. Level the pickers to the same plane, and work out any surface irregularities.
2. Lack of proper contact between transfer roller and glue roll.	2. Adjust the glue roll or transfer roller to provide a flush meeting.
3. Uneven transfer of labeling glue. glue.	3. (a) If the glue is too viscous to transfer evenly, dilute with a thin mixture of the same glue. (b) Adjust the scraper to provide a uniform clearance of the glue roll.
4. Use of labels having uneven absorption qualities.	4. (a) Use a suction-type labeling gum, or a fast-tacking, slow-drying glue. (b) Humidify the labels. (c) Switch to a uniformly absorbent label stock.
5. No thorough incorporation of water into the glue.	5. (a) Add water slowly and mix into glue thoroughly before placing in the glue pot. (b) When diluting glue already in the pot of the machine, use a thin glue mixture and stir in well.

Tearing of Labels in the Machine

Causes	Recommendations
1. Accumulation of glue on the transfer rollers, pickers, prongs, wipers, etc.	1. (a) Wash all gluing parts with warm water. (b) Dilute the glue as prescribed by the supplier. (c) Keep the machine in steady operation, or clean all gluing parts if labeling is interrupted.
2. Inadequate distribution of glue film to one or both picker fin-	2. (a) Adjust pickers to the same plane or level, so that

Causes	Recommendations
gers or plates.	both will receive identical glue films. (b) Adjust scraper to feed a slightly heavier film.
3. Faulty adjustment of label hopper unit.	3. (a) Adjust prongs so that label stack rides freely. (b) See that there is sufficient spring tension on the label follower, and that follower plate presses labels evenly but not too tightly against prong hooks. (c) Slant and smooth label prong hooks with fine emery cloth.
4. Engaging of labels by wipers.	4. Clean the wipers with warm water to remove accumulated and hardened glue.

FALLING OF LABELS OFF BOTTLES

Causes	Recommendations
1. Application of too much glue.	1. (a) Dilute the glue as instructed and tighten the scraper close to glue roll. (b) Test a label glue that dries to a plastic film.
2. Use of the wrong type of glue for nature or condition of the labeling surface.	2. If labeling greasy containers, or stippled bottles, consult- your adhesive supplier for a suitable glue.
3. Use of hard-sized resistant paper stock.	3. (a) Use a tacky, slow-drying label glue, or a gum that adheres by suction instead of penetration. (b) Change to a softer, more absorbent stock.
4. Remoistening of label glue on bottles stored in a damp location.	4. Use a non-hygroscopic (moisture-resistant) labeling adhesive.
5. Partial drying of glue on labels	5. (a) Apply labels to bottles

Causes	*Recommendations*
before they are applied to bottles (most often experienced in delayed table gumming operations).	promptly after gluing. (b) Use slower-drying labeling adhesive.
6. Inadequate glue application (rare).	6. (a) See that film control unit is properly adjusted. (b) Refill glue pot with properly diluted glue (incorporate prescribed amount of water thoroughly).

DISCOLORATION OR STAINING OF LABELS

Causes	*Recommendations*
1. Use of label stocks colored with water-sensitive or alkali-sensitive dyes.	1. (a) If plain water causes discoloration, consult the label supplier, as all commonly used labeling glues are of an aqueous nature. (b) If the color does not change with plain water but does with the type of glue you are using, test a chemically neutral grade.
2. Use of thin, translucent labels.	2. (a) Test a light-colored or transparent labeling adhesive. (b) Change to heavier, more opaque label stock.
3. Use of highly absorbent labels.	3. (a) Test a cohesive labeling gum that relies for adhesion on suction rather than penetration. (b) Change to a moderately hard sized and calendered stock.
4. The falling of oil or grease on bottles in filling or capping machines.	4. Clean fillers and cappers, and make necessary repairs or adjustments to prevent a recurrence.
5. Use of glue contaminated by foreign matter.	5. See *Molding of Labeled Bottles.*

Causes	*Recommendations*
6. Use of hygroscopic (moisture-absorbent) adhesive.	6. Switch to a non-hygroscopic (moisture-resistant) adhesive.
7. Application of too much glue.	7. (a) Clean wipers, brushes, and pressure pads. (b) Tighten the scraper close to the glue roll.

SMEARING OF BOTTLES OR LABELS

Causes	*Recommendations*
1. Application of too much glue.	1 Tighten the scraper close to the glue roll.
2. Excessive pressure of flap type wipers.	2. Lighten the pressure slightly, making sure that enough is exerted to prevent labels from curling away after application.
3. Use of a glue that dries to a cloudy film.	3. Test adhesives that dry to a clear film.
4. Slipping of labels.	4. See *Crooked Labeling.*
5. Use of freshly printed labels.	5. Age labels until inks are perfectly dry.
6. Use of poor quality ink in printing labels.	6. If ink smears on labels with normal handling, consult label supplier.

CROOKED LABELING

Causes	*Recommendations*
1. Uneven application of glue to labels.	1. (a) True the scraper or twin rolls to regulate a uniform glue film. (b) Check that pickers are on the same plane and smooth out any hollow spots on the face of pickers. (c) Check that transfer roller meets glue roll flush (see machinery instruction book for adjustments). (d) Use pickers of proper size for labels being run.

Causes	*Recommendations*
2. Excessive glue application.	2. Tighten the scraper close to the glue roll.
3. Lack of firm hold of labels by grip-finger.	3. Adjust grip-finger lever to exert just enough pressure to prevent wipers from shifting position of labels on the bottles.
4. Uneven pressure of flap type wipers.	4. Check wiper action to be sure of equalization of pressure at both ends of label.
5. Rough or gummed-up prongs on label holder.	5. Clean label holder prongs with a damp cloth and smooth gripper hooks with fine emery cloth.
6. Improper registration of labels.	6. (a) Align label holder to bring it in proper position and, if more than one, in proper relation to each other. (b) Square off bottle-rest mechanism.
7. Use of glue with inadequate tack.	7. (a) Run the glue as concentrated as possible. (b) Test glues with higher initial tack.

SCUFFLING OF LABELS WHEN NESTING

Causes	*Recommendations*
1. Application of too much glue.	1. Dilute the glue to the extent prescribed by the manufacturer, and apply the thinnest even film you can regulate by close scraper adjustment.
2. Use of a slow-setting glue.	2. Usually the setting speed of the glue can be increased by running at a higher concentration and applying sparingly, but if this fails, consult your adhesive supplier regarding faster-setting grades.
3. Scraping of freshly labeled bot-	3. (a) Check that the dimensions

Causes	*Recommendations*
tles while nesting.	of the bottle nest are correct for the size of the bottle. (b) If nesting by hand, slide bottles into compartments so that labeled area does not contact partitions or lower the case so the bottles can be dropped straight into the nest. (c) If nesting by machine, use proper grid, funnel, and guiding fingers for size of container being packed. Clean and adjust guiding fingers of boxing unit to prevent their engaging the labels. Replace broken guiding fingers.
4. Faulty wiper action.	4. Adjust wipers to apply proper pressure slightly beyond all edges of the labels.
5. Wrinkling or blistering of labels.	5. See *Wrinkling or Blistering of Labels.*

LOSS OF LABELS IN WATER COOLERS

Causes	*Recommendations*
1. Too high temperature of water in the cooler.	1. (a) Add ice more frequently or adjust the refrigerating unit to maintain a lower temperature. (b) Drain water-coolers, if left unattended. (c) Clean the cooling unit frequently and remove labels which have soaked off bottles.
2. Use of a water-absorbent label stock.	2. The face of bottle labels subject to liquid refrigeration should be rendered water-resistant by sizing, solid ink

Causes	*Recommendations*
	backgrounds, varnish coatings, etc., (e.g., foil-laminated labels are highly water-repellent).
3. Use of stippled or other decoratively molded bottles.	3. For best results, use a bottle providing a smooth labeling surface.
4. Use of a readily water-soluble adhesive.	4. Switch to a cold-water-resisant labeling glue.
5. Application of too much glue.	5. To ensure a close contact between the label and the bottle, regulate a very thin, even glue application.
6. Inadequate glue area on labels.	6. Increase glue area of labels by adjusting or substituting broader-faced pickers.

Difficulty in Removing Labels

(Of interest chiefly to soft drink and beer bottlers)

Causes	*Recommendations*
1. Faulty soaker action.	1. (a) Have soaker inspected and overhauled. (b) See that a sufficiently high water temperature and caustic concentration are maintained (must comply with state regulations).
2. Use of a warm-water-resistant glue.	2. Change to a more heat-sensitive type of glue.
3. Use of a highly water-repellent label.	3. (a) Test other cleaning compounds. (b) Consider the possibility of substituting a less resistant type of label stock, bearing in mind the greater importance of retaining labels in the field than of simplifying their removal in the plant.
4. Application of glue to an excessive label area.	4. Reduce glue area on labels by adjusting the pickers or by

Causes	*Recommendations*
	substituting pickers with a narrower gluing surface, making sure, however, to avoid extremes which would sacrifice field efficiency.

FOAMING IN THE SOAKER

Causes	*Recommendations*
1. Formation of soaps from reaction of bottle cleaning solution with (a) contamination on bottles (b) sizing, printing, and finishing of labels.	1. (a) Provide an efficient prerinse. (b) Dump label basket or screens frequently. (c) Change the soaker solution more often. (d) Test other cleaning agents. (e) Test lower caustic concentrations and water temperatures (not below legal minimums). (f) Test other types of labels.
2. Application of too much glue to labels.	2. Tighten scraper of bottle labeling machine.
3. Mechanical reasons such as excessive pressure of circulation pump, leaks on intake side of pump, insufficient tank height to provide room for blanket of foam.	3. Consult soaker manufacturer regarding practical correctives.
4. Use of soft water.	4. There is no direct corrective for this, as artificial means of hardening water seldom prove successful.
5. Fact that solution level in tank is below intake level.	5. Fill tank to proper level and adjust caustic to compensate.

MOLDING OF LABELED BOTTLES

Causes	*Recommendations*
1. Spillage of bottle contents at filling machine.	1. (a) Repair filling machine to ensure its proper functioning. (b) Investigate causes of

Causes	*Recommendations*
	forced stops at filler.
2. Contamination of the labeling glue.	2. (a) Clean all metal gluing parts of the machine (glue pot, roll, scraper, pickers, etc.), as well as mixing paddles, scoops, glue pails, etc., with boiling water and antiseptic. Rubber rollers and wipers should be washed with luke-warm water. (b) Use only clean water for diluting the glue. (c) Keep glue barrel covered when not in use. (d) Consult glue supplier if trouble persists.
3. Storage of labeled bottles in a damp location.	3. Store labeled bottles in a clean, dry location.
4. Inefficient cleaning action of bottle soaker.	4. Clean the soaker with the frequency prescribed by the manufacturer of the unit. See also *Foaming in the Soaker*.

Label Pasting Machine*

This machine applies any type of adhesive to labels, tags, markers or any form of printed matter that must be placed on package or product by hand.

It is operated by an electric motor and the labels are fed in to rollers which place the right amount of adhesive on the label with no excess to smear and dirty the package. This means a saving of ordinarily wasted glue.

Very small labels up to those 8 in. wide are handled with equal ease, and they may be of paper, cloth or cellophane to be placed on paper, wood, cloth, cellophane, glass or metal packages. Gummed stock costs are saved.

One operator feeding the machine can keep several others busy

*The following information is printed by courtesy of Alsop Engineering Corporation.

attaching labels or one person can do the whole job.

A continuous strip label would be gummed at the rate of nearly 100 ft. per min. A 2 in. label in 1/10 sec., a 6 in. label in 1/3 sec., depending upon the thickness of glue and type of label.

RECEIVING AND SHIPPING

Receiving

An entry should be made of all shipments received, as soon as received. Any evidence of damage should be noted. Containers should be weighed and any difference from the shipper's weight should be recorded. Comparison should be made with a copy of the purchase order checking the amount, packing, shipping expense, or other items.

Packing for Shipment*

Cartons

Regardless of the care with which corrugated boxes are designed and manufactured, they are no stronger than their seal. For proper protection they must be closed tightly and securely. That is the only way to insure against loss and damage to merchandise in transit.

Transportation systems exercise normal care in handling so that merchandise is safe, but they cannot do your sealing. Proper sealing is a job to be done only by manufacturers' shipping departments. It is a responsibility to customers; it is part of your service and selling program.

The first saving effected by proper sealing results from the reduction in damaged and lost merchandise. Proper sealing utilizes to the fullest degree the protection built into corrugated boxes, enables them to do the job for which they are designed, and delivers merchandise in as perfect condition as when it left the factory. Customers receive goods as ordered, without delay, reordering, or other bothersome circumstances. Good-will is retained. Costs are cut.

The second saving which results from proper sealing is one of materials. With a volume of shipping, hundreds of dollars can be lost in the use of excess tape, staples, wire, glue and other materials squandered by hit-or-miss sealing methods. Proper sealing methods save

---

*The following information is reprinted by courtesy of The Hinde & Dauch Paper Co.

money for you.

With properly standardized sealing methods there is invariably an appreciable saving in labor time. Unnecessary operations are eliminated, each worker knows the most efficient and quickest way to do a good job. Operations move even more rapidly as familiarity with these operations increases. Minutes saved add up to man-hours. Packaging moves along more efficiently in less time; merchandise receives better protection. You save money.

Finally, as is the case with most packaging operations, the proper way to seal is by far the easiset way. Once the habit of proper sealing is formed, corrugated shipping boxes automatically provide all the protection that is built into them. It takes less time to seal properly, it is much easier, it is more profitable. That is why it will pay every one who ships in corrugated boxes to review the sealing hints presented here, to adopt those methods most applicable to his packaging operations or methods.

Sealing with Adhesives

A quick-setting, firm-holding adhesive should be applied to the inner flaps with a wide, smooth, clean brush. The top area of the inner flap should be completely covered. "Spot" sealing is usually satisfactory for interplant shipments, and in those cases where boxes must be opened for inspection of contents prior to final shipment. After application of adhesive, top flaps should be folded down and pressed firmly together. Adjacent ends should abut with no open spaces. Apply firm steady pressure sufficiently long to insure a good set. In using automatic equipment, make sure it is adjusted so that final result is a squared-up box. Store the adhesive at normal room temperature and keep pans and brushes clean.

Sealing with Tape

Use a good quality, wide, gummed tape that meets all classification requirements. To seal, cover center of the box with a strip of tape extending 2-1/2 in. or more over ends of the box. Then cover end seams with strips extending 2-1/2 in. or more past corners on either side.

Measuring and cutting tape to length is simplified by an efficient machine with measuring and cut-off devices. Tape should be applied firmly, with steady pressure to assure continuous contact.

To counteract extreme drying-out of gummed tape supplies resulting from storage in rooms which are too hot or too dry, a small amount of vinegar should be added to the moistening water.

Sealing with Metal Staples or Stitches

This type of closure is specified in both Consolidated Freight and Official Express Classifications. Staples or stitches are of two types; the first is a continuous spool of wire from which machines draw the wire to form each stitch or staple; the second is a series of pre-formed staples. Both types may be applied by automatic machines, both give a secure, satisfactory closure. Boxes should be stitched across both ends and along the center seam.

Staples (or stitches) should be broad enough and long enough to clinch securely, and should be used in sufficient quantity, properly spaced. For freight shipments, rivets, stitches and staples on flaps must not be more than 2-1/2 in. apart; for express shipments, not more than 2 in. apart.

Sealing with Metal Straps or Wires

For export shipments, material packed in bales or bundles, merchandise which is unusually large or heavy, or when the value of contents justifies added precaution against pilferage, sealing with metal straps or wires is advisable.

For best results, use a good grade of strap or wire. Sealing equipment should be modern and adaptable, with devices to tension, seal and cut quickly, easily and accurately.

Straps should be drawn sufficiently tight to indent but not break corrugated board. It may be advisable to reinforce the edges of the box underneath the straps with pads or with metal protectors made for that purpose. The size of straps should be between 1/4 by 0.015 in. and 1/2 by 0.015 in. However, for very heavy or bulky packages, a size as large as 3/4 by 0.015 in. may be advisable.

While the four most commonly used sealing methods have been described as they best apply to certain styles of corrugated boxes, they are to a certain extent interchangeable. The degree of interchangeability depends upon contents, type of box, carrier and distance of shipment.

Whether you use only one or all four of the sealing methods here described, you will save money by studying railroad, express and postal regulations governing shipments, and by observing regulations imposed upon different types of packages. By sealing properly and in accordance with those regulations, your packages will not only be acceptable for shipment, but will earn for you the most economical shipping charges. For example, in shipping by express, savings can sometimes be effected by strapping or wiring together packages "forwarded by one shipper at the same time to one consignee."

Reference to three publications, *Consolidated Freight Classifications, Official Express Classifications* and *Postal Laws and Regulations,* will provide you with the necessary data for economical shipping by carrier standards. Copies of these publications should be kept handy in every shipping department and consulted whenever there is any doubt about proper classification.

Stacking and Loading Cartons

You save money by reducing loss on damaged merchandise when both stacking and loading operations are carried out efficiently.

With warehouse and freight space at a premium, it is imperative in the interests of economy that stacking and loading methods make full use of every cubic foot of storage space. A number of additional boxes in every car load may not appear in the profit and loss statement, but these boxes over a period of time add up to real money. A little care and planning in stacking and loading will expand storage space at no cost and provide free freight car and warehouse space.

Greater speed in loading and unloading from warehouse to freight cars and return is another by-product of efficient stacking and loading. As handling speed increases, you automatically increase storage space and reduce costs. The easiest, most simple way to increase that speed is to stack and load efficiently. It is a matter of methods, not machines.

Finally, by proper stacking and loading you gain full advantage of the designs and printing which characterize modern corrugated shipping boxes. Your stacks of merchandise become billboards, your freight loads become traveling salesmen. You advertise in transit.

Before loading freight cars or warehouse, pools of oil and water should be removed or covered with sheets of protective paper. Protruding nails should be removed or flattened so there is no danger of their ripping boxes and damaging merchandise.

More economical use of space is obtained and easier loading facilitated by estimating the number of boxes that available space will hold.

When loading boxes of uniform size, the easiest, quickest method of estimating is to place one row of boxes lengthwise in the space, then another row crosswise. Multiplying the number of boxes in the long row by those in the cross row and dividing this figure into the total number of boxes to be loaded will give the number of rows high you must stack to accommodate the boxes.

Conversely, multiplying the figures obtained by laying one row lengthwise, one row crosswise, and piling one stack as high as possible, quickly determines how many boxes a given space will contain.

When the inner-end flaps of boxes do not meet, and since outside flaps do meet, it is advisable to load the car with seams running crosswise, thus providing extra protection against end shock.

Each box should rest on two boxes below it. When the boxes do not fill the full width of the car, but the end box must not hang over more than 3 in. Likewise, incomplete layers of boxes should be held in place by an "incomplete layer brace" supporting crosswise rows of boxes. Incomplete layers remain in position, shifting of load and consequent damage to merchandise is prevented.

Tipping concentrates weight on one edge of the bottom box and can cause collapse of the entire stack with consequent damage to merchandise, confusion and jamming up of space. It can be avoided by stacking vertically so that the weight of each box is supported evenly by the box below, with the center of balance of the entire stack distributed to make full use of the corrugated boxes' ruggedness.

To minimize the danger of crushing bottom boxes, heavy loads can be distributed by interlocking rectangular-shaped boxes or staggering square boxes. In addition, these simple devices automatically *tie* stacks into stable units. To inter-lock, place one row of boxes end to end: the row above, side to side. The same effect is achieved by staggering square boxes so that boxes in the second row are centered at the juncture of the 2 boxes immediately below and above.

Both ease of identification and advertising value are attained by stacking boxes right side up so that both name and contents can be easily read.

As much as possible, the same size boxes should be stacked and loaded together in sections. So doing speeds handling, permits full, efficient utization of space and simplifies identification of merchandise.

In loading freight cars, operations are speeded by loading from ends to center tightly. Use of jacks will help keep loads tight against ends, while center bulkheads should be used to fill any space remaining in the center. These precautions will give you tightly packed loads with a minimum of shifting.

Boxes filling door space of freight cars should be protected with corrugated, fibre, chip-board or kraft paper so as to distribute load force against steel-strap door-bracing. For weather-stripping, use several thicknesses of heavy waterresistant paper and place next to steel strapping so that all boxes are protected against weather and dirt.

SLACK BARRELS*

Opening and Closing

Inspect both ends of the empty barrel before loading, and if the chimes (the ends of the barrel extending above the heads.) are the same thickness, take out the head with the least number of pieces.

If one chime is thick and the other chime is thin, take out the head in the end with the thinner chime. A thick chime usually denotes the bottom of the barrel.

If the barrel is to carry extra-heavy weight or is to be used for export, consult your local cooper.

After selecting the end of the empty barrel to be opened, turn it bottom side up and seal the bottom head. Fastening should be made by either of the following methods:

Drive nails through the hoop and staves horizontally into the head, spaced about 6 in. apart. Be sure that there is at least one nail at each end of each cant. Nails about the size of 3D common nails should be used.

Secure the head by stapling with a hand stapling machine. Staples should be not more than 8 in. apart, with at lesat one staple at each end of each cant.

Test the bilge hoops for tightness. If at all loose, drive them tight with hammer and redrive and replace hoop fasteners.

Release the top head hoop by tapping gently upward with a hammer until about half the hoop is above the edge of the chime. This will loosen the staves so that a light blow on one side of the head will drop it to the bottom of the barrel. The head then should be removed and laid in a clean place. Now drive the hoop down again level with the chime.

If a hickory catch ring is available, it can be driven down below the head hoop to take the place of the head hoop, which can be removed entirely.

Do not allow a barrel to stand for any length of time with head hoops removed. A barrel with the head removed may collect dust and dirt.

Closing One-piece Head

If a bag liner is used, read the instructions on bag liners in a following section. After the barrel is filled and the bag liner, if any, tied,

*The following information is reprinted by courtesy of The Associated Cooperage Industries of America, Inc.

replace the head, taking care that the liner is pressed down out of the croze so that the head will not pinch or puncture the liner.

Drive the head hoop upward about half its width, to release the staves. Place one edge of the head in croze and press or tap down lightly with a hammer until the head is seated in croze all around the barrel.

As the head slides into place, keep tapping top hoop with hammer gently around the barrel. Avoid driving hoop down tightly at any one place, but endeavor to bring it down evenly all around the circumference of the barrel.

When the head hoop is finally level with the end of the staves, the barrel is ready to have the hoops and head nailed or stapled.

Closing Multiple Head

If the head has several pieces, insert the middle cant (or piece) first, tapping the hoop down lightly. Then insert the larger side cant, saving the narrower side cant for the last.

If a side cant fills about one-half the barrle insert it first; then the middle cant, and the smaller side cant last. If there are only two pieces in a head, insert the wider cant first, and the narrower cant last.

While the cants are being placed, keep tapping the head hoop down all around the barrel until the head is snugly in place and the bead (or edge) of the hoop is level with the top of the chime. The barrel is now ready for nailing or stapling.

Inserting the Bag Liner

The bag liner is a bag that fits inside a slack barrel to protect the contents from moisture and sifting. It is best inserted with an inserting device, which consists of a cylinder on a handle, and which may be borrowed from liner manufacturers.

Place the inserting device into the bag liner as far as it will go. Then fold the ears of the bottom of the bag liner inwardly and insert the device and the bag liner into the upright barrel. Leaving the inserting device in the barrel, fold the ends of the liner that protrude loosely over the chime of the barrel. Do not fold the liner down tight against the top hoop, but leave a space of several inches between the fold of the liner and the chime of the barrel. The weight of the material used of fill the barrel will draw the liner down.

Now remove the inserter from the barrel, making sure that the liner lies smooth in the bottom of the barrel. The barrel is now ready to fill. After filling, gather the liner together over the contents, and force out the air. Tie the liner together as close to the *top* as possible. This is important. If you tie the liner near the contents, it will leave a large "ear" sticking up and the liner will billow.

You are now ready to head up the barrel. First, however, be sure that the liner is clear from the head, so you will not pinch it in the croze.

Proceed to head up the barrel and then nail or staple the head.

Hand stapling machines drive staples accurately through the hoop and stave into the head. They are more accurate than hand nailing, faster, and have less tendency to split heads. Staples may be extracted with a screw driver, and the heads removed intact. Stapling also prevents injuries on both the shipping and receiving ends, as the staples do not protrude on the inside of the package.

The following precautions are to be observed.

Do not knock a head in with an axe or hammer. You are damaging a valuable container.

Do not hammer and pound the hoops as hard as you can. Tap lightly and keep going around the barrel.

Do not try to insert bag liners without reading the instructions on bag liners.

Do not try to open and close barrels without the proper tools. Handy tools include a cooper's hammer, a hoop driver, a catch ring and a chime maul. A hand stapling machine is very valuable for stapling heads. If you use bag liners, you will need bag liner inserters. An investment of a few dollars in tools may save you many dollars in labor and in the resale value of your containers.

Handling and Storing

Because of its *double-arch* construction, the wooden barrel has rightfully been called *the container on wheels.* Due to its bilge, it rests on a small surface, reducing friction to a minimum. On its side, it responds readily to a push in any direction. Rolled, it requires little effort. On an incline, it moves of its own weight.

Never roll a barrel on the chime. Roll it on its bilge or side. To upend a loaded barrel, grasp it by the chime and rock it back and forth until you can rock it into an upright position.

Mechanical equipment for handling barrels includes hand trucks, power trucks, power conveyors, lifts, elevators and cranes. Inclined tracks, down which the barrel rolls of its own weight, are used in many industries.

Never store slack barrels on the bilge, whether empty or filled. Store them on end. Store them in a dry place, but not one that is overheated. Never tier barrels more than two or three tiers in height, depending on duration of storage and weight of contents.

If filled barrels are stored for a considerable length of time, check the bilge hoops occasionally to see if they are tight. Redrive them if loose, and fasten them just before shipment is made.

To lift a loaded barrel by hand, stand close to the barrel, feet close together, knees bent. Grasp the barrel firmly. Let the leg muscles do most of the work in the early stages of the lift, until the back and arm muscles assume the load.

For loading on shipboard, use chime hooks or slings. Use two barrels to a sling, whit an additional half turn around one of the barrels to prevent slipping. Never use loading nets.

Never let a loaded barrel drop, even for a few inches. When unloading from a car or truck, use a skid and ease the barrels down it. Do not permit them to smash into other barrels at the bottom.

Tight Barrels

Do not open a tight barrel to fill it unless absolutely necessary. Fill it through the bunghole when possible. If it is necessary to fill through the head, first select the head to be removed. With a piece of chalk, make a mark on the head and extend it up onto the chime, to mark the original position of the head.

Opening

Remove the top and second hoops by tapping them with a hoop driver and hammer. Loosen bilge hoop slightly if necessary. Push head firmly on side opposite the bunghole until it comes out of the croze. A head-puller can be used to remove the head.

Replace head hoop firmly. Keep head off the floor and out of dirt. Tighten quarter hoop slightly. Now drive all hoops firmly in place, using hoop driver and hammer, and tapping them evenly all around the barrel. When the hoops are tight, the barrel is ready for filling.

An alternative method of opening a barrel is to remove the top hoop only. Then loosen the quarter hoop and loosen the bilge hoop slightly if necessary.

Filling

Tight barrels are designed to be filled with liquids through the bunghole. Do not fill barrel full to the bunghole, but leave space for expansion of contents. When filled, insert the bung with the grain of bung running parallel to the grain of the bungastave or head. If filling through the open head, pour the product into the barrel, leaving room for expansion and clearance of head.

Closing

To replace the head, remove the head hoop and loosen second hoop slightly if necessary. Snap the head into the croze, lining up chalk marks. Be sure the chalk mark on the chime squarely meets the chalk mark on the head. Replace the hoops and drive them tight with hoop driver and hammer, working evenly around each hoop. Be sure all hoops are firmly driven.

Removing Bung

To remove the bung of a filled barrel, use a wooden bungstarter or wooden mallet. Tap sharply on the bung-staves on both sides of the bung until it pops out. Do not drive the bung into the barrel. If it is difficult to remove, use pick or chisel.

Emptying Contents

When removing liquids from a barrel through the bunghole, support the barrel on its bilge on a skid or rack so that the liquid may flow by gravity. If speedy flow is desired, a small vent hole can be made in one of the heads. If the vent is bored, do not let the chips get into the liquid. A satisfactory vent usually can be made with a 10D nail, driven in and then pulled out.

Avoid boring holes in barrels if possible. It lessens the reuse or resale value of the barrel. Let the cooper provide the necessary openings, properly plugged.

To empty the barrel through the head, set barrel up on one end, then bore an opening in the head diametrically opposite the bilge bung (that is, as far away from the bilge bung as possible). Insert a tapered spigot, faucet or gate. Lower the barrel so that it rests on its bilge, with the bilge bung on top. The bilge bung then can be loosened to provide a vent to speed the flow of the liquid through the faucet. In boring a bunghole in the head, bore it in the center of a head piece, never at a joint. The hole should be at least 1 in. away from the chime. Care should be taken that chips or shavings do not get into the product.

To pump a product from a barrel, stand the barrel on its head and bore an opening in the upper head for the pump.

Various devices are available for the filling and emptying of barrels. In many plants, fillers at the end of the production line pack products directly into the barrel. Spigots, faucets, gates and pumps are available for emptying barrels. Emptying machinery includes elevators and cranes, with tilting devices, which empty barrels into receptacles on the production line.

Storing

Never store a tight barrel on end. Store it on its bilge, or side always with the bungstave up.

The best method of storing tight barrels is on racks. These are available in wood or metal. Racks 2 barrels wide are best, with loading aisles from 4 to 6 ft. wide.

When storing without racks, tier barrels on the bilge on skids, not more than four high. Skids of 2×4-inch lumber are satisfactory. The inside edge of the skids should rest on the outside edge of the bilge hoops on both ends of the barrel. Tie skids together to prevent spreading.

Do not store barrels in the open, or in overheated warehouses. Inspect barrels periodically for leaking. If necessary, drive hoops tighter.

Plugging Holes

To plug worm holes in barrels, obtain from your cooper a quantity of shoe pegs (wooden plugs). Drive a sharp-pointed punch into the center of the worm hole approximately 3/8 in. deep, being careful not to go all the way through. Drive a plug into the hole until the plug is tight. Cut off the remainder of plug flush with the surface with a sharp knife or coopers' adze and sandpaper the end of the plug to level it off smooth.

Lining

Lining consists of coating the inside of the barrel with silicate of soda, paraffin, asphaltum, glue or other compounds to seal the pores of the wood and prevent the contents from leaking. Certain products require certain linings. In general, silicate of soda is used for oil-based products; paraffin for water-based products; and glue for alcohol-based products.

Usually, linings should be applied at a temperature of from 175 to 185°F. under pressure up to 5lb. per square inch. Pressures may be obtained by bunging up the barrel containing at least 2 gal. or more of hot lining compound to create vapor pressure. For the proper lining for your product, consult your cooper.

VI Marketing

INTRODUCING A PRODUCT

There are many known methods of introducing and selling chemical specialties. As a beginner with small capital, it is not possible or even desirable to force the issue by using more than one method at a time. Haste is best made slowly here.

Sampling is a favorite method of introducing a product. An attractive small package containing clear and complete instructions for use is made up for distribution. Of course, the larger number of samples distributed, the more people will be interested. You and your relatives can judiciously give samples to friends and other people whom you meet at home, business, clubs, churches, etc. Samples of products for home use can be left at the doors of prospective customers at a low cost. Care must be taken to check that such samples are delivered as directed and not disposed of in any other way.

Circulars sent to suitable prospects or small advertisements in a local newspaper, offering a free sample, are also used. People who write in for samples are known to use them to a greater extent than those to whom samples are given without any request on their part.

When considering the use of circular letters, advertising in newspaers or on the radio, as a means of interesting people to ask for samples, draw up a budget of how much money you have available for this purpose. Then select a printer who can help you in getting up an attractive, interesting circular and work together until you have a good circular. For your newspaper advertising, consult with someone of the newspaper staff who is experienced in writing good advertisements. Similarly, consult with a member of the staff of a radio station who can be helpful in producing a proper short script for use on the air. Do not be influenced too much by your desires or feelings. Try to approach the expenditure of money as an accountant does, by comparing how much you must spend against what experienced counsel tells you that you may expect in return. If you do this consistently, you will avoid the financial troubles which strangle the majority of small businesses.

SELLING

Demonstrations

Showing how well a product works is one of the best ways of selling. If a picture is worth more than a thousand words, then a demonstration (a living moving picture) is worth more than many thousands of words. Demonstrate you rproduct wherever you can, to individuals and groups; the more present, the more will buy. A demonstration will be far more effective if it is dramatic. The application of a little showmanship will greatly enhance the effects of any demonstration.

Recommendations

Word of mouth recommendations are very valuable. Ask satisfied users, where possible, to speak to others about your product. Get letters of recommendation, preferably from wellknown, important and leading individuals and companies. These serve to introduce you and your product.

Displays

Counter and window displays are effective when properly made up and placed. Counter displays are enhanced by illustrations showing the usefulness of the product. They only work for you when they are set up properly and placed in suitable situations. To make sure of this it is best to have them placed by the party who delivers them, otherwise they may be put aside and never used.

TYPES OF DISPLAYS

Win counter space from dealers and sales from consumers by merchandising with corrugated boxes. Properly planned display boxes provide panels for copy, show your product, are compact, give dealers a constant visual inventory.

To do an effective merchandising job, your package must stand out in *two* competitions. It must sell the dealer, must get across the thought that here is a box which is profitable to display; it must sell the consumer, must drive home the thought that here is the product to buy.

What is the most subtle, most efficient, most economical way to increase your unit of sale? Tie related items into a neat package and merchandise them as a natural group. In one example, the box transformed separate jars into a set of water colors. Customers were reminded and encouraged to buy the complete set. And they did!

Corrugated boxes can do this merchandising job by going into action with a visual story right at the scene-of-sale. Naturally, a compact, self-contained counter display, colorfully printed to spotlight the high points of your sales message is part of the merchandising force applied where and when it does the most good–makes the most sales. The "where" is on the counters of your outlets; the "when" is every day customers have an opportunity to pass, pause and purchase.

Dealers show a preference for case-lots which are easy to handle, which move quickly, which can be placed on counters to tell a pictorial story, which serve as visual inventory checkers. Meet those preferences with a box designed for the job. A box which combines shipping and display functions in one does a job for both you and the dealer. Easily packed, it is set up in a few seconds. No extra display accessories are needed.

If you sell low or medium-price items which must turn over fast, this may be the economical, efficient answer to your merchandising problem. It has already done an outstanding job for many manufacturers.

To introduce a new product or to inject a new sales appeal in an old product calls for merchandising to focus attention at the point-of-purchase. Set apart from competing items in a colorful corrugated box, your product gains a distinctive individuality which makes this job easier. Atmosphere to emphasize a particular quality of your product can be supplied by a properly designed corrugated box to make a deeper, more lasting impression on shoppers.

Corrugated boxes can provide the inexpensive flexibility and colorful design so necessary to merchandise a special offer or a seasonal sale. They enable you to pack a real sales punch into a short period of time, to stimulate immediate preferences, to work up dealer enthusiasm. Modern package designing is keyed to such lightning tactics, lends itself to utilization of colors and symbols which help make short selling seasons successful.

Properly designed corrugated boxes fit naturally into merchandising programs for special seasonal sales, for short special offer drives.

If a premium has a place in your merchandising program, consider the advantages gained by combining premium, product and sales message in an attention-getting display. You set the stage for both product and premium, link them closely in the minds of shoppers, put your premium to work selling *on location* where its *pull-ability* gets a real opportunity to demonstrate sales effectiveness.

For year 'round practical merchandising, use pre-packs to emphasize marketing efficiency and save on all handling operations. Merchandise pre-packed at the factory for delivery to ultimate consumer eliminates repacking by dealer, permits sale from sample, reduced breakage and soiling, and saves shelf and storage space.

In addition, your sales message, attractively printed on the package, stays with the product, goes into the home, remains on the "home storage unit." That builds repeat sales.

Another type of corrugated box not only merchandises your product, but also merchandises the use of your product. In this dual role it protects, displays, and gives the buyer high afterpurchase utility. You will win customers with a display that dramatizes the features of your product. You will earn good will with a good corrugated box which makes itself useful to your customers. You will be remembered if you merchandise with a convenient dual-use box, handy long after the sale has been made.

The basic rules for successful merchandising with corrugated boxes are: the box must be easy to pack; easy to set up; easy to use; and easily seen. These fundamentals are attained by building effectiveness, economy and practicality into one unit. A complicated set-up discourages dealer use. Too drab an appearance tickets your display for the rubbish heap. To strike a proper balance is your aim.

The artist has a definite place on the planning board when packages are chosen to play a part in merchandising, not just any artist, but a practical, packaging artist, trained in making the most of corrugated construction through the use of color and design.

A sales message can make or break the effectiveness of any merchandising program. What you say, and how you say it deserves caful consideration; these general check points will help you steer a straight course to copy effectiveness:

1. Approach. Key it to the purpose of your merchandising. Whatever objective you seek, whether the introduction of a new product or the promotion of a special sale, fit your copy to your porgram.

2. Brevity. Space is valuable. Messages must snap with flash thoughts. Every word must do its job quickly. Hence your copy should be analyzed. Are any words unnecessary? Does every thought expressed contribute to your sales story? Can the message be made more compact?

3. Information. Name, price, quantity, weight and other *specifics*

are an essential part of most package copy. Store traffic moves quickly. If your product is to move with that traffic, it must tell why.

When you adopt an unusual box for merchandising, you do so because it offers profit possibilities both for you and for your dealers. Be sure your dealers realize these advantages and know how to use that box. If your retail outlets are sold on your package, they will use it; more consumers will be sold on your product and your sales will increase.

Direct Mail

Certain products lend themselves to sale through direct mail, i.e. sales solicitations sent through the mails. Postage is usually the greatest expense here. The same amount of postage is paid whether the mailing piece is good or poor. Therefore it is essential that your mailing be studied from every angle to make it as good as possible.

Successful direct mail sales require expert attention and counsel. It is advisable to get the help of one experienced in this field, at least at the start. By doing this, you stand a much better chance of making good.

There are many variable factors which must be considered before attempting direct mail sales. First there is the envelope. How large is it to be? What type and color of paper is to be used? What will be printed on it and in what style of type? The same applies to the sales letter which will be enclosed in it and to a greater degree. It must be interesting enough to be read and stimulating enough to make the reader buy. This sounds very simple, but it requires hard work and mental alertness for proper fabrication.

Accurate records should be kept for all direct mail. A simple form is given below:

Date of mailing.
Articles offered in mailing.
Cost of mailing.
Number of orders received.
Value of orders received.

SALES LETTERS

Nine-tenths of all concerns with something to sell use letters to stimulate sales. Some do all their selling by mail; some use letters to pave the way for their salesmen; some to bring customers into their shops; some to keep dealers pushing their products; some to get inquiries that dealers or salesmen can follow up; some to revive inactive accounts.

But of all the millions of sales messages that daily choke the mails, only a small fraction rise above the dead level of mediocrity. The waste-basket is a yawning graveyard for most of them.

This is a dismal condition and a stupid one, because most of these failures might be successes if a few simple acid-tested principles were followed.

If a sales letter is so written that the writer can answer "yes" to the fourteen test questions that follow, it is pretty certain to bring results.

I. The Test of Strategy

1. Is the letter written from the viewpoint of the reader instead of the writer?
2. Does it get off to a good start in the very first sentence?
3. Is it short and packed full of meat, instead of long and wordy?
4. Is it free from tedious explanations?
5. Does it contain proof of facts, not mere assertions?
6. Does it make the right appeal to the class of people it addresses?
7. Has it originality of treatment?
8. Has it a good close that invites the prospect to take action?

II. The Test of Style

1. Is the letter clear, written so as to be understood easily, without mental effort on the part of the reader?
2. Has it proper emphasis and forceful expression?
3. Has it unity, logical sequence, coherence?
4. Is the message clipped of every superfluous word?

III. The Test of Form

1. Has the letter a high-class appearance?
2. Does it give the effect of an individual appeal rather than of a mass circular?

Let us consider each of these tests briefly and see where the average sales letter fails to make the grade.

I. Strategy

1. *Wrong Viewpoint* The most common fault of the ordinary sales letter is that it is written from the standpoint of the writer instead of from that of the reader or prospective buyer. The average business man forgets too often that no prospect has any real interest in him or in his proposition until he sees some direct effect upon his own welfare.

There is only one way to arrest the attention of a reader at the very outset and that is to set him and his interests in the foreground of the letter. To do this, the writer must visualize the prospect, determine what kind of man he is, his probable needs, his desires, his limitations and his prejudices. This diagnosis will enable the writer to find a point of contact between the prospect's self-interest and some aspect of the proposition, and then to discuss everything from the prospect's point of view. Here is an example of the right viewpoint—that of the reader.

Here is an opportunity to put on sale an article that will sell itself—the Pathex outfit—Camera, Projector and Accessories.

You don't tie up money in Pathex: you just keep turning it over and over. Every Pathex sold sells another. Pathex will bring customers to your store to whom you can sell other things, etc., etc.

Here is a letter, from another concern, composed entirely from the viewpoint of the writer instead of that of the reader.

Our concern is closely connected with the great—Company. This company has been known all over America for fifteen years as the leading—company. We have been able to make an arrangement by which a new invention—one of the most ingenious ever made—is placed upon the market. We offer you an opportunity to sell it, etc., etc.

2. *Wrong Start.* Arresting the prospect's attention and capturing his interest start with the opening sentence. Usually this sentence will determine whether or not the reader will be carried further into the writer's story.

Many letters begin with an introductory sentence that is so hackneyed and shop-worn that it is meaningless. "I want to call your attention to the most liberal offer ever made" is a sample of this kind of sentence.

Then there is the type of letter that begins with a striking but utterly irrelevant remark such as, "Noah had to wait six weeks for the rain to stop." Here the bid for attention is so forced that the reader senses the artificiality and is repelled. It is safer to start by saying something to which the reader will give spontaneous assent because it strikes him as interesting, true and important.

3. *Wordiness.* The mere length of a letter often turns the reader away from it. Just to look at it gives the reader a sense of effort, especially if it is not properly broken up into paragraphs. It is better to write a short letter that will be read than a long letter that will not be read.

When seeking requests for a booklet you need merely to write enough to arouse curiosity. A short letter will do this best. If you tell too much about your proposition, you will dull the edge of curiosity and kill the letter's pulling power.

If, on the other hand, you already have the inquiry and are working to turn it into an order, there is far less need to write a short letter.

You may use space to give complete information.

Some concerns use short letters to get inquiries and long letters to answer them on the ground that it takes few sentences to arouse interest, but a complete selling talk to get an order.

In brief, a sales letter should be long enough to accomplish its purpose, and not one word longer. In reviewing sales letters, the handiest tool to use is the blue pencil.

4. *Tiresome Explanations.* Many sales letters fail because they bore the reader with lengthy explanations. This fault is particularly common in letters concerning a technical product.

When you stop to think of it, a description of a product has slight selling value until you have awakened the prospect's interest in what the product will do for him. And if you get this interest, you will find that highly detailed explanations are unnecessary. The letter writer should make his descriptions so terse, graphic and specific, that the prospect's mind takes them in without effort and is in a receptive mood to hear reasons why he should buy.

Here is an example of such a description:

The Tyrod Hammer has a head drop-forged, heat-treated in three stages to give strength to the claws, toughness to the eye and hardness to the face. Its well-balanced handle is built of two pieces of selected hickory, double sealed by casein glue and hence water-proof. A steel rod that runs through the handle from the head adds great strength. The point of break, just back of the head in ordinary hammers, is thus eliminated, and the head cannot fly off. These points make the Tyrod far superior to any other hammer manufactured.

VS. 5. *Assertion Proof.* Even though a sales letter may measure up on the score of terse and graphic description, even though it may successfully attract our attention and awaken our interest, it may still fail to convince us. To accomplish this it should introduce real evidence or proof. Mere assertions, or boastful claims of superiority, pass over us like the idle wind. We can all recall sales letters studded with sentences like these:

We sell only the highest quality.
Ours is the best real estate proposition in Florida.
We offer you a chance to make BIG MONEY.

These are old familiar friends, floating in on every mail.They inspire little faith. They are mere assertions, with no evidence to back them up.

But when we are told the names and given the statements of people that have used goods or accepted a proposition with profit to themselves, conviction sets in and the letter accomplishes its purpose. Here is an example of a letter containing proofs:

Dealers like our proposition. We have added 900 dealers in the last 60 days.

Let us send you the names of dealers near you, no matter what your location. Write to them and they will tell you that their sales of our product were large at the start and have increased steadily every month through the year.

6. *Wrong Appeal.* It is surprising how many letters miss the point in appealing to a certain type of prospect. The problem of the letter writer is the same as that of the advertisement writer. If the latter is advertising women's shoes, he should realize that with most women style is a more potent appeal than price or wear. This may not be true of stockings, or of automobiles, but it happens to be true of women's shoes. In like manner, a sales letter in its appeal must be geared to the product and to the prospect. The service appeal, for example, is strong with both men and women, but the quality appeal is stronger to men than to women. Such appeals as style, profit, exclusiveness, sentiment and price require careful handling if they are to be capitalized to the fullest extent. It takes a good letter writer to use them properly.

Again, careful consideration may profitably be given to a variety of proper appeals to determine which is the strongest.

For example in selling Hooven automatic typewritten letters it may be proper to point out the size, the high character and capacity of the organization, and to emphasize the standards of promptness, or accuracy, and of general fitness. But the strongest appeal to business men would be on the question of profits. From this viewpoint the following is a good appeal.

Your fellow manufacturers are using Hooven letters in an increasing degree because experience has proved that these typewritten letters get inquiries at the lowest net cost. Hooven letters are worth more than others because they bring twice, or three times, or even four times, the returns of other letters. (Here might properly follow specific examples of profitable results.)

7. *Lack of Originality.* No one has ever been quite able to determine just what constitutes charm in a woman. Just as elusive is the quality of originality in a letter. We may not be able to prescribe a formula for making letters original, but we know instantly whether a letter possesses that subtle stamp of individuality. The only suggestions possible here are negative ones. When a writer composes a sales letter, he simply must turn his back on hackneyed, stale, commonplace expressions which give the message a stereotyped and conventional ring. He must get out of the rut that thousands of others have traveled and devise, if possible, some new way to get his message across. Sometimes this will take the form of a striking presentation, and if not too "smart,"

will yield surprising results.

The Delco-Light Company, for instance, selling small electric light and power plants to farmers, worked out such a letter. The conventional type of letter would have asked the farmer if he didn't wish the convenience of electric lighting and heating enjoyed by his city brethren, and then would have described the Delco-Light equipment, how it works and what it does. Instead of that, a letter was prepared with an actual copper penny cemented to the top. The letters began as follows:

> There is tremendous power back of this penny. It will buy enough kerosene for fuel, and enough lubricating oil to generate 200 watts of electricity, with a Delco-Light Plant.
>
> With 200 watts of electricity you can:
>
> Pump 235 gallons of water on a 22 foot lift
> Light one 20 watt lamp for 20 hours, or
> Light five 20 watt lamps for 2 hours
> Run a 9 inch desk fan for 5 hours
> Run a washing machine continually for 40 minutes
> Run an electric iron for 22 minutes
> Run a sewing machine for 2 hours.

Here is practically the whole story of the advantages of a product told in a very few words. No long introduction, no tedious description or long-winded explanation, no conventional treatment found in the usual run of sales letters. The sheer originality of this letter brought such big returns that it was used for three successive years before its pulling power weakened.

The encouraging thing about this is that almost any business has a story to tell which can be vivid and interesting if original thought is applied to the task. There is endless variety in words and in ideas, just as there is in music, and as long as creative minds are on earth, new and fresh music will be composed, and new and original sales letters will likewise be composed.

8. *Closing on a False Note.* Not long ago a sales manager was asked by a cub salesman what he ought to say after he had given a full presentation of his proposition. The answer was so self-evident that it stunned the salesman, "Ask him to buy." It is really pathetic to see how timidly some letter writers skirt around this issue in winding up their sales message. Unless your letter is purely educational propaganda, its main value lies in getting the prospect to do something that you want him to do. If you want an inquiry for further information, why not close with the strong, positive suggestion that he send for it, and make it easy for him to do so? If you want him to express

himself about your line, why not come right out and ask him to do so? If you want him to send you an order, you cannot make the closing suggestion too emphatic.

As a general thing, most sales letters seek some action on the part of the prospect, but few of them are definitely singleminded and positive enough in calling for this action.

Asking a man to let you know when a salesman may call to see him is usually a futile request. Even though you have aroused interest in your proposition, you will find that few men are willing to give a direct invitation for a salesman's call. In spite of this fact, hundreds of letters continue to make this weak closing suggestion.

A good close does not threaten a prospect with a salesman's call. It merely asks him to indicate if he wishes further information, and it makes it easy and natural for him to ask. It calls for as little effort on his part as possible. Often it does not even require him, for example, to find ink and paper to reply, or to call in a stenographer and dictate a letter. It suggests merely, "If you are interested, simply jot your initials on this letter and mail it back." Above all, it assures the prospect in a very casual but none the less effective way that he will be under no obligation if he replies.

II. Good Style

If letter writing is "salesmanship on paper," then the style of a letter corresponds to a salesman's personality. Just as the salesman reveals his personality in everything he says and does from the moment he makes his approach to the conclusion of the interview, so the style of a sales letter either advances or retards its effect in every paragraph, every sentence, every word chosen. Here again, the tests are simple, but rarely met.

1. *Clearness.* Many sales letters are not clear. To get their meaning requires mental effort on the part of the reader. Often he must read parts of a letter a second time before he can understand them. He naturally hates this work and ordinarily will not read letters that impose it on him.

Herbert Spencer said that the chief principle of good style is that it economizes effort on the part of the reader.

A good letter is easily understood. The sentences are short. It is broken up into paragraphs, and these are varied in length. Some will be only a line or two in length, some several lines in length.

A clear letter uses short words. It is easy and conversational in style, it reads as men talk. It is never stilted, learned or technical.

Lincoln's address at Gettysburg was only five minutes of simple, easily understood talk. Yet it is called by critics the greatest piece of prose ever written in America.

The same day Edward Everett spoke for two hours on the same topic. His speech was filled with long, scholarly words. Everett's speech is today almost forgotten. But Lincoln's speech lives on immortal. Lincoln could have written a good sales letter.

Here are some obvious contrasts in expression: "You can obtain the goods;" it is simpler and more effective to say, "You can get the goods." "Purchase an allotment"; it is better to say, "Buy a lot." "This machine assists you in your duties"; it is better to say, "This machine helps you do your work." "Indicate your selections", it is better to say, "Take your choice." "We are unable to discern"; it is better to say, "We cannot see."

2. *Wrong Emphasis.* Choice of words is vitally important. It is not, however, more so than emphasis. Proper emphasis of a word is gained by placing it in just the right place in a sentence.

The most emphatic place in a sentence or paragraph or letter is at the very end; the second most emphatic place is at the beginning. The least emphatic place is in the middle.

"We not only do your printing well, but we deliver it to you promptly," is a sentence taken from a sales letter. Here emphasis is placed on the word "promptly" by its position at the end of the sentence.

"Results, that is what our machine gives you." Here, by the location of the word "results," emphasis is given to it.

3. *Flabbiness.* Letters may be clear, correct, exact, without being forceful. They may be understandable without being impressive. Deliver us from flabby sales messages.

How can this fault be removed?

Concreteness adds force to statements. To say that a thing is valuable is to make a general statement, usually meaningless; but to say that it will increase the prospect's profits $9 a day, 300 days in the year, is to be concrete.

Figures of speech add force to style. To say, "Don't be held down," "Don't keep your nose on the grindstone," is far more forceful than to say, "Don't permit yourself to stay in a position where you have to work hard many hours a day without opportunity."

Short, sharp pointed sentences that contain figures of speech are far more forceful than long sentences without figures.

"His sentences are full of punch" is a figurative expression. They liken the power in a sentence to the blows of a prizefighter.

Slang is practically all made up of figures of speech. That is why it is so expressive.

We all use slang in conversation, because it makes talk more picturesque and forceful. A great deal of slang is vulgar, and should, of course, be avoided. But since slang is so forceful that practically everybody uses it in talk, why should not letters gain force through it? There are many slang phrases that are not vulgar, that have become correct through usage.

In striving for force and conciseness, we should not, however, be abrupt or curt. A certain amount of force may be gained by these qualities, it is true, but it is not the kind of force that leads to good will.

Let us by all means avoid peremptory commands. Such commands may seem to secure an immediate object for which the writer is striving, but to use them is bad business if they outrage the reader.

The words "please" and "thank you" are very important, especially in selling the policy of the house. Courtesy is really an element of force, because it makes the reader feel that you are treating him as an equal, and consequently makes him more responsive.

Personality and individuality in a letter are most vital elements of force. They are the qualities that make a letter stand out as distinctive among the many on a business man's desk. They make the reader feel that he is listening to a real, live person.

Force comes through intense conviction, and a writer's desire to drive his message deep into the consciousness of the reader. The letter of a really forceful man is not of the "rubber stamp" variety. The reader feels, as he reads it, that the writer is blazing his own way. They personality, the individuality of the writer, is apparent in every line he writes.

5. *Scattering Fire (Lack of Unity, Sequence, Coherence)*. Many sales letters produce the same effect as the firing of a rusty shot gun, loaded with miscellaneous slugs of every size, weight and shape. They scatter their charge all over a wide area, instead of sending it straight to the mark.

Such a letter is a mere collection of nondescript sentences, badly chosen and badly arranged. It confuses the reader; it gets nowhere in the pursuit of an order or inquiry.

A good sales letter has unity. Every part is carefully selected. Not a single sentence, not a single word, is put in that does not help to build up one central idea.

The arrangement of parts is logical, so that the reader's mind proceeds in a straight line through the letter to the very end. The sequence of sentences is correct.

Finally every sentence is closely connected in meaning with the sentence preceding. As a result, all the sentences hang together, and coherence is achieved.

III. Good Form

1. *High Class Appearance.* No one disputes the importance of a salesman's appearance as a factor in successful selling. If he is less than well-dressed he handicaps all his cleverness and force. What impression would you have of him if he appeared before you in a ready tied necktie and a celluloid collar? Both serve their purpose as well as, and far more economically than, a hand tied scarf and a linen collar. But we are strongly prejudiced against incorrect apparel.

It is the same with letters. No one can challenge the wonderful utility of letter processing and printing machines, but they have decided limitations when used to produce sales letters. A processed letter, even when the name and address are cleverly filled in, places the message under the same handicap as is a salesman with a slicked-up celluloid collar. A few people do not notice these things, but most of us do.

Custom stamps its unseen influence on our minds. As long as most of us can remember, the worth-while business communications coming to us have been dictated to a stenographer and written by her on a typewriter. When some one attempts to short-cut around this custom and sends us a processed communication with all the ear-marks of mass circulation, we give it about as much attention as we give a hand bill.

Is there any reason for this human prejudice? The full message is in the processed letter and that is the only thing that ought to count. But then, try to explain the celluloid collar! The fact is that we like the compliment of a typewritten letter just as we like to handle a fine piece of bond paper. We are unfavorably impressed by a palpably filled-in processed letter, just as we are by a salesman in incongruous dress. That is a human weakness that simply has to be catered to if we want our letter to be read instead of ignored.

It has become axiomatic among skilled promotion men that a letter which will produce 2% returns when processed will pull 5% returns when individually typed.

The successful salesman calls each prospect by his name, he speaks of his prospect's specific industry, of his particular local conditions.

2. *Individual Appeal.* Think of the letter campaigns that are perfectly

planned; lists selected with the utmost care, almost regardless of expense; well planned letterheads on high-class paper; the copy skillful and compelling; all these steps carried through with the greatest care, only to have the entire message fall flat because the mechanical presentation is so plainly a make-shift that every man addressed spots the letter at a glance as a mere circular.

The Little Schoolmaster of Printers' Ink speaks of this all too common practice as being a great deal like saving at the spigot and wasting at the bung. He goes on to say, "PEOPLE do have their feelings, and their little vanities, and occasionally a sense of proportion. Those things are all carefully taken into account when the advertising copy and the printed matter are under consideration—why are they not of equal importance in connection with a letter?"

Thinking of these things, realizing the far greater sales power of letters presented in a superior physical form, we may legitimately ask: why are not all sales letters typewritten instead of processed? Why are they not all given these marks of individuality?

The answer would be obvious if it were necessary to have a stenographer type all these letters. But with the invention of the automatic typewriter, which turns out genuine typewritten letters at high speed, each letter with all the marks of individuality, and none of the marks of mass circulars, the answer is not so obvious. Because the cost of typing a thousand letters for mailing is not so much greater than that of processing, whereas the results in returns have been demonstrated to be from two to three times as great, and often incredibly greater.

A sensible man does not hesitate to invest a few more dollars in a mailing effort if by so doing he reaps several hundred dollars additional profit. That is the only reason he spends two cents in postage, instead of a cent and a half—he knows that first-class mail gets more favorable attention. Similarly, individually typed letters have the call over the imitation variety.

In compiling a mailing list for this direct-mail sales letter, since nearly every family has a car, names selected from telephone books can be used. Special lists of names of automobile owners may be rented from many concerns. The latter have hundreds of lists of names of specialized prospects and buyers that can be used for selling anything by direct mail.

It is best to always first test out a list by making a trial of 1000 to 2000 names. If this pays, then 5000 may be tried next, then 10,000, and so on. If a mailing does not pay, it may be due to the product, circular, list used, time or other variable. Studying each one of these factors is the difference between success and failure in selling by mail.

Example of a Good Sales Letter

H. Bryane Polish Co.
22 Negrin Avenue
Choiceville, Va.
January 5, 1942

Dear Automobile Owner,

How many hours have you spent polishing your car? Have you always been satisfied with your results?

As an owner of a car like yours I would like to answer these questions by seeing the many extra hours that I have put in and the far from satisfactory results.

Today, I am glad to say, that I only spend about half the time and get a much better and *permanent* polish. Do you want to know how it is done? It's really very simple. All you have to do is to get some Trutone, the latest development in this field.

Trutone differs from all other polishes in that it contains a new synthetic wax which enables anyone to polish his car in half the time and leaves it as bright as a new car. The polish, because of the synthetic wax, does not pick up dust. What is more, it does not "rain-spot" and lasts longer than ordinary polishes. Trutone is more expensive, but a little goes a long way. Try it and you will swear by it. Your money back if it doesn't please you—100%.

You can't be the first to use Trutone but don't be the last.

Very truly yours
H. Bryane Polish Co.
H. Byrane

Use this coupon for special trial offer

H. Byrane Polish Co. *1/2 Price Special Offer*
22 Negrin Ave.
Choiceville, Va.

Gentlemen:

Please send me one pint of Trutone (regular price $1.00) for the special introductory price of 50¢. It is understood that you will return my money in full if it does not do what you say.

Enclosed please find

Stamps ☐ Very truly yours
Coin ☐ Name
Check ☐ Street
Money-order ☐ City State

Quantity Purchasers

Mail-order houses and chain-stores will buy meritorious or unique products brought to their attention, even though they are new and relatively unknown. These houses expect and get very large discounts because they can sell huge quantities through their mail-order catalogs or counter displays. It is usually profitable to sell them because no selling or advertising expense is incurred and they pay promptly.

Sales Agents and Salesmen

Sales agents are individuals or companies who sell one or more related products to jobbers and large and small users. Since they are already selling to these outlets they can economically handle one or more additional related products. Thus a sales agent who is selling equipments, parts or accessories to garages, service-stations, etc., is properly situated to sell an automobile polish. Similarly a food broker selling canned goods, sugar or flour could take on a new food specialty. Suitable sales agents may be located by consulting specific trade papers or trade directories.

Good salesmen are hard to find. Good salesmen must be treated fairly and be paid the highest possible commission to keep them satisfied, particularly at the start, when sales are hard to make.

Jobbers and Wholesalers

Jobbers and wholesalers, who sell to retailers, will seldom take on a new item. They are primarily order takers who supply the wants of retailers who, in turn, order what their customers want or need. Accordingly, no attempt should be made to sell them until a retail demand is present. The exception to this is that sometimes jobbers and wholesalers can be sold on a unique useful specialty if they are given an especially large discount. In such a case they undertake to introduce and sell a product through special efforts.

Exporters

Some products lend themselves to sales in foreign countries. Since exporting is a complicated business and ties up money for longer periods than domestic sales, it is not desirable that a new business engage in this field directly. There are many exporters, however, who buy and pay here for merchandise which they sell abroad. They tell you how to pack for export and attend to all other details.

Certain exporters may ask for exclusive representation for one or more countries. In certain cases they may be entitled to such future protection for extra efforts and expense which they put forth in introducing a new product.

Exclusive Sales Representation

When asked for exclusive sales representation for a specific territory, you should weigh carefully what the sales-representative *definitely guarantees* he will sell within a certain trial period against what you will lose if he does not fulfill his promise. It is wise not to give exclusive representation for more than 6 months for a limited territory. If satisfactory sales do not develop, then this arrangement should be terminated automatically.

Publicity

News items, if written in an interesting manner, are published free by many trade-papers and some small newspapers. They should be neatly typed and double-spaced. The example that follows is typical:

For release Sept. 8, 1945

After many experiments and numerous tests, a new type of automobile polish has been developed. It is based on a new synthetic wax recently produced by American chemists. This polish, called Trutone, saves half the time usually required and produces a lustrous polish that does not pick up dust, does not rain-spot and will not injure the finish of any car.

Trutone is being manufactured by the H. Bryane Polish Co., 22 Negrin Avenue, Choiceville, Va.

NEW PRODUCT DEVELOPMENT

The initial idea for a new product may come from divers sources. It may come from a user, dealer, salesman, friend, magazine or newspaper, etc. A customer may ask for a specific product as a rust-remover or a special adhesive comparable to what he is buying. Or he may have a problem for which a new product must be made. For example, he may be concerned about rust-formation in cracks and crevices in a metal unit. Perhaps a special detergent or penetrating protective film will solve his problem. Because what he is now using is often difficult to obtain, e.g. an imported product, or is too expensive, e.g. a patented or monopoly product, he may want a substitute. Finally, a product may be developed to fill a latent need. This may require considerable selling to introduce it, e.g. an antiseptic spray (not for insects) to kill bacteria.

Most people are conservative in that they prefer to continue using a product which gives satisfactory service. Some, however, are alert and are continually seeking improved products. Others are receptive to new improved products, when the latter are presented properly.

1. *Appearance*

   a) Color

   Color may be changed by the use of different grades (darker or lighter) ingredients, bleaching, or by the addition of a dye, lake, pigment or chemical that will effect a chemical change with a resultant change in color.

   b) Clarity or opacity

   Clarity can be effected in a liquid by decanting after standing to permit settling, centrifuging, or filtering; or by using purer or specially treated raw materials.

   c) Consistency

   A product may be changed from a solid or paste to a liquid by means of water, solvents, emulsification, or by the addition of another ingredient that causes liquefaction. Conversely, the consistency of a liquid may be increased by dissolving one or more substances in it or treating it by chemicals, heat or pressure.

   d) Form

   The forms in which a solid product can be produced are cakes, bars, lumps, sheets, granules, crystals, beads, flakes, powders, etc. Each of these can be in different sizes as desired.

2. *Application or use*

   a) Ease in use

   The trend is toward lightening labor, both at home and in industry. Thus, any product that can be applied more easily is desired. This is exemplified by the shift to non-rubbing floor polishes from the older rubbing types.

   b) Saving in time

   Similarly any product that will effect a saving in time finds a ready acceptance. Thus a drug powder that dissolves instantly is more desirable than one which requires a longer period for solution. Paints which dry quickly have taken over a large portion of this market.

   c) Economy in cost and amounts used

   A product that is lower in price or of which a lesser quantity can be used to effect a saving meets very little sales resistance.

   d) Absence of harmful effect on containers, or materials with which it comes in contact

Certain products corrode or discolor containers or materials with which they come in contact when being used. Tincture of iodine is an example of this. That is why many prefer to use other antiseptics (even though less potent) to avoid such effects. Carbolic acid (phenol) suffers from similar disadvantages and, in addition, is toxic and corrosive to plants, animals and humans in cetain concentrations. Accordingly, it is displaced by other more modern antiseptics.

e) Better odor, taste or feel

It is admitted that the odors of certain products are disagreeable, yet often nothing is done about it. Cover odors (industrial perfumes) and preservatives to prevent bacterial decomposition are freely available to cover up or change undesirable odors.

Undesirable tastes, too, can be corrected by the judicious use of sugar, saccharin, salt, spices, essential oils and flavors.

The feel of a product can usually be altered where it is objectionable. Thus a rough or harsh surface may be sheared, ground, or treated with a lubricant or coating to make it more agreeable.

f) Uniformity

Uniformity of the contents of an individual container and as compared to other similar containers is important. Different sizes of particles of different ingredients will cause formation of different layers because of shaking during shipping and storage. Unless each package is thoroughly shaken or mixed, before use, the results obtained may not be satisfactory.

g) Greater durability

Increasing durability or life of service through the use of a new product is an important factor. Thus if a flame-proofing for clothing or curtains can be made to withstand a number of washings or dry cleanings, it will be far more desirable than one which disappears on its first cleaning.

3. *Reduction in hazards*

a) Fireproof qualities
b) Non-explosive qualities
c) Non-toxic qualities
d) Non-corrosive qualities

Any reduction in danger to life or property is a worth while accomplishment. An inflammable, explosive spot or stain remover may be made non-inflammable by emulsifying or replacing the solvent with one that is non-inflammable or non-explosive. In a shoe dye containing toxic nitrobenzol, the latter should be replaced by a non-toxic solvent. A tile cleaner containing volatile hydrochloric acid, the fumes of which will corrode metal fixtures nearby, could utilize, instead, a non-volatile acid salt such as sodium bisulfate.

4. *Stability on aging*
   a) No precipitation, separation or bacterial decomposition
   b) No marked change in appearance, odor, taste, consistency or feel
   c) No marked change in use

   Some products deteriorate with age and under certain conditions to which they may be subjected at an early stage in their life. Thus exposure to heat, cold, sunlight, air, electrical discharges, shaking, etc., may produce undesirable changes during shipment, storage or on dealers' or consumers' shelves.

5. *Change in any of the following properties which may be desirable*
   a) Acidity or alkalinity
   b) Density
   c) Viscosity
   d) Freezing or melting point
   e) Evaporation rate
   f) Boiling-range
   g) Solubility
   h) Hardness
   i) Tenacity or cohesion
   j) Plasticity or ductility
   k) Flexibility
   l) Elasticity
   m) Stickiness
   n) Slipperiness
   o) Drying qualities
   p) Feel or "hand"
   q) Hygroscopicity or efflorescence
   r) Edibility
   s) Effect of heat, light or pressure on
   t) Effect of solvents on
   u) Interaction of ingredients
   v) Emulsifiability

w) Dispersing or suspending properties
x) Adsorption or absorption
y) Heating or lighting power
z) Electrical or thermal conductivity

6. *Concentration or elimination of inactive ingredients, effecting a saving in*
   Freight and trucking charges
   Storage space
   Labor and power in handling and compounding
   Size, weight, and cost of consumer packages

   Where a product contains a large proportion of inert materials such as water, solvents or dry fillers, their elimination may effect the above savings.

7. *Repackaging*
   a) Appearance
   b) Utility
   c) Re-use of container

   Repackaging may make a unit more desirable in appearance through a change in size, shape or color. Its utility may be enhanced by making it easier to handle, store and less subject to breakage or deterioration. The container may also be redesigned so as to make it more usable when empty.

8. *Additional factors often making a new product desirable*
   a) Better availability
   b) Greater uniformity
   c) Freedom from legal restrictions

Study of New Opportunities

Product Story

Whether the new product involves a disinfectant, or a new type of hand lotion or whatever it may be, it is realized that to make any progress at all, all of the people who are to concern themselves with the product need to know what is really under discussion, and so it must be the responsibility of some member of the organization to write up as fully as possible the type of product, package, article or process which he has under discussion, which he wants to submit for consideration.

In developing product stories, each manufacturer's problem is different, but in the main similar. The why of the product and the service it is to perform need to be fully explained. With such an explanation properly prepared, it will not be too difficult to make a quick check, not necessarily a final check, of how much consideration the product or process should receive.

Competitive Factors to Be Studied and Outlined

Assuming that the product story that has been written up looks fairly promising, or promising enough to warrant further study, it can then be decided whether additional information not available within the organization should be sought.

Are there similar products on the market? How are they marketed? What are their advantages and disadvantages? In what price range do they sell? How are they distributed? Are there outstanding weaknesses in the competitive products which might be met by either an improvement in manufacture or improvement in sales procedure or distribution? There are other questions; these are only indicative.

Consumer Market and Merchandising Survey

If further information is needed, it is desirable to make studies of the marketing opportunity, both with the consumer and the trade. Whatever type of survey the facts in hand warrant should be made to explore further the product opportunity.

Product Research and Product Development

Having decided certain specifications in the particular product or process under consideration which cannot be met from information in hand, a definite line of research and study should be carried out in the laboratory or by an outside expert. The work will only be progressive if it is made quite clear exactly what is wanted. Otherwise, much time will be wasted in floundering or in following up paths of investigation and thought not related to the objective.

Laboratory Product Tests

If the product or process is one that can be subject to testing, and most of them are, whatever tests are necessary to assure that the product meets the specifications should be carefully established and the product should be tested against these requirements caarefully before going further.

Are Outside Laboratories Needed for Further Check?

In some circumstances, it may be desirable to test the results found in one's own laboratory in an outside laboratory, where independent views may add further improvement or safeguard against spending funds on a product in which serious flaws may later be discovered.

Can It Be Tested by the User?

If it is product that can be adapted to consumer tests before marketing, it may be desirable and extremely helpful later on to submit products to sufficient users to explore further its acceptability and utility.

The extensiveness of all this testing will be determined by the very nature of the product itself and the price class in which it is expected to find a market.

Storage, Spoilage Tests

It may be that a product can be developed, tested by the laboratory and checked by ultimate consumers all within so short a time that the effects of storage or spoilage as the result of particular processing may not be revealed. This is particularly true in food products, although it may not apply to many others. It is important, therefore, that storage and spoilage tests be conducted to be sure taht no serious mistakes will be made.

Packaging Specified

If you can get this far and you still feel favorably inclined toward the idea, the proper packing is next determined.

Packaging Tested

In too many instances new types of packages, which appeal to the eye and are very attractive to display, may be found too expensive to use, either because of the labor of setting up the packages, filling and sealing, or because such packages may not stand up under ordinary usage. It is our purpose to test all such packages to see if they will carry the products for which they are intended.

Process Approved

Though this subject can be taken up earlier, depending on the nature of the product under study, if it involves a new type of processing or manufacturing it will be important to consider whether the process which has been developed can be carried out. Consideration should be given to whether additional required equipment or required plant changes will be more expensive than the product will warrant.

Provided, however, that the product is one that is allied to the lines being manufactured, it will be possible to specify such modifications as may be necessary in order to approve the process in principle.

Production Line Plan

It may be that the product is one not possible on present production lines. With the process and package identified, and with a rough estimate of quantities that might be sold, a production line can be planned on paper. At this point, the number of people who will be required in processing the article can be specified.

Product Cost Estimates

With the previous information available, it will be possible to prepare an estimate of what the product will cost in given quantities.

Price and Gross Margin Settled

If the product has lasted on the idea list this long, and you have developed pretty closely your ideas as to what price you can get for the product, you are ready to consider it nearly in the final stage of planning.

You can determine what the gross margin will be in the product and consider whether it will be worth while.

Questions involving the cost of any new equipment, plant changes, changes in production line, etc. can be weighed against the probable profit opportunity of the product.

Preparation of Marketing Plan and Financial Data

If it looks promising, a marketing plan can then be sketched that will involve certain expenditures for either advertising or sales work and a recommendation based on the profit possibilities can be prepared.

Equipment Ordered, Production and Distribution Started

It then remains to decide whether the equipment can be ordered and when the product can be actually started on its way to manufacture and distribution.

For recording the various stages of the development and marketing of a new product, the following chart can be used. The headings of the various columns should correspond to the titles of the preceding sections.

You will note, by referring to the chart, that *Product A* has completed all of the necessary work up to and including Step 14. *Product B* has a number of blank spaces in the course of its progress representing the extent of information, work or progress yet to be made. *Product C* has had little or no work done on it.

In the secondary and larger white space underneath the broken bar brief memoranda that are needed to guide the person responsible for the program in his follow-up at regular intervals can be noted.

New Product Checklist*

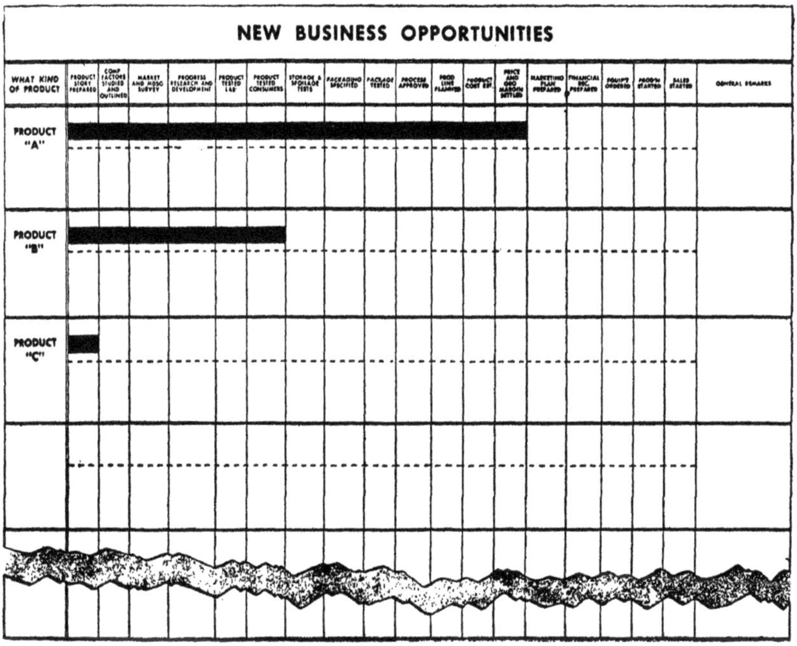

A manufacturer who introduces a new product—no matter how good or how useful it may be—without having clearly thought out the course which he will follow to move that product from his factory into the hands of the ultimate consumer, can find himself in very serious difficulties. His position is like that of the man who built a magnificent boat in his garage, only to discover when it was completed that he could not get it through the doors. A great many excellent consumer products have died in factories because their manufacturers had not figured out in advance how to get the product out-of-doors—that is, how to *sell* it successfully as well as make it.

This check list for the introduction of new consumer products is offered as a working tool to the manufacturer *who wants to make sure that he has thought his merchandising and selling problems through* before he launches a new product on its way. And the manufacturer who calls his executives together and goes over it, line by line, with his pencil in his hand, will get the greatest possible benefit from it.

**Economic Series No. 41, United States Dept. of Commerce, 1945.*

The check list cannot answer all of the marketing and merchandising problems which you may have, but careful analysis of it will be like inspecting your entire marketing procedure with an X-ray machine. When you get through with it, you will know exactly how carefully you have charted your course, and how much more thought or investigation will be required of you before you can be sure that your new product will be launched into the competitive arena with the best possible chance of being successful.

The check list is divided into six parts:

1. The users of the product.
2. Channels of distribution.
3. Competition.
4. Manufacturer's price policy.
5. Selling and promotion.
6. Legal and related problems.

Each of these six parts contains a brief introduction, followed by a list of specific questions. It is suggested that you read it all first, and let others in your organization read it also. Then sit down, with responsible executives in your organization, and, pencil in hand, go through the questions, item by item.

Some of the questions can be answered with figures; some with a "Yes" or a "No"; others with a sentence or two. In still other instances, you may not find it worthwhile or desirable to put the full answer to the question down; but in all such instances, you will know, after considering the question, whether you really know the answer to that question or not.

Then, when you have completed the detailed checking of questions in the main body of the book, transfer your answers to the summary check list at the back. On this summary check list you will find three columns: one to indicate that you have a satisfactory answer to the question posed; another to indicate that this question requires additional information before it can be answered, and a third to show that the question is not applicable to your product.

When the summary check has been completed, you will have at your fingertips a brief "control sheet" which will show you exactly how well you have equipped your product for its competitive battle in the market place. And you will know exactly what further thought and study you must devote to the problem.

I. *The Users of the Product*

The *easy* way to "define" the market for your product is to say that

"Millions of people will want it," or "Every housewife needs something like this," or "Every one of America's 200,000,000 people will be a prospect for this product."

That is the *easy* way to convince yourself that you can create an unlimited demand for what you want to sell. But it is never true!

Every product's potential sales are limited by a number of factors. Some can be sold only to young women; others can be sold only to automobile owners; still others will be bought only in warm climates, and not in cold.

Furthermore, every product's potential is limited by the product itself, by its price, by the use to which it may be put, and by a variety of factors over which the manufacturer himself has no control. To give a very simple example: Most of us consider tooth brushes and tooth pastes as items that are "universally" used. Yet obviously they are not used by some 2 million infants who do not yet have teeth, nor to any significant extent by the several million oldsters who have lost theirs. And, of course, a large part of the remaining population—which has teeth and should take care of them—never buy brushes or paste. And among users of tooth brushes and dentifrices, usage is by no means evenly distributed. Some groups use two to three times as much of these products each year as do other groups in the population.

There are all kinds of other factors that affect the sale of products—some of them obvious, some of them difficult to determine—but equally vital to the successful marketing of any product.

In many cases even successful products might be much more successful if their marketing problems were in sharper focus. Take the case of a manufacturer of high-class shoes, selling for over $20 a pair. This manufacturer had always assumed that the bulk of his business came from wealthy men and women, since the price of his shoes is considerably higher than the average, and as a result he has always directed the bulk of selling and marketing activity at the "class market."

Yet an analysis of 16,000 retail sales in New York proved conclusively that only a small fraction of his business came from the highest-income districts of the city, and that his sales—per capita—were just about as good in the low-income neighborhoods as in the highest-income neighborhoods! As a result of this study, he changed his whole approach to the New York market, with satisfactory results.

These brief examples are sufficient to indicate that generalizations about the size, location, or character of the possible market for any product are dangerous, and that the only sound, sensible way to determine the possible market for a product is to test, secure the facts, and consider all of the factors involved.

Fortunately, once you are satisfied that the product is properly designed to meet specific consumer needs, this is usually a fairly simple task, requiring nothing more than a little investigation and some thought. If you will get out your pencil and check off the items on the following list, for instance, you will be able to determine quickly the answers to this vital question:

"Am I *sure* that I know the size, location, and characteristics of the market for my product?"

Check List I. The Users of the Product

1. What types of consumers will use the product?
   (Are they housewives, school children, farmers, salesmen, mechanics, animals, or what?)

   ____

   ____

2. How many potential customers are there?
3. What factors cut down the size of the total market?
   (Enter numerical estimates in the spaces provided.)
   a) How many live in sections of the country where they will not use the product? ______
   b) How many are in the wrong age group to use the product? ______
   c) How many are likely to be poor prospects because your product is too expensive or too cheap for them?
      (To what income classes will your product appeal?) ______
   d) How many are likely to be out of the market because they already have similar products giving satisfactory service? ______
   e) How many live in such a manner that they cannot use the product? (E.g. if you live in an apartment, you are not a good prospect for a lawn mower.) ______
   f) How many would not be prospects because of religious or racial reasons? ______
   g) Are there any other market-limiting factors peculiar to your product? (List and enter number affected.)

   ____

   ____

   ____

   ____

   Total No.3 ______
4. How many prospects are left for your product?
   (No. 2 minus No. 3) ______
5. Where do the potential customers live?
   a) Are they scattered throughout the country? ______
   b) Are they limited to particular sections or States? ______

c) What size community do they live in and what are their buying habits?
   i) Cities, and buy at:
      Neighborhood stores ______
      Downtown stores ______
   ii) Small towns ______
   iii) Farms, and buy:
      In town ______
      By mail ______
   d) Are there climatic or employment factors in their location that you should investigate? (E.g. is your product designed only for cold climates or for workers in steel mills?) ______

6. Will the price of the product meet the requirements of your logical prospects? ______
7. Will the price of your product compare favorably with existing products of this kind and with similar products which may be introduced shortly? (See Sections III and IV PP.000 and 000.) ______
8. What is the present consumption of products of this type? ______
9. Is the market for this product likely to change in size during the next 2, 5, or 10 years?
   a) Expand ______
   b) Contract ______
   c) No change ______
10. How often will consumers buy this product? (Is it a durable goods items which will last several years or something in everyday use which needs replacing every few days or weeks?)

____

____

11. Will the product sell evenly throughout the year or will the bulk of the sales be concentrated at certain seasons?
   a) Evenly ______
   b) Seasonally ______
12. What features of the product appeal most to consumers? (Rank in the order of importance.)
   a) Cost of operation ______
   b) Durability ______
   c) Ease of repair ______
   d) Price ______
   e) Service and maintenance ______
   f) Style ______
   g) Trade-in value ______
   h) Other ______
13. What are consumer prejudices, if any, in regard to:
   a) Cost of operation? ______________________
   b) Durability?______________________________

c) Ease of repair? ______
d) Price? ______
e) Service and maintenance? ______
f) Style? ______
g) Trade-in value? ______
h) Other? ______

14. Are products of this kind usually bougth by consumers:
a) For cash? ______
b) On open account credit? ______
c) On a time payment plan? ______
d) On an installed basis? ______
e) With the expectation of a service guarantee? ______

15. If bought on an installed basis, will:
a) Cost of installation be included in the price of the product? ______
b) Installation of the product require additional expense for the consumer? (E.g. rewiring for an electric range) ______

16. Does your product have any possiblilities for industrial use? ______
a) If so, what? ______

______

II. *Channels of Distribution*

It is no exaggeration to say that the success of almost any consumer product depends to an exceptionally large degree on the efficiency of the distribution system which the manufacturer of that product is able to build.

No product has ever been successful unless it flowed smoothly and profitably from the factory through the various channels of distribution into the hands of the ultimate consumers. And no phase of the introduction of a new consumer product is more important than the establishment of proper and efficient channels through which the goods can flow to the ultimate user.

The distribution system in the United States is so highly developed that any manufacturer has a wide variety of possible methods of getting his goods into consumers' hands. This infinite variety of choices makes his sales problem much more simple, and at the same time more complex. If there were only one or two possible ways in which to distribute the product, it would be a simple matter to decide on which to use; but since there are almost any number of paths and combinations of paths that can be followed, the choice of the right paths becomes a matter of very careful study and intelligent thought.

It is vital to remember, too, that the only reason for a distribution system at all is to *make sales to the final user,* and that therefore the system of distribution selected should be the one system or combination of systems which is most likely to accomplish this end with the least difficulty and with the greatest expectation of profit.

Too often manufacturers try to set up their system the way they "think it ought to be," instead of the way it *must* be to make sales, particularly if they have other items in their lines, related or unrelated to the new product. Occasionally, it is possible for a manufacturer with an outstanding product to disregard existing channels and the buying habits of consumers, and yet be successful; but generally *it is much smarter, much more effective, and much quicker to fit distribution as nearly as possible into the existing purchasing habits of consumers and into well-established trade channels.*

Naturally, it is impossible for anyone to tell you the "ideal" method of distributing your product without a close study of the product and all your problems. But if you will actually check through the questions in this section, you will probably gain a much clearer idea of the many things you must know and consider before you come to a final conclusion on this very vital subject.

Check List II. Channels of Distribution

1. Through what channel or channels are consumers accustomed to buying products of this kind?
   a) From retail stores? ______
   b) By mail from catalog? ______
   c) From house-to-house salesmen or demonstrators? ______
   d) From service outlets? (E.g. plumbers, auto or radio repair shops) ______
   e) Others (e.g. direct from ads, book clubs, or cooperatives ______
   ____

2. Put this question to yourself and your friends and associates: "If *you* were going to buy a product like this, from what outlets would you prefer to make your purchase?"
   ____
   ____

3. If the method by which you expect to get the product into consumers' hands is not the same as the preferences stated above, what is the reason for the difference?
   a) Difficulty in securing dealers? ______
   b) Does not follow pattern used for other products in the line? ______
   c) Requires too much time to secure market coverage? ______
   d) Competition too great? ______
   e) Any other? ______________________________ ______
   ____

4. Are the reasons indicated in question 3 important enough to justify

your decision to market your product in a way that does not coincide with the way consumers prefer to buy products of this kind? ______

5. Assuming that you will distribute your product through retailers, what kind of retail stores will sell it? (E.g. grocers, department stores, hardware stores, gift shops, or what?)

____

____

6. How much do you know about the operations of this type of retailer?
   a) How many such stores are there? ______
   b) Where are they located? ____________________

   ____

   ____

   c) What percent of total sales volume in this field is done by:
      i) Independent stores? ______
      ii) Corporate chains? ______
      iii) Voluntary chains? ______
      iv) Manufacturer-owned stores? ______
      v) Consumer cooperatives? ______
      vi) Others? ____________________ ______

      ____

   d) On what basis do your competitors usually sell products of this kind to retailers?
      i) Exclusive franchise? ______
      ii) Selected distribution? ______
      iii) General distribution? ______
   e) If on an exclusive basis, do dealers expect to be protected against competition from other retail outlets in their town or neighborhood? ______
   f) Can you sell your product through both independent and chain stores? ______
   g) At what season of the year will retailers be most likely to buy your product?

   ____

7. What is the best and easiest method of selling to these retailers?
   a) Through established wholesalers or jobbers? ______
   b) Through exclusive distributors? ______
   c) Through your own factory-controlled sales force? ______
   d) Through a combination of the above? ______
   e) Others? ________________________ ______

   ____

8. How many wholesalers, jobbers, or other intermediate distributing organizations are there which can logically handle your product? ______
9. What is the best and easiest method of selling to them?
   a) Through salesmen, demonstrators, or technical personnel working out of the factory? ______

b) Through branch offices of the factory? ______

c) Through brokers, sales agents, or others? ______________ ______

d) Do you know the wholesale trade practices or discounts, allowances, billings, credit, and warehousing which you will have to meet? ______

10. If products like your are usually installed or have a service guarantee, will you conform? ______

11. If so, have you arranged for the service and installation through the distributive channels? ______

III. Competition

It is obvious, of course, that a major portion of the success or failure you achieve with your product will depend on the competition you meet—its size, extent, and intelligence.

In very rare instances, products have been introduced which have been so revolutionary, so extremely important, or so clearly an improvement over everything previously in existence, that it was possible to say, truly, the product had no competition. One might point to the development of a successful means of making a chocolate coating adhere to ice cream (as in the original Eskimo Pie), or the development of nylon yarn, as typical examples of products which had no competition initially. Yet even in these cases elements of competition soon developed.

The chances of developing a new product which the public needs and wants, and which is completely and wholly different from any other product, are extremely small, and therefore it can be accepted that your new product will enter a competitive market in which it will have to win the battle for orders against stiff, intelligent competition. It will have to be made better, or cost less, or be sold more aggressively and intelligently—preferably all three—in order to win a profitable share of the market.

Since these things are true, it is apparent that *you must know your competition* and must be able to beat it or meet it in the market place. So, while the questions in this part of the check list are few and simple, it will pay you to *know* the answers to them as surely as you know the answers to the other questions in this list.

Check List III. Competition

1. What competition will your product face? List:

2. What is the reputation of:
   a) Leading competitive firms? ______________________ ______
   ____
   ____
   b) Leading competitive brands? ______________________ ______
   ____
   ____
3. Are manufacturers likely to enter the field with products similar to yours? ______
   a) If so, what is their reputation? ______________________ ______
   ____
   ____
4. Can any competitor bring out a seriously competitive item quickly? ______
5. Will marketing your product cause competitors to give you additional or keener competition on your established products, if any? ______
6. How does your company stand in relation to competitors in the field to be served by the new product?
   ____
   a) Is your name and reputation already established in the minds of consumers and retailers with relation to:
      i) Similar products? ______
      ii) A general reputation for quality? ______
      iii) A general reputation for dependability, stability, honesty, and fair dealings? ______
7. In the light of your own standing or lack of reputation in the market for your new product, is it best to trade on your company name in introducing the product, or would it be easier and more satisfactory to build up a separate name for the product itself?
   ____
   ____
8. Can your product compete favorably with similar products already on the market?
   a) As to price? ______
   b) As to quality? ______
   c) As to style? ______
   d) As to service? ______
9. To what degree does nonproduct competition exist for your item? (E.g. commercial laundries for washing machines.)
   ____
   ____

IV. Manufacturer's Price Policy

The soundness of the price policy you establish for your product will be the pivot on which the success of your marketing efforts will turn. For that reason, every step in the establishment of a definite and concrete price policy should be carefully considered.

Certain steps in the establishment of a satisfactory price policy cannot be avoided. For one thing, it is obvious that retail prices *must* be in line with competition for the same quality, service, or styling, and that discounts and allowances must be in line with the established practices in the field. Therefore, a thorough knowledge of trade practices and customs in the field served by the product is a first essential in the establishment of a workable pricing policy.

It is likely that, in attempting to enter a new field, or to enter an old field with a new product, you will find it necessary to set up a schedule of discounts and allowances to wholesalers, retailers, and others which will make it attractive to them to add your line; but it should also be borne in mind that many a manufacturer has foundered on the rocks of a system of distributive discounts which has been so large that it has defeated its own purpose.

Unless resale price maintenance is practiced (and this practice, now possible under the laws of many States, should be carefully investigated), an overgenerous scale of discounts can often lead to price-cutting practices under which it becomes virtually impossible for anyone to make any money on your product. If this occurs, the product becomes a commercial football, with results that can be disastrous.

The three most important items, then, in setting a price policy may be said to be these:

1. Be certain that your scale of prices, discounts, and allowances is, at least, in line with long-standing trade practices.

2. Be certain that discounts and allowances are such that each factor in the distributive system, all the way from factory to retailer, not only can, but *should* make a satisfactory profit.

3. Be certain that a proper relation exists between all of the distributive elements who may handle your product, so that no units or groups in your distributive set-up will have so large a price advantage over other units that it is impossible for some wholesalers or retailers to compete successfully with others.

Incidentally, it is well to bear in mind, when setting up your price policy, that clearly-defined Federal laws prohibit the establishment of any price policy which discriminates between individual buyers. You can, of course, provide larger discounts for larger buyers, give special concessions for merchandising service, etc., but in each case these special discounts and concessions must be available on an equal basis to all buyers who meet the specifications you set up.

You should also remember that a price policy embraces more than prices, discounts, and allowances. It embraces everything that goes into the establishment of a true price, including especially such items as credit and collection policies; whether goods are sold f.o.b. factory or on a delivered basis; consignment policies; policy on returned goods, cancellation of orders, and credit for damaged or unsatisfactory goods.

All of the questions in this particular section of the check list can be answered with a simple "Yes." or "No." If you have developed a carefully planned, intelligent price policy, the answer to every question will be "Yes." If one or more questions must be answered "No," you will have clear-cut warning that further thought and attention must be given to these particular questions.

Check List IV. Manufacturer's Price Policy

1. Do you know, in general, what your price policy will be on this product? ______
2. Have you determined what classes of customers will be entitled to discounts and allowances? ______
3. Have you determined the schedule of discounts and allowances to:
    a) Exclusive distributors? ______
    b) Wholesalers? ______
    c) Retailers? ______
    d) Other distributors? ______
4. After considering trade customs in the field for your product, have you determined policy with respect to:
    a) Pricing point? (E.g. f.o.b. factory; local warehouse; delivered) ______
    b) Credit? ______
    c) Credit facilities? ______
    d) Collections? ______
    e) Returned goods? ______
    f) Consignment? ______
    g) Order cancellation? ______
    h) Damaged or unsatisfactory goods? ______
    i) Use of minimum retail prices through price maintenance agreements? ______
5. Have you considered insurance and transportation costs as well as costs of manufacture and selling in determining your price? ______

V. The Sales Program

Having determined the market for your product, the channels of distribution which you will use, the effects of competition, and the price policies which will govern your operation, it now is possible to outline a sales, advertising, and promotion program which is logically and carefully designed to fit the specific needs of this product.

Too much stress cannot be laid upon the wisdom of tailoring your sales, promotional, and advertising efforts to the particular sales problem and the particular product at hand. It may be that your existing sales and promotional set-up can handle the new product with little or no difficulty; but it may also be true that a greatly changed or an entirely independent sales and promotion program will be required for the new product.

It is impossible to generalize on this subject, and only a careful examination and evaluation of all the important facts will provide the proper answer. But what is all-important is that you have a clear realization of the problems involved, and that every effort be made to to make your selling program conform to the needs of the product, rather than attempt to fit this set-up into your existing system, at the expense of destroying its effectiveness.

While sales and promotional problems are very closely related, and should be dovetailed carefully, it may add clarity of thought to consider each separately.

A. The Sales Force. In many instances, the addition of a new consumer product may be extremely simple from the standpoint of sales organization. This is particularly true if the new product is sold through the same distributive channels and to the same people as existing products, in which case the new product merely becomes an addition to the line which the salesman normally sells.

In many cases, however, this simple solution of the sales problem is not possible, because the new product is sold to a different group of buyers, to different people in the buyer's organization, or requires a different type of selling and consequently a different type of salesman.

A simple example of this is the development of drug store lines by several manufacturers whose primary business has been concentrated in grocery stores. Even though these two types of stores seem rather closely related in merchandising methods, grocery store product manufacturers usually find it necessary and desirable to set up completely different sales staffs to sell their products through drug stores, since the buying habits of these types of retail outlets are considerably different, the distributive organizations in the field operate differently, and merchandising and promotional activities must be especially geared to the practices in each field.

Similarly, some products require a sales force which is technically trained to deal with distibutors, while others may require no such training at all; some staple products which are marketed unchanged year after year may be successfully sold by salesmen who have little contact with the factory, while other products may require constant re-education and re-orientation of salesmen, in order to keep them informed of the advantages of new developments, new models, etc.; some products may permit successful sales operations with the use of a small sales force covering a very large territory, while other products may require a much larger force, working relatively small territories intensively; some products may require personal attention from the manufacturer's salesmen all through the distributive process and perhaps even into the hands of the consumer, while in the case of other products the manufacturer's stet salesman may beable to complete his work when he makes his initial sale to a wholesaler or jobber.

Similarly, the system of salesmen's compensation should be carefully studied, whether the company already has a sales force or is about to establish one, because the method of compensation which may be perfectly satisfactory for salesmen selling one line of goods may be comoletely unsatisfactory for salesmen selling a different line of goods.

In general, it should be emphasized again that the manufacturer introducing a consumer product should have adequate knowledge of normal sales processes in the field in which he intends to compete, and should not assume, without investigation, that his existing set-up, or one which he "believes" would be correct, will be most satisfactory for the particular product the sales force will be required to sell.

Check List V. The Sales Program

A. The Sales Force

1. If you have other products, can your existing sales setup handle the new item without harming the sale of the others? ______
2. If you already have a sales department selling other products, will any changes or additions be necessary? ______
   a) Will it be necessary to:
      i) Increase the sales force? ______
      ii) Decrease the size of territories? ______
      iii) Do both? ______
   b) Will it be necessary to set up additional supervision or control in the sales department? ______
   c) Will it be necessary or advisable to change existing methods of compensating salesmen? ______
   d) Will the salesmen require special training to sell the new product? ______
   e) If the product is a seasonal seller, will its sales peak coincide with that of others of your products and thus add to your problems, or will it come at a different time and thus fill in the valleys?

____
____
____

3. If a separate sales force is required:
   a) Will a type of salesman which differs from your present type be needed? ______
   b) Will the men require special or technical training? ______
   c) Will the entire sales force work out of the home office? ______
   d) Do you know how many salesmen will be needed? ______
      i) Have you determined their compensation? ______
      ii) Have you worked out territory assignment and routing through the field? ______
   e) Have you given special thought to the "missionary" selling they may have to do for introductory purposes? ______

B. Advertising and Sales Promotion. Before any consideration is given to sales promotion or advertising activities, the management should definitely establish the principle that all decisions on advertising and other sales promotion should be made with the same objective efficiency that applies to the selection of production machinerey for the factory. No matter how enthusiastic the manufacturer may be over the possibilities of his new product, he must never forget that all sales promotion is designed to help sell the product and for no other purpose. Advertising and sales promotion work, like credit or purchasing, is the work of specialists. Overruling the judgment of a trained sales promotion executive in order to substitute plans based on individual prejudice or personal desires of other executives can be extremely costly, and in some cases disastrous.

Advertising and sales promotion activities are largely intangible, but that does not mean that they are not subject to the operation of sound basic principles. Furthermore, all activities along this line must be considered in relation to the whole marketing problem—advertising and sales promotion are not things apart, but are integral parts of the selling mechanism, and must be coordinated carefully with all the other operations of the company.

Your promotion program, like your sales organization, cannot be completely successful if it is developed arbitrarily—it must be logically and carefully designed to fit the specific needs of the product, and to dovetail as perfectly as possible with all your manufacturing, marketing, and selling operations.

Because sales promotion and advertising are largely intangible, it is wise to give the most careful consideration to them, to make sure that their objectives are clearly fixed, that the policies which are set up hew closely to the line of these objectives, and that nothing is done which does not fit logically and efficiently into the entire merchandising operation.

Furthermore, it should be realized that advertising and sales promotion perform no miracles, and that a successful program requires a great deal more than the purchase of a "big name" network radio show or beautiful four-color pages in the leading magazines. Well-conceived and intelligently executed promotion and advertising plans have been of inestimable value in increasing the sale of hundreds of products, but on the other hand, poorly conceived and ill-advised efforts along this line can have the most disastrous effects on a business.

Sales promotion and advertising may be likened to air and artillery support for ground troops. Their function is commonly to "soften up" resistance and to make it easier for the ground troops (that is, the salesmen) to accomplish their objective. Thus sales promotion and advertising must not only be intelligently conceived and executed, but they must be perfectly coordinated with all the other factors in the sales army. They must be aimed at the right targets, they must be big and effective enough and sufficiently sustained to do the job assigned to them, and—perhaps most important of all—they must be used in exactly the right places and at the right time.

Let us consider individually each of these prerequisites of successful promotion. Aiming advertising at the "right target"—i.e. the logical customer—would seem simple enough; yet, to hit the right target is frequently a major problem. One example of this is the case of an important manufacturer of a woman's product, who conducted a very popular and "highly successful" network T.V. program. That is, the program was highly successful in reaching a wide audience and providing that audience with entertainment. But unfortunately it did not result in many sales because it missed the target. Research showed that most of the prospects for this product were young women of the working-girl type, whereas the T.V. program got the bulk of its listeners from the ranks of somewhat older, married women. These women enjoyed the program, but had little interest in the product.

Similarly, sales promotion must be large enough and sustained long enough to do the job required of it. Some years ago a manufacturer developed a new household product and decided to test its marketability in a single city. He set up a finely executed and well-coordinated promotion and sales campaign in that city, and with the aid of very attractive introductory offers, succeeded in achieving good distribution and making a satisfactory volume of initial sales. But at the end of 3 months, his repeat sales were negligible, whereupon he discontinued the promotion and sales efforts, and prepared to write the new product off as a failure.

Fortunately, however, he was induced to make a check among housewives who had made initial purchases in the city, to determine why they did not like the product and had not repurchased it. This check quickly demonstrated that those who had purchased the product liked it very much, and that the sole reason for the small volume of repeat business was that the "use cycle" of the product had been miscalculated. There was only occasional use for the product in most homes, and it had not yet been reordered because the original purchase had not been used up. Subsequent introductory campaigns, extended 60 days beyond the 3-month period of the original test, indicated satisfactory repeat business and enabled the manufacturer to establish a successful product firmly in market after market.

The matter of making certain that the effort is expended in exactly the right places and at the right time is equally important. The list of advertising campaigns which have been abysmal failures becuase they were launched before adequate retail distribution had been secured is an extremely long one. No advertising at all is far better than advertising which is so poorly timed that it induces the prospect to ask for the product in his retail store, only to discover that the retailer does not carry it. In such cases—and there have been hundreds of them—the damage done is almost immeasurable.

It is easily seen, therefore, that intelligent planning for promotion and advertising consists of a lot more than merely appropriating a sum of money to be spent. Yet no one need be overwhelmed by the difficulties and possible pitfalls involved.

There is a simple way to set up an advertising and promotional program, and to determine how much money should be invested in it. That simple way is to start with a definite objective, or a number of definite objectives, and then, like the military commander, assign each part of the "fighting forces" to the task they can do best.

For example: One of your first objectives in connection with your new product is to get it in the hands of a sufficient number of outlets, in other words, to get adequate distribution. If you have worked out the preceding sections of this check list carefully, you know how many such retailers there are, where they are located, and their size. You also have a pretty good idea of how many of these retailers you will have to sell in order to have adequate retail distribution. You know, too, the extent of your "ground forces" (your own salesmen, and the salesmen of wholesalers and distributors) available to makes the sales.

It is now comparatively easy to gage the extent of promotion and advertising support these "ground forces" will need to do the job most effectively, quickly, and economically. You are then in a position to formulate your promotion and advertising campaign to accomplish this specific objective. You may decide that advertising in the trade papers reaching these retailers, sending out direct mail, exhibiting at trade shows, and similar activities are called for.

Having laid out the plan of campaign, the necessary budget almost sets itself. Instead of picking an arbitrary promotion and advertising budget out of the air, you start at the other end, determining the objective first, then detailing the steps involved, and finally estimating the cost. Of course, you will probably find that the "ideal" campaign will cost more than you are able to spend, and that a compromise must be reached. You must then re-evaluate each item in your list, cut or trim the least important, and finally arrive at a program within your financial means and reasonable from the standpoint of possible returns, but still adequate to perform the specific task or tasks which you want performed.

Operating in this manner, with each step in the promotional and advertising operation given a definite objective, this phase of the company operation becomes as scientific, and as amenable to evaluation of results achieved, as your sales or your production operations.

It must be borne in mind, however, that initial costs—for selling, advertising, and promotion—will probably be extraordinarily high in introducing a new product. For products which have fast turnover and are bought over and over again, it is not unusual for promotion expense during the introductory period to be as high as, or even higher than, the total sales volume. And even in the case of consumer capital goods, which have a long life, the cost of introduction to the consumer market can reasonably be expected to be very high.

How high these costs should be and how long these extraordinary costs are justified are matters for individual determination in each case, but they can be determined, and the potential value of each new user can also be determined. On one product selling for $1, it may be well worth while to spend $2 or more to gain a single new customer; on another $1 product, it may be unprofitable to spend as much as 50 cents to gain acceptance in the introductory period.

One other point should be covered in connection with promotion and advertising, and that is the possible employment and selection of an advertising agency. You may already have an agency, and this agency may be perfectly competent to help you with the new product—or it may not. Whether to employ an agency for the first time, to extend your present agency relationships to the new product, or whether to engage a different agency for it, should be a matter of careful consideration. It should be remembered, however, that advertising and sales promotion are specialized fields requiring specialized talents.

B. Advertising and Sales Promotion

1. Have you an existing advertising and sales promotion set-up which can handle the new product without harming its efforts on behalf of existing products?
   a) If so, how much additional personnel will be needed? ______
   b) If not, are you planning to set up separate advertising and sales promotion sections or departments for the new product? ______
2. Have you an existing advertising agency connection which will be able to handle the new product? ______
3. Have you decided on the details of the advertising program? ______
   a) What advertising and promotional support, if any, will be given locally to:
      i) Distributors? ______
      ii) Wholesalers? ______
      iii) Retailers? ______
   b) Who will bear the costs of this local advertising and promotional support, and in what proportion?
   c) How will you control such assistance so as to make certain that it is used most efficiently for the benefit of the product?
   d) What approach will you use in advertising your product? (See I-12, P.139.)

____
____
____
____

4. Are you familiar with the general sales promotion practices followed by distributors in reselling in this field? ______
5. Are you familiar with the sales promotion and advertising practices of other manufacturers in this field? ______
6. Have you determined what type and how much sales promotional assistance you will give to your own salesmen? ______
7. Are you planning an organized publicity campaign (as distinguished from advertising) to help make your new product familiar to the public? ______
8. Will your promotion and advertising budget be determined on the basis of:
   a) The cost of attaining a definite objective? ______
   b) A certain percentage of estimated sales? ______
   c) An arbitrary sum? ______
   d) Some other system? ______________________________
   ____

Publicity and Advertising

A new small business may not be able to consider much paid advertising, if any. Products for industrial and institutional use can get publicity, if properly presented, that often is more valuable in producing sales than paid advertising. Trade journals regularly run items on wen products. To take advantage of this the following method should be used. Prepare a news item, as below, and send to New Product Editor, *name of journal*:

News Item Date
(For Immediate release)

A new cleaner for aluminum window frames has been developed and is being marketed under the name of "Brite-Lume." This is a non-inflammable, odorless clear liquid that removes dull surface layer from aluminum and leaves a bright weather-resistant film. "Brite-Lume" is manufactured and sold by Brite-Lume, Inc., 340 Dahlia St., Bruton, Md.

When you are ready to advertise and have a definite budget for it, confer with a representative of the newspaper or magazine, who will help you with type, style and wording of your ad. When and if your advertising budget reaches a fair sum you should work though an advertising agency which can be very helpful. Additional publicity can be obtained by writing a general article that mentions your product and company. This can also include a photograph of your plant, compounding and laboratory areas. This article should be newsworthy. For example, as the manufacturer of the aluminum cleaner you could write about the causes of aluminum tarnish, older unsatisfactory products and how this new product was developed.

VI. Legal and Related Problems

In addition to the many phases of the introduction of a new consumer product we have already discussed, there are a number of very important questions the manufacturer must be able to answer satisfactorily before going ahead with a new product. In general, these are problems that should be studied by the company's legal counsel, because they deal primarily with patents and trade-marks, legislation affecting marketing, taxes, etc.

You are probably reasonably familiar with patent and trademark legislation, although it will obviously pay you to make certain that you do not infringe on any other manufacturer's rights in this field. But perhaps the most important legal difficulties to avoid are those which have to do with marketing; and these are likely to be less familiar to you, because in so many instances they differ greatly from State to State and even from locality to locality.

In the national field, the principal statutes and regulations by which the marketer in interstate commerce must abide are antitrust laws, including the Robinson-Patman and Miller-Tydings Acts, the Federal Food, Drug, and Cosmetic Act, and the Federal Trade Commission Act. It will pay you well to have your legal counsel examine your product, your labeling and advertising, your sales agreements, price policies, etc., in the light of these statutes and State Fair Trade and Pure Food Laws.

In the state and local field, laws, ordinances, and regulations which circumscribe the field of operation are too numerous to detail, but mention of a few of them will give you an idea of the possible difficulties which may be encountered unless great care is taken.

Premiums and certain types of contests in connection with the sale of merchandise, while legal in most states, are prohibited in some, and material dealing with such premiums or contests must be kept out of the prohibited states, or must contain a specific disclaimer, saying that the premium or contest offer is not good in such states.

In the case of alcoholic beverages, and a number of other products, state laws vary widely on what may be said and shown in advertising and display material; and in many States, certain types of products, such as drug products, may be sold only in certain types of retail outlets.

A substantial number of cities throughout the country have adopted ordinances circumscribing the activities of house-to-house canvassers, and frequently requiring a license. In other states, restrictive laws have been passed which may tend to cut down or eliminate sales of out-of-state products competing with dominant local industries.

We are all familiar with the fact that health laws, building codes, and sanitary codes are in operation in most, if not all, cities. In many instances, these laws prohibit the use of certain types of products in construction or in homes.

Check List VI. Legal and Related Problems

1. Is the new product patentable? ______
2. Is its trade-mark protected? ______
3. Are all claims to royalties or other indemnities settled? ______
4. Do royalties limit the market for the product? ______
5. Is there anything in the product, its labeling or advertising which may cause you to become involved in a possible violation of a Federal, State, or local statute or ordinance? ______
6. Is the nature of the product such that regulations, trade agreements, etc., may restrict or prohibit its sales in certain areas? ______
7. Is there anything in your pricing policies, trade practices, or selling setup which might involve a violation of Federal, State, or local statutes or ordinances? ______
8. Have local licensing and tax problems been considered? ______
9. Have you determined how the product should be prepared for shipment? ______
10. Have you investigated the relative merits of the various freight carriers available for moving your product to the market? (E.g., rates, accessibility, territory served, handling of perishables, time in transit, etc.) ______
11. Will it be desirable to insure your product while in transit to the market? ______
12. If so, have you made the necessary arrangements for insurance coverage? ______
13. Are there any special wartime, and possibly postwar regulations, which affect your product? ______

13A. Are there any embagoes on shipments to certain foreign countries? ______

14. Are there any labor or union regulations which might affect the manufacture or distribution of your product? ______
15. Are there other problems, peculiar to your product, that should be considered? ______

Summary Check List

When you have completed your consideration of the questions on the check lists closing each of the six preceding sections, use the following summary check list to record your progress in finding satisfactory answers. When you know the answer to a question or have worked out a satisfactory solution, place a check mark in the first column. If you find that you need additional information before you can arrive at a satisfactory answer, place a check mark in the second column. If the question is not applicable to your product, place a check mark in the third column.

The list will then seve as a guide to the degree to which you have satisfied yourself in surveying the factors which influence the successful introduction of a new consumer product. (Note: The numbering system used is identical with that of the six section check lists and it will be noted that in some instances intermediate questions have been omitted where another single question adequately summarizes the problem involved.)

I. The Users of the Product	Have satisfactory answer	Need more information	Question not applicable
1. What types of consumers will use the product?			
4. How many potential prospects are there?			
5. Where do they live?			
6. Will the price of your product meet their requirements?			
7. Will your product's price compare with existing products and with similar products which may be introduced?			
9. Is the market likely to expand or contract in the next 2, 5, or 10 years?			
10. How often will consumers buy the product?			
11. Will the product sell evenly throughout the year, or seasonally?			
12. What features of the product appeal most to consumers?			
14. Are the products of this kind usually bought on time payments—on installed basis—with the expectation of service?			
15. If bought on an installed basis, will the price include cost of installation?			
II. Channels of Distribution			
1. Where and how are consumers accus-			

tomed to buying products of this kind?
3. If you expect to get your product into consumers' hands in some way other than through the usual channels, are the reasons for this decision valid ones?
5. Assuming that you will distribute your product through retailers, what kind of retail stores will sell it?
6. How much do you know about the operations of this type of retailer?
7. What is the best and easiest method of selling to these retailers?
8. How many wholesalers, jobbers, etc., are there who can logically handle your product?
9. What is the best and easiest method of selling to them?
10. If products like yours are usually installed or have the expectation of a service guarantee, will you conform?

III. Competition

1. What competition will your product face?
2. What is the reputation of competitive products?
3. Are manufacturers likely to enter the field with similar products?
4. Can any competitor bring out a seriously competitive item quickly?
5. Will marketing your new product cause competitors to give you additional or keener competition on your regular line?
6. How does your company stand in relation to competitors in the field to be served by the new product?
7. Will you use your company name on the new product, or will you build up a brand name for it?
8. Can your product compete favorably on a price basis with similar products already on the market?

IV. Manufacturer's Price Policy

1. Do you know what your price policy will be?

IV. Manufacturer's Price Policy-Continued

	Have satisfactory answer	Need more information	Question not applicable
2. Have you determined who will be entitled to discounts and allowances?			
3. Have you set up a discount and allowance schedule for distributors, wholesalers, retailers, and others?			
4. Have you decided on your f.o.b. net pricing point and policies affecting credit, collections, returned goods, consignment, order cancellation, and retail price maintenance?			
5. Have you considered insurance and transportation costs in determining your price?			

V. The Sales Program

(A) The Sales Force—

1. Can your existing sales setup handle the new product?
2. If so, will changes or additions be necessary?
3. If a separate sales force will be required, have you determined how to set it up, how many and what kind of men you will need, how to pay them, etc.?

(B) Advertising and Sales Promotion—

1. Have you an existing promotion and advertising setup which can handle the new product?
2. Have you a satisfactory advertising agency connection, or are you planning one for the new product?
3. Have you decided on the details of the advertising program?
4. Are you familiar with sales promotion pratices followed by distributors in this field?
5. Are you familiar with sales promotion and advertising practices of competitors?
6. Have you determined what type and how much sales promotion assistance you will give your own salesmen?
7. Are you planning an organized publicity campaign (as distinguished from advertising)?

8. Will your promotion and advertising budget be based on (a) cost of attaining a definite objective; (b) percentage of estimated asles; (c) an arbitrary sum; (d) some other system?

VI. Legal and Related Problems

1. Is the new product patentable?
2. Is its trade-mark protected?
3. Are all claims to royalties or other indemnities settled?
4. Do royalties lmiit the market for the product?
5. Is there anything in the product, its labeling or advertising which may cause you to become involved in a possible violation of a Federal, State, or local statute or ordinance?
6. Will codes, trade agreements, etc., restrict its sale in certain areas?
7. Is there anything in your pricing policies, trade practices, or selling set-up that might involve a violation of Federal or other statute or ordinance?
8. Have local tax and other problems been considered?
9. Have all transportation problems been considered?
13. Are there any special war or postwar regulations which affect your product?
14. Are there any labor or union regulations which might affect your product?
15. Are there any other problems, peculiar to your product, that should be considered?

VII General Business Principles

CAPITAL

Some can start business on a shoe-string and become successful and others starting with thousands of dollars fail. How is it done? There is no simple rule for success. The following principles have helped others:

1. Do not spend any appreciable amount of money or make substantial commitments before studying and conferring with experts.
2. Limit the amount to be spent for equipment, materials, services, etc. to a definite part of your capital. Do not exceed this for any reason.
3. Do not be tempted to put on a false front. This costs money and is soon found out.
4. Do not accept orders for amounts larger than your capital permits.
5. Do not borrow unless you are *sure* that you can repay the loan, when it comes due, from profits or earnings.
6. Do not hope that certain things will happen within a certain time. If *you* cannot make them happen, they just will not happen.
7. Do not extend credit to friends, large users, or others unless you are certain that they can and will pay bills when due.
8. Instead of extending credit, allow a cash discount on bills paid on presentation or within ten days.
9. Do not tie up cash by buying large quantities in order to get a lower price. Do not speculate by buying for future delivery.

Some of the *do nots* above may be hard to adhere to at times, but they may be the difference between success and failure.

At such time that you are sure of customer approval of your product and that it is profitable to make and sell, then and only then should outside financial assistance be considered.

Before turning to outside sources for capital, see what can be borrowed on your life insurance, by way of mortgages on your real estate or any other unused assets. If equipment or a truck is needed, the seller will usually sell on convenient time payments. Money can often be borrowed on stock and accounts receivable.

It is not usually desirable to sell an interest in a business to an outsider who puts in money without services (a *silent partner*). If, however, a partner can contribute real ability in addition to money, he should be considered. If additional capital is needed, it should be obtained from a bank or individual who is paid interest only. If your business is sound (no matter how small) and your character is good, your bank will try to help you. That is their business.

Going to loan-sharks (individuals or companies that give loans to undesirable business risks at high interest rates) is not advised. It is better to do a small business for a longer time and grow from profits than to jeopardize one's future by trying to grow too rapidly.

If you cannot meet payments when due, then you will be through as a business man.

Duties of a Business Manager

A business manager must be able to accept responsibility and make proper decisions promptly. On small matters, off-hand decisions may be made. On important matters, time should be taken for careful study and consultation with others to avoid serious errors in judgment. You must be able to say yes or no and mean it.

You must be alert and a good mixer. You must get around and not confine yourself to your office, factory and customers only. You must make many friends and learn from every possible source. You must devote much more time and work much harder and more intelligently than ever before. The degree and speed of your success will depend entirely on your effort.

Of course an individual may be an excellent "inside" man, excelling in production and management, but a very poor mixer and salesman. Both of these are needed in most businesses. It may be necessary to hire the corresponding type or, if desirable, to take him in as a partner or stockholder.

BUSINESS ORGANIZATION

If you do business under your own name, no preliminary organization is needed. To do business under another (trade) name, e.g. Delrox Products, it is necesary to file a certificate of registration at the county clerk's or other designated recording office. Such registration merely identifies you legally as the owner. It does not relieve you of personal liability as a corporation does. A general partnership makes each partner personally liable for all partnership debts and obligations.

In considering whether to operate a business as a corporation, you must understand the nature of such an organization. Primarily, it is a form of enterprise in which a person may invest a given sum in the business without becoming personally liable for its indebtedness beyond the amount invested.

To form a corporation, you and two or more other incorporators or prospective stockholders apply for a certificate of incorporation, or charter, and the application is accepted by the appropriate state authority. After all legal requirements are fulfilled, the corporation is authorized to own and manage property, and to conduct the type of business specified in the charter. You turn over property or money capital to the corporation and get back in return shares of stock which represent your interest in the corporation and are readily transferable.

Although the small corporation offers many advantages, such as limited liability, central management, and transferability of ownership of shares, certain practical drawbacks must be taken into account. One is the cost of incorporating. This, with subsequent state franchise, capital stock, and stock transfer taxes, may be considerable. Another is the federal income tax, which is levied against the corporation and again upon the dividend received by each stockholder.

Another disadvantage of corporations is the frequent difficulty in obtaining bank or other credit. The creditor knows that he cannot proceed against the individual stockholders to satisfy a debt thats beyond the ability of the corporation to pay from its own assets.

INSURANCE

There are seven kinds of insurance that are likely to find a place in any business: life, guaranty and fidelity, casualty, public liability, fire, product liability, and workmen's compensation.

Life Insurance must be considered by independent proprietors and partners whose principal reliance for support is in the business. This is not only for the protection of their successors, in case of death or disability, but to enable them to continue the business so that it may be sold as a going concern, or to bring about its orderly dissolution without undue loss.

A practice that has become common in recent years is that of taking out life insurance for the benefit of the business on the partner or corporate officer whose personal service is deemed essential to the welfare of the business.

Guaranty and Fidelity Insurance is used for bonding employees handling funds, for contract performance bonds, and various other purposes.

Casualty Insurance is used to cover the risk of damage to property, such as plate glass, boilers, machinery, fixtures, and buildings.

Public Liability Insurance covers damage or injury to outsiders lawfully on the premises or who are injured or damaged by the owner's truck or other equipment used away from the premises. It is an important safeguard for the shopkeeper, since his action in opening a shop may be construed as an invitation to the public to enter with the assurance of safety.

Fire Insurance is always important since fire is an everpresent hazard.

Product Liability Insurance is very important. This should cover any damage caused to person or property by the purchaser of a product if used according to directions.

Workmen's Compensation or employer's liability insurance is required by law in most states, when a certain minimum number of workmen, usually three or more, are employed, unless the employer can show financial ability to pay compensation awarded workmen injured or killed in the course of their work.

Other forms of insurance, burglary, automobile, travel, and credit, may be necessary or advisable.

Insurance is a complex subject, with unexpected rules and exceptions. Accordingly, it is wise to consult an insurance representative who has a good reputation in the community. His advice should be valuable on all questions relating to insurance.

BUSINESS LAW

While you need not be a lawyer or retain a lawyer, on starting a small business you should be aware of the following general legal-business principles:

1. All agreements and contracts are to be in writing, without any verbal understandings. They should be witnessed when possible.
2. All orders for buying and selling should specify all particulars as to quantity, quality, packing, price, terms, delivery, etc.

3. No guarantees or promises are to be made which you cannot or will not fulfill.

4. Local, state or national regulations regarding the manufacture, shipping or sale of your product are to be studied to avoid any violation.

5. A product is not to be marketed which, if used in accordance with your instructions, may cause injury or damage to people, animals, clothing or any article of value.

6. A reliable lawyer is to be consulted when there is any doubt or unfamiliarity about a problem having legal aspects.

Before entering upon a business it may be necessary for you to comply with certain preliminary requirements. These vary with the state and city, according to the type of business. For example, city, county, or state may require a permit or license to engage in certain occupations, and failure to obtain it, or to pay the fee or tax, may subject the owner to a penalty.

If workmen are employed to operate any type of machinery, the owner must comply with state and local factory regulations. Elevators, scales, or other applicances may require inspection certificates. Occupation of a building for business purposes may require approval of zoning authorities. Also, the premises may have to be approved for safety and sanitation.

SOCIAL SECURITY

There are two separate social security programs.

Workers covered by one program are not necessarily covered by the other. Both programs, however, identify the worker's account by the social security account number which appears on his social security card.

Old-age and survivors insurance provides monthly payments to an insured worker and his family when he reaches 65 or older and stops work, or provides payments to his family after he dies. It is operated by the federal government through the Social Security Board. The same rules for payment of benefits apply to all persons who are covered by it, regardless of the state in which they live.

Unemployment compensation, known also as unemployment insurance, provides weekly payments to an insured worker who loses his job and cannot find another immediately.

It is operated by each state government through its unemployment compensation agency or commission. Each state has its own unemployment compensation law and its own rules for payment of benefits.

In the case of old age and survivor insurance, workers and employers share the cost. The employer takes a certain amount from each dollar of a worker's pay, adds an equal amount from company funds, and sends the total amount to the federal government. In the case of unemployment compensation, however, the employer pays tax contributions of about 3% of his pay roll, except as the rate is modified by law.

LABOR RELATIONS

This area of business operations is more or less circumscribed by union treatment of employee relations. If you determine to enter business in a community, where there is a dominance of mass-production operations, you should contact the responsible representatives of the unions concerned in your chosen field. You should get in touch with these individuals because a preparatory word will always bear fruit in final dealings.

If you confine yourself to small operations, your employee relations will, in all probability, be highly personal. If an employee is affiliated with a union, then it is likely that there will be conditions of employment that may affect the practices in your shop, and these should be ascertained from the employee or his union representative.

Should you, however, enter into mass production or large-scale production, you will find that your employee relations are to a large extent governed by the National Labor Relations Act (Wagner Act), and decisions of the National Labor Relations Board that have been rendered from time to time under that act. Familiarize yourself, therefore, through your advisers, on just what phases of the act and its functions apply to your contemplated type of operation.

In your employee relations, as mentioned, you will find your activities governed to a large extent, in their security phases, by workmen's compensation laws of the state in which you choose to operate, and the various aspects of social security laws of our federal government.

The total number of employees you may hire before subjecting yourself to the incidence of any of these laws differs among the several states. In any event, irrespective of the state in which you operate, you will be required to comply with all phases of the social security and workmen's compensation laws. Therefore, it would be well for you to determine the requirements of the laws of the state in which you contemplate activity.

Nearly all social security laws require some pay-roll deductions and payment by the employer, and, in a few states, the employer is further required to make deductions from the employee for employee contributions required by state statutes.

Remember that you can catch many more flies with molasses than with vinegar. It is recommended, therefore, that you contact personally the agencies handling employee relations just as you do your landlord, your banker, the local town officials, or any other persons whose good will you value.

GOOD WILL

Good will is the foundation of all successful businesses. We all know that it is useful, but many of us do not try to build it all the time. It is important that we build good will not only with our customers but with everyone with whom we come into contact—with suppliers, employees, neighbors, etc. This will often work out most advantageously. When materials are scarce the seller will prefer to supply those who pay promptly, do not make unjust claims and treat their salesmen courteously. They often can give advance notice of intended price changes, scarcities, or other valuable information.

Fair treatment of employees will result in better and greater production, less spoilage and damage, and countless other advantages. Courteous treatment of truckmen, mail-carriers, electric service men, etc., will not only result in better service but may produce additional recommendation or use of your products.

Keeping the good will of neighbors involves a little work and expense but is always worth while. Attention should be given to keeping dirt, odors and smoke at a minimum.

Good will is built with customers by not making any claims or statements that cannot be fulfilled completely; by making prompt shipment; supplying uniform products that arrive and remain in good condition; by offering the same prices and terms to all without any exceptions; by being courteous, obliging, and rendering any reasonable service.

RENTING VS. BUYING

A new business, if possible, should rent rather than buy space. A few years of experience will put you in a better position to decide when and where to buy a factory.

Even though a factory be bought, it cannot be considered rent-free. Charges such as interest on mortgage, taxes, maintenance, etc., must be taken care of. The tying up of capital to the amount paid in cash must also be considered. Factory buildings must often be sold at a great loss when times are bad.

Leasing has many advantages. The initial investment is lower, leaving you more cash for equipment and operating costs. There is less probability of needing a bank loan. You enjoy freedom from such ownership risks as fires, tornado, or value changes.

In rental payments, it is just as important to avoid extravagance as it is to avoid miserliness. A man is proud of the business he is establishing and of the personal mechanical skill upon which he counts to make that business a howling success. He naturally wants to dress it up to reflect his confidence and pride.

However, this attitude can be overdone from the viewpoint of dollar-and-cents income. Remember, a reasonable appearance of economy is no liability to a new business. People are essentially conservative and will mistrust a new business which has spent too much on flash.

Rent has to be paid every month whether or not the business is making a profit. Therefore, rent is going to be one of your major items of fixed expense. Do not cut down a limited working capital by obligating yourself to pay a discouragingly large rent bill every month. High rent may not seem important during a period of good business, a prosperity cycle, but it may strangle your business if a slump comes along before you have been able to accumulate adequate reserves.

Ordinarily you would build only if space requirements were not obtainable or no existing plant could be bought on favorable terms. If you are to buy, your financial position must, of course, permit the investment. Keep in mind possible needs for plant expansion as the business grows. The time to check on whether or not sufficient space for expansion is available is before you buy or lease.

Many sectional or demountable small buildings are available at times. If one can be bought at a bargain, and if it is more suitable for your purpose than any available building which can be leased, then its purchase should certainly be given careful consideration.

Location

The proper location of a factory is one that requires serious study in order to get the greatest economic advantage and also to avoid possible dangers. Of course, if the start is very small, location is not too important and moving is not very costly. Considerations that are important for a medium-sized plant will be mentioned here. Those considerations that are not of great moment can be disregarded.

Distance from Home

The nearer the factory is to your home, the less time and energy will be wasted in traveling back and forth. Since you will be working longer hours, this becomes doubly important.

City, Suburbs, or Country

The advantages of the city are chiefly those of transportation, saving of time and the convenience in your calling on customers and their coming to you. Similarly, your employees will find it advantageous. Shipping and receiving time of materials, containers, etc., is usually better. Postal deliveries are also better.

The disadvantages of a city are higher rent, special taxes, manufacturing restrictions, and minor matters, such as noise, dirt, poorer light, etc.

Raw Material Source

In most small businesses, it is not imperative to locate near the source of raw material to avoid excessive transportation charges. If, however, such charges are proportionately large and can be eliminated by proper location, then it is in order to locate as near the raw material source as possible.

Finished Product Delivery

If your finished product is bulky and most of it is to be delivered to one place, then location near this place would effect reduction in carrying expense.

Hazards

When inflammable materials are used, special tanks, containers, ventilators, etc. are required in certain places that may cause unexpected outlays. Locating next to a hazardous business will increase your insurance and danger. Odors or loud noises produced by your factory may be obnoxious to your neighbors, or theirs to you and your employees. In such cases, it is often desirable to locate in the country in order to avoid creating a nuisance.

Public Nuisances

Air and water pollution are two nuisances that must be carefully studied and avoided. Different municipalities have different regulations. What is permitted in a previous town or in the same town at a later date, may not be allowed in a new place or time. Installations to absorb smoke, noxious fumes and objectionable liquid or solid wastes can be very costly.

Water

When large quantities of water are required, location beside a stream or where good well water can be obtained will effect a saving against buying city water.

General Considerations

Many a location which looks perfect on a map and seems to satisfy nearly every requirement is actually unsuitable. For this reason it is necessary to see the plant site and its possibliities with almost X-ray vision.

Is there any "catch" to the property you are planning to lease or buy? Specifically, are sewer connections available? Is the road paved so that access is possible at all times? Is there ample police and fire protection? How near is the property to a fireplug? If you buy, is the title clear? Before making a considerable financial investment in the community, it is important to know just what your tax dollars are going to buy that will be of direct value to you.

Is the town generally favorable to the establishment of new businesses? The leading citizens of some towns resent the introduction of new businesses and try to block the issuance of a license to a stranger. The old-timers control everything, own most of the property, and run the town council. Keep out of such a town and save your bank roll, your courage, and your optimism. Some other community nearby may be a better place in which to establish a new business.

The plant itself is discussed fully later. But before you consider any building, be sure that electricity, gas, and water will be available at all hours and seasons in the amounts required. Is ample pressure maintained at all times? Do not take anything for granted. Find out whether or not the water charge is included with the rent; if not included, learn whether water is sold at a flat rate or on a metered basis. In some places water rates are extremely high.

The cost of commercial electric current may be a measure for determining how much of the shop equipment should be electric-powered. It may be necessary to buy a Diesel generator set to insure ample current for the shop's power and lighting requirements.

Locations To Be Avoided

In a previous paragraph you were advised not to set up in a town that is unfriendly to new business. Within each town or city will be found areas which are undesirable and detrimental to the success of a particular business.

Most cities have blighted areas, usually old run-down residential districts on the fringes of the commercial section. In these districts hoodlums, petty thieves, moochers, panhandlers, and hoboes may be numerous. The presence of such persons not only complicates the plant-protection problems by day and night but tends to keep timid customers away.

Petty thievery might make it impossible to store bulky materials outside the shop; tools disappear too often. The disadvantages and losses might amount to more than could be saved on rent.

It may be necessary to have a watchman at all times. Even this may not always prevent broken windows. Locating near a railroad or certain factories may be detrimental because of heavy smoke, fumes, vibration, etc.

Older cities are often situated in the lowlands at the fork of two rivers. Some sections of those cities may be imperfectly protected against flooding in time of high water. The record of a location with respect to the inconvenience and damage caused by floods should be investigated. Some entire cities are below the level of the river levees and suffer periodically from uncertainties due to weather conditions. Even in cities which are not in the river bottoms poor surface drainage and lack of storm sewers can lead to damage from heavy rain or high water.

It may be well to discuss matters with the Chamber of Commerce, Board of Health, police, church, bank and insurance agent to try to get additional useful information about your proposed location.

Type of Building and Layout

Industrial vs. Commercial Buildings

A factory or industrial type of structure is preferable to space in a business block or commercial type of building. Each type has been developed to best meet the requirements of the kind of business that it is intended to house.

Floors

A concrete floor may itself be a sufficient foundation for light machinery if the foundation bolts are set in neat cement. Check the thickness of the concrete in the floor. Drill a hole through it if it can be measured in no other manner. Do not trust heavy loads on it unless it has a safe margin of strength. In comparison with concrete, wooden floors, whether of planks or blocks, are not as strong, durable, stable, or level. Also, while wood is softer and less tiring to the feet, it is likely to become oil-soaked and then the wood floor is extremely inflammable.

Ordinarily a ground-floor location is preferable, as it lends itself better to the provision of proper foundations for machinery. Ample footings are especially important in the case of machine tools. They are assembled units and their parts can be thrown out of alignment by unequal lagging. Poor alignment creates strains and vibrations which make accurate work impossible.

Level, nonsagging floors are also necessary. Since these conditions may not be found on upper floors, except in the case of a steel and concrete building, your first effort should be centered upon finding a ground floor location either without basement, or with only a part basement in which the heating plant is installed.

Ceiling

Next, examine the ceiling. If it has exposed steel beams that is all the better. To unload equipment a chain block or hoist is required. This hoist should hang from a beam strong enough to carry safely any load it will be called upon to support. Such structural features are not generally found in a commercial business block, and laboriously sliding or rolling heavy loads off or on a truck can add up to a lot of expense. Such loading also slows down work, and takes men off more important tasks.

A ceiling height of 13 or more feet is desirable. This allows for setting up mixing kettles on legs or a stand so that a drum can be placed underneath for filling. It may also be useful to build a platform around the kettle near the top, from which to load it. A man should be able to stand upright on this platform.

Single-Floor or Multi-floor Buildings

Single-floor factories are usually operated more efficiently than those having more than one floor. The exception to this is when the upper floor is used for an office or laboratory. In a single-floor unit, there is no need for an elevator or hoist and everything can be easily seen by the manager. Any difficulties or stoppages can be attended to promptly. There is no time lost by the manager and employees in shifting materials or containers from one floor to another. Also, second floors, unless specially reinforced, will not safely carry heavy machinery and stock of materials and finished products.

Yard Space

A certain amount of yard space, or vacant land, adjoining the factory, is very useful for storing empty drums, carboys and waste.

Light and Air

A light and airy factory is desirable because it tends to increase efficiency and reduce waste and hazards.

Windows and Skylights

Windows are very important and are likely to be found properly placed, or can be easily installed, in the industrial type of structure. Have plenty of windows and keep them clean, not just for the sake of appearances and good houskeeping, but also because cleanliness will pay dividends in many ways. Good light from windows will cut down your electric bills, lessen the possibility of accidents that could result in loss of time as well as injury to an employee, and help to prevent mistakes in measurements.

If the window and skylight area take up less than the equivalent of 30% of the floor area, the daylight inside the room is insufficient.

Actual measurements have demonstrated that vertical windows lose half their efficiency if left unwashed for 6 months. If they are sloping, the light loss is much greater and might be as much as 83% in 6 months. And, remember, some three-fourths of that dirt collects on the inside of the window. A washing is especially desirable in the fall of the year when the days become darker and shorter.

Electrical Facilities

When the equipment has been installed, call in the lighting-fixture salesman with his intensity meter to insure that all general lighting and special job lighting will be properly placed and in the right amounts and intensities. Ecxellent artificial lighting costs about one cent per hour per workman. It is poor economy to work in badly lighted premises.

Most of the electricity generated in the United States is for use as 3-phase, 60-cycle, 220-volt current, or, if only two wires are brought in, it is single-phase, 60-cycle, 110 volt. Check with the local public utility company, however, to insure that 3-phase current is available for the shop's power needs. To discourage theft of bulbs, use 220-volt lighting current, if possible, because it takes bulbs that cannot be used in homes wired for 110 volts.

The machinery and equipment will have to be provided with motors suitable for the kind of electric crurent available. Some sections of cities are supplied only with direct current, and families moving into such areas find their electrical gadgets useless unless they happen to have universal-type motors. Heating devices can be used on either alternating or direct current, but motors (above fractional horsepower, universal-wound) must be supplied with the current for which the rotors are wound. When you are buying second-hand machinery it is especially important to determine whether the motors are wound for the proper current.

If inflammable vapors will be present see to it that all lights, switches, motors, etc., are of the explosion-proof type. Steam lines should be insulated, to avoid heat losses. Water lines should be so placed as to minimize freezing risks during very cold weather. If acids or other corrosives will go into the drain, the latter should be of acid-proof ceramic, lead or other suitable alloy. All steam lines and steam-heated process vessels should be of proper construction so as to be safely used at the steam pressure required.

Space Requirements

The first considerations in the choice of a building, once its location and structural characteristics have been determined, are the amount and layout of the floor surface and the available headroom or interior space. A cramped interior tends to restrict the amount of work which can be carried on. Crowding also prevents the natural growth of the business. It makes good housekeeping difficult, if not impossible.

By all means provide plenty of space, both horizontally and vertically. Allow for too much rather than too little. Avoid the loss, both of time and money, in having to move unfinished jobs around to make space for new or rush ones.

The size of the entrances is also of prime importance. Can the largest truck used for delivery purposes by your suppliers drive inside to be unloaded? Can large cases brought to the shop be trucked inside and easily unloaded without excessive hand labor? Can the door be opened easily by one man for entry of trucks and is there a smaller door for persons to enter or leave?

Large entrances also supplement the lighting and ventilation facilities, and offer easier escape routes in case of fire, explosion, or other unusual hazards.

Fire Hazards

If the space occupied is only a portion of a larger building, are there fire doors and fire-resisting walls between your space and the rest of the building? Make certain that your fire insurance rate will not be excessively high because of a neighboring occupant's type of business. In case of doubt have an experienced insurance agent check the building and advise what the insurance will amount to and what, if anything, can be done to reduce the rate. Small technicalities, easily corrected, sometimes increase insurance rates excessively.

Aisles

Wide aisles are essential along all traffic lanes in a wellmanaged factory. Often the aisle edges are painted a light color, and employees are prohibited from stacking anything in this space. Work is speeded up, accidents reduced, and damage to work lessened if hand trucks operate only on aisles marked for them. Do not skimp on passageways when estimating the floor area required for the factory at the start and as it grows. Shafts, housings, pillars, braces, or other dangerous projections which extend into the aisle space and which cannot be removed should be painted in attention-compelling colors. Yellow and black bands painted on such obstructions will prevent collisions. Use yellow striping also to mark the edges of open pits, elevator wells, and so on.

Heating and Ventilation

Heat is not usually included with the rent in a factory type of building. However, a heating plant might have been installed by the owner since it is to his advantage that the space be satisfactory to the tenant. If heating equipment is to be installed by the tenant, a choice must be made between the various types of systems available, hot water, steam, vapor, or unit heater.

Usually the heating system is designed to combat the average low outdoor temperature rather than extreme low temperatures. It is not likely to be damaged by being overworked during short spells of unusually cold weather. Temperatures that should be maintained for efficient working conditions are approximately 65° for the factory and 70° for the office. Extremely cold days seldom exceed 5% of the heating season; cold weather averages 20%, mild weather 75%.

Health and comfort demand a proper circulation of air in summer and winter. Ventilation cannot safely be left to chance alone, especially when there are noxious fumes in the air. In some places, such as engine and boiler rooms, the air has to be changed every few minutes. In the average shop about 1,200 cu. ft of air are needed per person per hour to provide proper air sanitation and air cleanliness. Installation of a hood and blower to eliminate dust and fumes at their source may provide sufficient air turnover for the whole shop. Humidity is best maintained at 50% but can range safely from 35 to 80% in ordinary shops.

If you content yourself with poor ventilation in order to "save money", you lose the price of good ventilation many times over in lessened output by your employees. If natural ventilation through doors, windows, and skylights is not dependable, it should be supplemented with exhaust fans and, if necessary, with ducts which will reach the bad spots.

Lease Provisions

The discussions up to this point have been mainly concerned with the choice of a satisfactory building, without consideration of the conditions under which it should be leased. There has been some mention of making certain what facilities, water supply, toilet connections to sewer, plant protection, are included in consideration of the rental payments agreed upon.

After making a thorough investigation of the various sites and buildings available, and arriving at a decision to lease a certain property, the principal matter remaining to be settled is the period of the lease. Should rent be paid from month to month with the right to move out on short notice? Such an arrangement might be a boomerang.

What if the property owner, seeing that the new business is successful, decides to double the rent? He may think that you cannot afford to take time off the job to find a new location and then go to all the expense of tearing down shop and installing the machinery, equipment, and supplies elsewhere. If his rent gouge succeeds, he may raise the amount again later on. He can even dispossess you on very short notice to rent the place, possibly to a competitor who ascribes the success of your business to its good location.

If the location has been carefully selected and the rights and obligations of owner and tenant are fully and clearly agreed upon, in writing, you need have no hesitation in signing a lease for one year or longer. You should reserve the right to renew it for a similar period by giving notice some 60 days before the end of the period. A signed lease will be one thing settled, one less matter to worry about. The rent may never again be as low as it was when that first lease was signed, so make the best of it.

There are few good reasons for you to sign a lease for more than one year. However, it may happen that the building selected will require extensive alterations. The alterations can be paid for by the property owner if he has an occupant who promises to lease the place for a term of years. No such concession can logically be demanded on a short-term rental or lease agreement. On some long-term leases, provisions are included which permit the rental rate to be increased at stated intervals. The amount of the raise may be arbitrary, or may depend on the profits of the business, or some locally recognized index of business activity.

Responsibility for maintenance of repairs to the building, the roof, plumbing, floors, walls, and other parts should be clearly agreed upon. The owner and the tenant may have opposing points of view on this subject and any differences should be brought into the open and fully discussed.

Plant Layout

The factory as a whole is a production machine and must be suitable for the work it is to turn out. To make certain that it is right for your work, a simple production analysis is necessary.

Consider what will be the flow of work through the factory; the processes through which the material must pass and in what order; the number and sizes of machines or equipment required; and the volume of business which must be handled in order to insure the financial success of the venture. If the layout is clumsy and amateurish your production machine will function in an awkward and inefficient manner. A well-planned flow of work will decrease the labor cost, shorten hauls between machines, cut down the time of work in process, lessen space requirements, facilitate inspection, and encourage such neatness and orderliness as can reasonably be expected in a factory.

It is preferable to have all production activities on one floor. If a work space of serveral floors is under consideration, remember that elevators cost money, time, and power, supervision will be more difficult, machine foundations more complicated, and illumination less efficient.

The position of the truck entrance to the shop is also important. Can the delivery trucks reach the storage room without interfering with work in progress? If you have an inspection bench, is it set in a spot where there is good natural light? Can additional windows be installed economically where necessary? Will the lease permit the installation of more windows?

Suppose a factory in the neighborhood offers a subcontract? Is the plant layout such that the material can flow naturally from machine to machine or process to process? If some extra equipment is required, will space be available in which to install it? Hindsight is of little use when such emergencies come up. But with foresight these emergencies become opportunities for unexpected profits.

A good method for checking layout is to draw, to fairly exact scale on a large sheet of paper, the floor plan of the shop you intend to lease. Show also the size and location of doors, windows, and partitions. Make it like the house plans which home magazines picture. It is not difficult.

Then, to the same scale, cut out of heavy paper the plan of each piece of machinery, equipment, workbench, storage space, or other space-occupying item. Arrange them on the floor plan for efficient flow of work, taking best advantage of natural light and keeping in mind the various other factors mentioned in this chapter. Shift again and again until you make certain that ample space will be available. There is no substitute for this method of checking proposed floor plans. The largest industries use this procedure.

Measure to see if there are plenty of aisle spaces. Indicate on your plan where the aisles must be to insure the easy flow of work from one bench, machine, or station to another through the shop. Is there enough space left for expansion as more equipment and working space are needed?

This simple planning removes all guesswork. You know definitely whether or not the building will be satisfactory. Some minor compromises will have to be made to fit the flow sheet to the available building, but if any major disadvantages are indicated it is best to face them before the lease has been signed or a building purchased.

It has been mentioned previously that space outside of the building is essential to the successful shop. Is there sufficient area to take care of parking, an irreducibly small waste pile, a stock of drums, cases, etc., which cannot be stored inside? Is there space in the street or alley for delivery trucks to back in easily?

While looking around the outside of the proposed location glance at the windows. How easily can the shop be broken into? Some of the windows may need bars which you can make and install at little expense.

Adequate drainage of the yard is important. If it is not paved, perhaps a thick layer of cinders, properly graded, will be sufficient. If the yard is in good condition, some jobs can be done outdoors in good weather.

How does the building look from the outside? Some paint might be a good investment and would not displease the property owner either.

Make it easy to keep the plant surroundings clean and attractive.

CREDITS AND COLLECTIONS

The need for the small businessman to extend credit to his customers raises a very important issue. An enormous number of businesses have failed because they were too lenient in granting credit. Knowing this, you will probably be tempted to adopt a no credit policy. However, such a course would limit your field of business opportunities to such an extent that you would never be able to earn more than a very meager living, if that.

The assumption that you must take losses if you extend credit is false. Some losses may occur, and probably will, but these are usually due to unforeseen circumstances. Where credit is extended when there is a serious doubt of collecting, you are not extending credit. You are gambling in an effort to enlarge your business. This type of risk-taking is common practice but is one which you should not attempt with limited capital. Nor should you ever do it without first giving careful thought to the possible results.

When properly and carefully applied, credit can be a profitable business builder. There is no necessity for taking wild chances. Always remember that, as a small operator, you can be in close contact with the activities of your customers merely by keeping your eyes and ears open. What may appear on the surface to be a risk may be perfectly safe when all the facts are known. And, of course, the reverse is also true.

You should decide upon a credit program before you become actively engaged in business. The extent of the program will be limited by the amount of working capital at hand, that is, money you have available to pay such expenses as materials, labor, and monthly maintenance bills. In other words, you must consider, before granting credit, how long you can operate with the money left over after setting up business.

After leaving a substantial margin for safety, you can then estimate the amount of credit which you are in a position to handle. It would be wishful thinking to count on getting loans from your bank or from other sources to supplement your working capital for several months, at least. By carefully estimating your ability to handle credit, you can determine your limit. You will not be turning away profitable business by refusing to accept any terms except cash, nor will you be expanding beyond your depth.

The next step is to determine how to extend this reasonable credit to the best advantage. During conversations with prospective customers, you will have discussed the question of prices, volume of business to be expected, and other details. Although you may not have discussed credit terms at great length, you will have some idea of what to expect.

Trade customs vary in different localities and individuals have different ways of doing buisness. It would certainly be foolish to insist on extending credit if the customer prefers to pay cash, or to insist on a monthly settlement if he prefers paying his account weekly.

The less credit it is necessary to extend to get an order, the greater is your opportunity to take on more business. Your working capital is just as much a tool as is a motor or mixing tank. Work it hard but do not make it do needless work.

By carefully estimating the amounts each existing account will require and taking the turnover, which means frequency of settlement, into consideration, you can compute the funds still available for business expansion. If an account can be expected to run $1,000 a month and weekly settlement is made, it will take approximately $250 capital to handle it.

Do not estimate too closely. A settlement within 30 days usually me ans it will take 45 days before the check is deposited and cleared. However, profits tend to offset this lag as your business operations progress. In other words, the $1,000 account does not represent a $1,000 expenditure since it includes profits.

Learn the credit practices existing in each field and be guided by them. Generally speaking, particularly lenient terms should be offset by higher profits. If profits are low and price competition is keen, credit terms will usually be restricted.

Government contracts involve practically no risk from the standpoint of credit. However, the problem of having your capital tied up for a relatively extended time makes it imperative that you consider the advantages and disadvantages of such business. You may be limiting the amount of business you can do with private firms, and, incidentally, government contracts usually do not bring automatic repeat business. On the other hand, a completed government contract is very often adequate security for a bank loan if the government agency is willing to accept your assignment.

Manufacturing firms will differ as to credit risk, and the advantages or disadvantages of handling their business will depend upon the type of operation in which they are engaged.

These are only some of the very general practices. You will have to learn the individual credit habits of each customer with whom you do business.

The investigation of a prospective customer's credit standing is a most important and difficult problem. Before a banker loans money, he requires a financial statement, and he does not hesitate to ask for any other pertinent information he may deem necessary. After a thorough investigation, he has a complete knowledge of you and your business. A supply house will follow a similar procedure before extending credit.

One source of information on prospective credit customers is your banker. He is in a position to know the financial status of practically every businessman in his community and can often help you to get information on firms in other towns. However, this source of information can only be used after you have established bank connections and your banker looks upon you as one of his regular customers. You can readily see the advantage in becoming well acquainted at your bank as soon as possible.

Direct information can also be obtained for a moderate fee from local credit investigation associations. In many instances the firm you are investigating may be a small corporation or a partnership in which one or two people have the controlling interest. In this case, information on the men personally may be indicative of their methods of running their business. If John Jones does not pay the butcher and grocer, it is quite possible that the John Jones Manufacturing Company will not pay you.

In the case of unincorporated firms, the owners may have other assets in the form of property, other businesses, or securities which strengthen an otherwise weak-appearing credit.

Much of this type of information can be obtained directly from a credit report.

After you have collected all the available information, you must make your decision. The application of sound judgment and good business sense is involved here. There are several pertinent points to keep in mind in making this decision.

First examine the moral risk. All information pertaining to the subject's paying habits and previous dealings should be considered. Remember, no matter how strong his ability to pay, unless he has the habit of paying promptly, he will be troublesome.

Next, examine his ability to pay. Two classifications may be used in analyzing ability to pay, the primary and the secondary. The first group consists of businesses that are being operated at a profit, and for which the prospects seem good. In other words, you grant credit to such a firm because it is a profitable, going concern, and the work that you are doing for it will help build its profits.

The second group is composed of credit risks which have other security besides the business, such as property owned, or endorsements or guarantees by substantial backers. The credit in this case is being extended more on the worth of the owners or backers than on the profit-making ability of the firm. Although such a customer is usually dependable, settlement of his account is often slow and sometimes requires costly legal action. The ability to pay arising from a profitably operated business is very definitely the stronger of the two.

All credit information gathered on a customer should be put in writing and kept on file where it will be readily accessible for future reference. Add to this any information you may obtain through contact with him.

A word of warning is in order at this point. Be particularly careful of business that is too easily obtained. If the prospective customer requires very little selling before he agrees to give you an order, it is quite possible that his credit has been canceled by others because he has failed to pay his bills.

In obtaining credit information, there are several precautions that need to be recalled. Never quote the source, or discuss the information you have obtained. The average businessman considers his credit above reproach and resents prying inquiries.

Avoid telling the prospective customer that credit cannot be extended until an investigation has been made. No matter what he may say, he will not like it.

Large business houses usually have definite requirements before extending credit. If their sales are lagging, they will relax the requirements. If they are getting all the business they can handle, they will make the requirements more strict. Nevertheless, their credit program is essentially impersonal and inflexible. This policy is a great boon to the small businessman who has difficulty in competing on price alone with his larger brothers. Because he is closer to his customers and because the extension of credit is a personal matter, he can do things which the large operator is unable to do safely. Make wise use of this very definite advantage.

Overextension of Credit

A great deal can be said about overextension of credit. Its causes are numerous and its results disastrous. In fact it probably is the chief reason why many new businesses have failed. The two main contributing causes of such overextension are carelessness and poor judgment. Carelessness can be eliminated by being forewarned and ready. Good judgment is a necessary asset for the successful operation of a business.

The first indication of overextension of credit will be a cramped cash position. You will find that you are unable to take on new business and will begin to lag in the payment of bills. It will amaze you how quickly this will get around to your creditors and how quickly they will begin to limit your credit. A great deal of time, effort, and energy can be wasted in trying to operate with insufficient funds.

To avoid trouble, carefully examine the practices which lead to overextension of credit. It can usually be traced either to a poorly constructed credit program or to the failure to follow a well-planned one. A laxity in the extension and collection of accounts is usually evident.

Overextension of credit does not necessarily mean that the accounts which have been opened are not satisfactory from a credit point of view. You may have been expanding the business faster than the existing capital justifies. If so, you would do well to discuss the matter with your bank and obtain additional working capital, if possible.

All too often, however, your trouble may lie in allowing the mental attitude of "he's good for it" take the place of sound judgment. As a result, you acquire accounts that are not readily collectible and then you fail to put forth the thought and effort necessary for their collection.

Another frequent mistake is to allow profits to remain in accounts receivable. This practice is likely to develop when the collection program is governed entirely by the needs of the business. In other words, no attempt is made to collect until it is seen that cash will be needed for such necessities as rent, pay rolls, and materials. You then call on your creditors as though they had become your bank account.

The danger of such a system is that you contact the persons whose accounts are most readily collectible, and allow the slower or less conscientious customers to slide. Since your main objective is to raise cash quickly you naturally turn to the sources where you have the best prospects for getting it. As a result, you eventually have on your hands a large accumulation of past-due accounts which are difficult to liquidate. In time you find your business in a very precarious position.

The first evidence of overextended credit is to be seen in your bank account. Therefore, if you consistently try to keep your bank balance at a reasonably high figure, you will be on the right track.

However, the extension of credit operates so quickly and so easily that methods other than maintaining a good bank balance are necessary to control it. One of the best is a periodic analysis of account by delinquency. The more frequently this checkup is made, the closer will be the control. Such an analysis can be made by listing the dollar value of past-due accounts into groups. The accounts may be grouped according to the length of time they are overdue, as 15 to 30 days, 30 to 45 days, 45 to 60 days, and over 60 days. Of course, the past-due classification will depend on the general credit program and the credit practice in the community.

This examination of your accounts should be made at least three times a month. It will quickly show whether they are getting out of hand. A steady increase in the over-60-day group means trouble.

Keep in mind that an analysis made on June 30 is more truly comparable with an analysis as of the preceding May 30 than with one made on June 15. The reason is obvious: Between the first and the fifteenth you can normally expect more remittances from your customers, whereas the analysis of the thirtieth would include many accounts from which you would receive remittances in the following week.

On the more serious past-due accounts, those over 30 days or over 45 days, depending on the credit customs of your locality, it is suggested that a further analysis be made. These accounts should be listed, by name, on a columnar work sheet. The first column should show the total amount of the account; the second column, the amount you have received on the account in the past 30 days; and the third column, the amount received in the 30-day period preceding the last.

The value of this sort of analysis is easily seen. An account on which you are receiving money in regular amounts is not as serious, from a credit point of view, as one on which you have received no money for some time. However, it is still a problem from the point of view of tying up your working capital, and should not be encouraged.

If these analyses are made frequently and used intelligently to guide you in keeping your accounts in order, they will be a great aid in preventing overextension of credit.

Avoid, so far as possible, advancing credit to small occasional customers. To attempt to watch and control a large number of small accounts is a full-time job. There is a tendency on the part of the occasional customer to leave his account unpaid until he wants to place another order. Then, if he does not have the money available to pay you, he will order from a competitor. The result is that you not only have the inconvenience of handling the account, but actually lose business. There are several methods of discouraging this practice. A good-natured, businesslike manner is one of the best. Don't be afraid to ask for your money.

However, you may find it necessary to advance some credit to small customers to expand your business, and thus build up a larger number of accounts. In this case, it would be well to make your past-due account analyses by number of accounts as well as by dollar value. It takes a lot more time and effort to collect $100 from ten different people than from one. By comparing the number of accounts as well as the dollar value in your analysis, you are able to determine which class of customer is causing the trouble.

Another helpful guide is the relation of the volume of outstanding accounts to cash. Such a relation will vary during the month but a definite trend will be indicated. Watch this ratio carefully; then your bank account will not get so dangerously low as to cause you to go on a collection spree.

If, through carelessness and neglect, the accounts are allowed to get out of hand, the remedies are neither easy nor productive of quick results.

Collection Methods

The effectiveness of your collection method will determine the credit that you are able to extend. No one can afford to give credit unless he is able to collect later.

That a well-organized collection program can be outstandingly successful is shown in the operations of small-loan companies. These companies loan money in relatively small amounts to people who, from the point of view of income or collateral, are the lowest extreme of possible credit risks. Yet the losses of these companies average less than 1/2%!

It is not recommended that you study the methods of the small-loan companies and attempt to adopt them for your own use. Those collection methods must be utilized that best suit your business and the types of debtors you will have. Different methods have to be used with different customers and the more versatile and resourceful you are, the better will be the results.

Learn the trade customs of your competitor so that your collection program will be neither too strict nor too lenient. Remember that extension and collection of credit are a part of the service rendered. Find out the needs and habits of your customers and be guided by them. Collecting money is an easy job if it is properly approached.

Your mental attitude is an important factor. Do not be afraid to ask for your money when it is due. Be genial, but forceful. Make up your mind that you are going to get your money. Your customer will sense this determination and forget any intention he might have had to stall you.

When you call to collect from an important customer, do not be overawed by a haughty secretary and a mahogany desk. Be self-respecting and businesslike in appearance and manner and you will get results.

Never be apologetic in asking for a settlement, for you are then placed on the defensive. Do not tell a long story about needing the money to pay the rent or meet the pay roll. The reason you want the money is that it is due you.

Courtesy always pays, especially if it is instinctive. Nobody likes a grouch. Surly belligerence, usually caused by a subconscious fear that the customer will get the best of the interview, gets you nowhere. Moreover, such an approach is likely to jeopardize the future of your business.

The suggestions given above are not intended to imply that you will have great difficulty in collecting. Be systematic in sending your bills and you will find that the great majority of your customers will remit within a reasonable time after receiving your statement. These accounts cause you no trouble from the collection angle.

In the rendering of your monthly statements, keep two things in mind. They should be accurate, and should be sent out promptly. Everyone resents getting a statement for an amount already paid or for a wrong amount. The necessity of sending out corrected statements sometimes leads to embarrassment.

You can hardly expect your customers to pay you promptly if you send out your statements haphazardly. The average concern receives statements shortly after the first of the month, verifies them, and issues checks in payment. If your statement comes in several days or a week later, the chances are it will not be paid until the following month.

A reasonable time should be allowed after the mailing of statements before any follow-up is made. Even if prompt attention is given your statement, a period of a week or 10 days may elapse before the remittance reaches you. The routine of verifying the statement, issuing the check, and mailing takes time. Before mailing the statements, consider whether or not it would be best to collect by personal calls. The disadvantages of calls are the time consumed and the fact that some customers may feel they are being dunned.

On the other hand, a personal call permits you to get your money several days sooner. It also gives you an opportunity to become better acquainted with your customers. And you will be able to find out if there are complaints or suggestions for bettering your work and service.

Remember that your customers will not remain with you automatically. They must constantly be solicited. Inasmuch as your work is of a manufacturing nature, statements will occasionally have to be adjusted for unsatisfactory merchandise and adjusting can be kept more to your advantage by a personal call.

Another point to be remembered is that some of your customers may be extending credit. Thus, their bank accounts are at the lowest point on the first of the month, and will remain so for several days until their monthly receipts start coming in. If you appear for a check on the first or second day of the month, you may not get it.

Find out when your various customers are in the best position to pay, and contact them at that time. You will find your relationships will be much more pleasant.

The slow accounts should definitely be followed by a personal call. Use of the telephone is often satisfactory, but past-due notices are hardly appropriate. Do not let thoughtless and crude collection methods drive good business to your competitors.

You may have few bad accounts but those you do have should be collected because accounts written off come directly out of your profit.

There is real difficulty in recognizing an account that has gone "sour," and distinguishing it from a good account which has temporarily slowed down owing to some setback. Ability to differentiate between these two is largely a matter of good business judgment.

The first thing to do is to study carefully the credit information which you have accumulated on the customer. Ask yourself the reasons for the extension of credit in the first place Have they changed materially?

The chances are that the account is one on which no credit report was obtained. In other words, the account was opened on a hunch. It would be well, in this case, to call immediately for a credit report.

These reports usually show the names of others who have made similar requests, and the dates. If a number of your debtor's other creditors have made recent inquiries, you can be sure you are not the only one who is not being paid.

Businessmen who have put themselves into a tight spot by failure to meet their bills often adopt methods of defense that differ according to their temperament. One of the most common is the evasive type. He will never be home or at his plant when you call. When contacted, he will endeavor to give you a check which may be worthless. However, if you see the customer during banking hours, you can verify the worth of the check. Eventually you will get payment from the evasive type of customer if you persist.

Other types of customers are the arrogant or indifferent ones. These are classed together because their attitudes arise from similar basic backgrounds. They usually have been through this before and know all the tricks. They will tax your disposition to the breaking point, and endeavor to arouse your temper.

Be careful of your legal position. These types are quick to bring action. Attempt to get a check. If it is worthless, it is a definite legal weapon.

It might be well to attempt to obtain a promissory note. Be careful of an offer of a high interest rate as this is in violation of the usury law.

The subject of checks bears discussion. Whenever possible, avoid depositing checks that may bounce back. Bad checks put your bank to a lot of trouble and give your banker a poor impression of your customers. You may want to borrow money sometime, and your accounts receivable are one of your most important assets.

Moreover, if your account is low and your bank has let you draw on uncollected funds, a bad check may cause some of your own checks to be returned, in which event your standing with the bank will be greatly damaged. If you are dubious about a check, cash it directly at the customer's bank. In general, it is poor practice to accept post-dated checks, but you may find that some of your good substantial customers will post-date every check 4 or 5 days for no particular reason. This habit is usually a hang-over from early days when their capital was not adequate to meet business needs.

Often, when presentation of the statement is made shortly after the first of the month, the customer will not want to give you a check until he has received some returns on his own accounts. To accept a post-dated check under these conditions is satisfactory. It saves you a return trip.

The subject of handling bad accounts would not be complete without discussing legal action. A common practice is to turn over a number of bad accounts to a lawyer to collect on a percentage basis. Even then the results are not always satisfactory.

Avoid court action as a general practice. When it seems necessary consult a good lawyer. If, after other means have failed, you feel you must resort to legal action, make it count. Attach the customer's property or assets. Follow through.

BUSINESS GROWTH

A small business making a fair profit is far better than a large business operating at a loss. It is not wise to expand too rapidly. It is better to grow slowly each year, without taking great risks, than to attempt too rapid growth. Do not be jealous of the size of another's business. In time you too can grow. Do not be too concerned with dollar volume. If you build a good foundation and conduct your business on a profitable basis, it is sure to grow. The first few years of a business are critical. At this time your mistakes will be made on a small scale. You are learning. Later, when you have learned from your experience and have built a smooth-running organization you can gradually increase your size with little or no danger.

Keep as much of your profits as possible in the business. Keep your personal drawings to a minimum. In this way, your capital will grow and enable you to extend your operations.

FINANCING

From the day you open and as long as you stay in business you will have dealings with a bank. Therefore, it is wise to choose one that will best serve your needs from the standpoint of convenience and services offered.

After you make your choice, your first contact with your bank should be by introduction. If you have a friend or business acquaintance with good connections there, so much the better. You will be asked for considerable information on your background as well as your business setup. Give full details. If there is an unfavorable incident in your background, give the necessary facts without reservation.

On your initial visit and for some time after your business is established, your dealings with the bank will be confined to your checking account. Keep a sizable balance; it will get your credit off to a good start.

The best way to build a good reputation with your bank is to operate in a businesslike manner. This means you must pay your bills promptly and keep your accounts receivable collected.

Bills can be paid promptly only if you do not overbuy, if you accept no credit except on regular trade terms, and if you avoid expanding faster than your capital and ability warrant. In other words, if you purchase materials on the basis of amounts required for filling specific orders, your investment will be quickly converted into finished orders which will bring you either cash or accounts receivable.

As to accounts receivable, efficient collection will provide cash with which to pay your bills and thus relieve you of the necessity of accepting extended terms. Furthermore, your banker will consider proper control of your accounts receivable as a sound asset on which loans may be based. To him, machinery and equipment have only junk value, and inventories of materials are little better. Even aside from his opinion, these particular assets have a limited market in normal times.

At whatever time you need bank credit, your banker will consider you a poor risk if you have availed yourself of too many sources of credit or if you have extended too much credit. He will realize that in either or both cases your working capital is strained beyond your capacity.

Here is an illustration. You fill orders with materials which you have not paid for and which have been processed on machines you are buying on the installment basis. If you extend credit too freely, you will run into serious difficulties. On the surface, your business will seem to be growing by leaps and bounds, but if you are unable to collect in some cases you will learn too late that you might better have turned down the orders in the first place.

The day will come when you will need more capital if your business is to grow. For instance, a profitable business deal will be offered but you will be unable to accept it without financial aid.

Here is a time when you cannot afford to be overconservative. By all means, go to your banker and get his advice before turning down such an opportunity. You may be surprised to find that, in spite of your limited capital and other assets, he will consider you a good risk.

Before approaching a bank for your first loan, decide on the exact amount you want to borrow, and have definite, sound reasons for requesting it. Your first loan and each succeeding one should be sufficiently substantial to take care of your specific need.

At this point in your business career, your manner of approach is especially important. It must be both businesslike and confident. Remember, your banker will base his judgment on your mental attitude, background, assets, the growth of your business as indicated by operating statements, and the reasons you want to borrow. If you are convinced in your own mind that your position is sound, your manner should reflect that conviction. If it does, your banker is bound to be favorably impressed.

Banks grant credit with extreme caution. The banker has trained himself to make decisions on facts and not on emotions. This conservatism is necessary. The banker owes a debt to the community and he knows it. He realizes that the bank is an integral part of the community. If the bank fails, or even if it is suspected of loose credit practices, the whole economic structure of the community may be affected. In contrast, failure of your business for lack of credit would be unfortunate for all directly concerned but relatively unimportant from the general point of view.

The bank is interested in helping responsible borrowers. It makes its profits chiefly through loaning money. Other services are offered as a convenience to its customers. Commercial banks take deposits, handle checking accounts, discount notes, handle foreign exchange, aid in credit investigations, and give advice on many other financial affairs. Such services provide little revenue.

Since loans are the chief source of his profit, the banker is always willing to listen to a good loan proposition. Try to see things from his point of view. First and foremost, he must be sure the money will be returned. This money is not his. It belongs to the depositors and stockholders. So, regardless of profits, the money must be returned. A bank does not sell money; it loans it. The second thought in the banker's mind is *when* the money will be returned. You are not the only customer. There are others to be taken care of, and their wants must be anticipated. When the money will be returned is very important to the banker.

With these basic thoughts in mind, we can proceed to an examination of details. A common conception of banking principles is that if the prospective borrower's ratio of current assets (those which can quickly be turned into cash) is two to one over current liabilities (debts which must be met within the next 60 or 90 days) he is eligible for a bank loan. Or to put it another way, if he has $2 in his pocket, and owes only $1, he is a good bank risk. Generally speaking, the two-to-one ratio is an indicator. No bank will loan money to anyone to pull him out of difficulties with other creditors. Remember this fact, and avoid embarrassment and trouble later.

Banks will loan money for temporary needs only and must know when the money will be returned. They are not in a position to loan capital for long periods of time. Neither will they loan money for speculation or promotional purposes where there is very meager capital invested by the would-be borrower. Many businessmen feel that their ability, time, and ideas should warrant the financial backing of a bank. They do not want partners with capital because partnership would mean splitting possible profits. Instead, they feel that the bank should put up the money. If the venture prospers, the bank gets its 6% for the period the money is used; if it fails, the bank loses. The borrower, however, stands to gain all but the 6% or to lose nothing. Consider this sort of proposition from the banker's point of view and see how ridiculous it sounds.

In seeking bank credit you will be required to fill out several forms asking for detailed information. Be sure that you answer each question thoroughly and honestly. The first and most essential step of a credit investigation is to uncover misstatements and discrepancies in your answers.

There are definite advantages in establishing bank credit. In the first place, your business will have certain seasonal fluctuations. Also, there will be special demands from time to time which you will not be able to meet with your normal profits.

In other words, no business can have several thousand dollars lying idle to take care of all emergencies. If you waited until sufficient profits were accumulated before making expansions of any sort, the growth of your business would be so slow that it would never really get anywhere.

Business involves a certain amount of gambling. By forethought and planning, the items of chance are curtailed as much as possible but they can never be entirely eliminated. Your banker provides the funds with which to expand soundly so that your basic capital will produce the largest possible profits. But he will depend on this basic capital to cover the element of chance.

As said previously, your banker should also be considered in the light of an adviser. Because of his varied knowledge of all business activity in the community, his opinion is of great value in formulating future plans, buying property, expanding business activity, and in providing supplementary services to your business. His advice and aid in cash planning will be invaluable, and although it may be conservative, it will certainly be basically sound.

Your first encounter with the need for obtaining credit will probably arrive before you are fully aware of it. It will be with a finance company or discount house when you purchase equipment for your factory. Financing of this sort is usually handled in one of two ways, depending on the plan used by the factory or dealer from whom the equipment is being purchased.

Under one plan, your note is retained and collected by the factory, the money being furnished by the finance company which accepts collateral trust notes as security. Although the finance company's money is being used, your dealings will be entirely with the firm from whom the equipment was purchased. With the other plan, the note is immediately discounted with the finance company, in which case the company handles the collection.

The unfortunate part of either plan is that the notes must be paid or you may find yourself operating a factory without equipment. The equipment is direct collateral and may be repossessed if the notes are defaulted. The same finance company may also be handling the notes on your car, truck, refrigerator, radio, or washing machine, and on the Federal Housing improvement loan for the new roofing job on your home.

The terms offered on the purchase of machinery are usually 1 to 3 years, although longer terms are sometimes given on the purchase of expensive pieces. The interest rates charged are relatively low, and the down payments required usually depend on the demand for the machinery. If the firm has no difficulty in disposing of it, the down payments will usually run 20 to 30%. A highly competitive market will push this down to 10%, even lower in some cases.

But do not base your ability to purchase on the cash available for a down payment. Carefully consider the monthly notes you will face. Then estimate conservatively the amount of monthly business you must do to make enough profits to meet these obligations. The notes will be a fixed expense regardless of how much or how little business is done.

Do not be oversold on equipment merely because the credit terms seem easy. Instead of asking yourself, "Do I need it?" ask, "Can I get along for a while without it?" Remember, a small down payment means a higher monthly note.

Nor does your problem end with selecting equipment and agreeing on terms. State laws differ in regard to installment selling. Some require a chattel mortgage form. Others use a conditional sales contract. Whatever the form of the legal instrument, it will be a promise to pay a certain amount each month on a definite date.

Decide which time of the month will be most satisfactory for you to pay. Obviously, it might be embarrassing to have the contract dated the same day the rent is due. It would also be foolish to have it dated the latter part of the month, if you anticipate having considerable cash tied up in accounts receivable at that time. Also, bills for material will come due on the 10th.

When several pieces of machinery are purchased at one time, it is customary to cover the whole purchase with one note. However, a single contract may or may not be convenient or wise.

You may have purchased a machine which at the outset you thought would be needed. Later you find it cannot be used for the required work or that your investment is not warranted. It may be that you would like to keep the machine but just cannot afford to.

If there are separate notes for each unit, it will be possible to pay off the finance company in full for the machine in question. Then you can sell it, and thus eliminate a portion of your monthly cash outlay. If all the machines are covered by one note, none can be sold until the entire note is paid in full.

On the other hand, if you have only one note, bookkeeping will be simplified.

These are just two points. Consider all angles before signing. The finance company will not want to make any changes in a note after it is discounted. They will expect you to fulfill the terms of the agreement exactly as it is written.

Incidentally, read the contract before signing it. Never sign a blank contract.

Supply-house credit is a matter which should be given due attention. The supply house will furnish the materials used by your shop in turning out the finished products. Credit will be needed as a matter of convenience.

If materials are being purchased several times monthly, it would be a burden to write a check in payment of each delivery. At the outset such handling might be advisable, but as the business get under way a more satisfactory arrangement should be made.

The supplier will require information similar to that requested by the bank. Forms will have to be filled out and questions answered. Since the supply house makes its profits from sales, its credit program will be developed to increase them in every way possible. For example, such suppliers often offer exceptional terms during slack seasons.

Terms are usually 2% discount 10 days, 30 days net. Statements are sent out at the end of June for that month's purchases. If you pay by the 10th of July you will get a cash discount, usually 2%. Otherwise, the full amount of the bill is due on the 30th.

When you start in business, you will probably be asked to remit by the 10th and no cash discount will be allowed. After the supply house has had some experience with you, the regular trade terms will be allowed.

While competition is a controlling factor, credit customs vary in different localities. Terms of 60 to 90 days are frequently offered as an inducement to buy. Unfortunately, this type of inducement usually leads to overstocking or purchasing unneeded materials, and encourages lax paying habits.

If you are operating on a limited capital, accepting such generous terms is extermely questionable. Never forget that an important part of the formula for sound business practice is "pay as you go," as much as possible.

Planning for Cash Requirements

Planning for the cash requirements of the business can be divided into 2 classes: (1) Planning for long-time or future requirements, which involves the need for reserve funds, and (2) planning for current needs.

There is no infallible rule to follow in planning for future requirements. Indeed, it is difficult to estimate accurately how much cash should be laid aside from profits for the replacement of machinery as it wears out or becomes obsolete.

On the other hand, reserve funds for taxes can be estimated so that an amount of cash approximately equivalent to the need will be available when the taxes come due. If you will establish a reserve fund for taxes you will avoid drawing on the cash balance on hand at an inopportune time.

Reserves must be kept in the form of cash or readily marketable securities. Of the two, cash is greatly preferred. You will have sufficient problems to keep the business going properly without attempting to keep up with stock market quotations.

The cash for this reserve should be kept in a separate bank account to avoid its being absorbed by the business. It should not be invested in real estate, nor should it be borrowed by the business for meeting occasional current needs.

Generally speaking, it seems a bit far-fetched for a new venture to consider putting away part of profits for future needs. At least for the first year, you will have your hands full providing cash for current needs.

But do not lose sight of the fact that a reserve fund should be started as soon as possible. A good plan is to deposit 10% of the profits in an account for contingencies. Make sure that this fund is reserved for emergency purposes only. If it is allowed to grow over a period of years you will have a very fine "spare tire" to help you survive the bad periods that come to every business.

It is not suggested that you attempt to decide mathematically how much you will have to save monthly in order to replace your machinery at the end of an estimated 10 years or so. In the first place, there is no way of knowing what the price of a similar machine will be at some future date. New developments may call for replacement sooner than you expect. On the other hand, through intelligent care your machine may last twice as long as you first anticipated and still be up-to-date.

In the final analysis the best way to provide for future needs is to put away a certain amount of your profits; 10% has been suggested. Perhaps your profits will warrant a larger amount. On the other hand, it is quite possible that in the first few years of business 5% will be the maximum you can spare. Even if it is only 1%, put it away. It is your insurance against future difficulties.

Planning for current cash needs can be done with a fair degree of accuracy, and your success as a businessman will depend largely on your ability to do so. The average successful small businessman has trained his mind so well that he has only to know his current bank balance. From constantly being on the alert to have adequate funds available at needed times, cash planning has become second nature to him.

An executive of a large business, however, is unable to keep all of the details in his mind. He must plan on paper. Although in your venture you can hardly adopt all big business methods, it is highly recommended that you do your cash planning for current needs on paper.

Here is a suggested method, involving the use of a large columned pad. Across the top, head the columns with your fixed cash requirements, rent, notes on machinery, bank notes, etc. Leave a column for subtotaling, and head the remaining ones with such variable cash requirements as accounts payable, electricity, materials, and pay roll. Careful thought should be given to making up this first analysis so that nothing is omitted.

In the left-hand column, list the dates vertically. In the rent column you will put the amount of rent on the date it is payable. The analysis may well be made for 60-day periods. Total the amounts across to the subtotal column.

It is now apparent at a glance just how much cash will be needed each day for fixed cash outlays. As bills are received, or as it becomes evident what the variable amounts will be, enter them and total across. You will then know at least a week in advance just how much cash you will need.

Taking June as an example, your subtotal for fixed outlays for June 10 is $200. On June 2, you receive a statement for materials amounting to $500 which should be paid on June 10. You check your other variable items immediately and find that none will fall due on this date. So, your total cash requirement for June 10 is $700. You know this on June 2 and have a week to plan for it.

If you find that variable cash demands can be estimated with some degree of accuracy, you will be able to determine cash requirements several weeks in advance.

By carefully checking your analysis of accounts receivable, you are in a position to know well ahead of time what your cash position will be.

Assume, for example, that your 60-day cash-planning sheet for May-June indicates a heavy demand during the first week of June. It may be caused by any number of reasons, among them negligence and carelessness on your part. Whatever the cause, you will know about it in May, a month in advance, and can then decide what must be done.

Perhaps a bank loan is in order. Your banker is consulted and is told that $1,000 will be needed in the following month. If you are poised and confident, the banker will be favorably impressed and, other things being equal, the chances of getting the loan are fairly certain. What might have proved a serious blow to your credit standing has been avoided.

The subject of withdrawals of funds for personal use should be mentioned. The type of business organization, whether corporation, partnership, or single proprietor, will have a bearing on this matter.

In case of a corporation, you will draw a fixed salary and the profits will be paid in the form of dividends, or will be left in the business as surplus. In a partnership, profits are divided according to agreement, and careful records must be kept of the partners' drawing accounts to make proper settlement. In the case of the single proprietor, probably the best system is to decide how much you will need for personal living and to draw this amount each week. In this way, a proper balance is maintained, and profits can be accurately determined.

Along this same line is the question of what to do with surplus earnings. Many a small businessman has worked hard and intelligently, putting in long hours and living modestly. Eventually his business starts producing profits above the needs of normal expansion.

When you arrive at this profitable position in your career, turn to your friend, the banker, for advice. He can help you invest your profits in a sound manner.

BUYING MATERIALS AND EQUIPMENT

Equipment

Makers of equipment have engineers who will be glad to study your problem and make the proper recommendations. It is imperative that you disclose to them complete information about your raw materials, processes, finished products and customer requirements. Only with such complete information can they really help you. If you can afford to get the advice of a qualified engineer, do so by all means. He can guide you not only in the selection of equipment and accessories but also in setting them up so that they will produce efficiently and avoid any hazards.

Second-hand equipment has its advantages and disadvantages. Its only advantage is lower initial cost. Cost of maintenance and repair parts are usually higher. Used equipment should only be bought from a reliable dealer.

No equipment, new or used, should be bought unless the seller suitably guarantees, in writing, its proper performance under the conditions required by you. If possible, where there is any doubt of its performance, arrangements should be made to try it out.

Most equipment can be bought on terms, the seller retaining a chattel mortgage for the unpaid balance.

A list of sources of equipment is given in the appendix.

Raw Materials

For small quantities of raw materials, it may be necessary for you to buy from a wholesale druggist or chemical supply dealer. When you are ready to buy in barrels, drums or larger units you may find it advantageous to buy from a manufacturer's agent or directly from the manufacturer.

A comprehensive listing of suppliers of chemicals, drugs and allied materials is given in the *Buyers' Guidebook Number of Chemical Week* published yearly by the latter at P.O.Box 386, Times Sq. Sta., New York, N.Y., 10036, and the O.P.D. Green Book, Oil, Paint & Drug Reporter, 100 Church St., New York, N.Y. 10007.

Importance of Prudent Buying

If after working in the factory all week, calling on prospects, managing the business, and studying the trade magazines at night, the owner looks forward pleasurably to the day salesmen come around, so that he can become the purchaser for a change, he is probably headed for a spree of overbuying.

At such a time it would be well to go to the stock room, and take a good look at the mistakes made in moments of rash buying, on the impulse to be a "big shot" and impress some salesman, in moods of over-enthusiasm regarding an anticipated big order (the one that got away), or an order given so as to bring its shipping weight up to that required by the carrier's minimum charge.

These mistakes are easily recognized, looking back at them, but hard to avoid at the time they are made. Checking on them is the first step in learning to avoid letting pressure of any sort influence what should be a cold and calculating procedure, the compliing of a stock order covering minimum amounts of indispensable materials and operating, maintenance, and office supplies.

In his budget a man sets down the many items that are a drain on his funds, rent, power, heat, water, raw materials, supplies, labor, and so on. Every dollar spent for any of these items has to be earned back by the shop and carry a profit along with it.

Take a look at the notes due next week on equipment purchases. Count up the bills coming due the first of the month, the rent, the sign painter, light and power, stationery, pay roll. Get yourself down to earth before making out that order! Figure out not only what you want to buy but, even more important, what you can do without.

While buying on a hand-to-mouth basis, it is important to patronize a firm that makes frequent deliveries and has fairly complete stocks. Stick closely to the supplier who can furnish the greater part of your requirements. Such a policy will entitle you to expect more consideration from him, reduce the number of checks to be made out each month and provide better justification for the extension of more favorable open-account privileges as time goes on.

Remember that to get good service from a supplier your business must be worth having. Spread thin among many, it will not be attractive to anyone. After all, mutual confidence between supplier and purchaser is the basis of all successful business relations.

How To Get Best Prices and Discounts

Advice on obtaining best prices could as well be headed, "Buy large quantities in order to get best trade discounts." But quantity buying is contrary to the viewpoint which places emphasis on keeping purchases small in size while the business is new and untried and working capital must be turned over as rapidly as possible. So far as quantity discounts are concerned, the owner of a new business, unless he has ample funds, must content himself with avoiding extra charges for broken packages. A penalty for buying in what would be practically retail quantities is something to guard against.

Use a calling card to establish your identity as a businessman every time you make a purchase, and ask what your discount is. If you are told there is no discount for a purchase of that size, ask then and there what the discounts are for larger lots, and record the information for future use. Possibly the amount you planned to buy is just a little under the first breaking point and by increasing the order slightly you may receive a profitable discount.

Ordinarily the clerk will not give you this information on discounts unless you ask him directly. If his attitude is not co-operative or if his authority to quote discounts is questionable, ask to be waited on by the department head. Make this man's acquaintance. Then, if at any time you find it necessary to telephone orders, or to send a messenger for them, he will recognize the account and there will be no delay or argument at the counter about getting lowest prices.

Buying at wholesale usually necessitates purchasing a quantity not less than a fixed minimum in order to get a price below retail, and discounts often vary, according to quantity bought.

There is usually a reason for the quantity discounts being as they are. In the case of a certain chemical, for instance, the price on 10-lb lots compared with 100-lb shipments was exceptionally high. Investigation disclosed the former was based on express shipments from Philadelphia (halfway across the country), whereas the rates for the latter were determined by the freight rate, which was low on 100-lb shipments. You can avoid much ill will over seemingly exorbitant prices by making a point of finding out and recording systematically the various discounts, penalties, or other price factors for every separate item you purchase.

Until you can buy in quantity, your best discount will be for cash. Sometimes an even higher one is given for cash with order. A simple, profitable rule to make is; never pass up a cash discount. Even the low rate of 1/2% 10 days net 30 days is the equivalent of 18% per year interest.

Aid From Salesmen

The value of maintaining cordial relations with the salesmen who call at your shop cannot be emphasized too strongly. Your own personal calls to get trade will help you discriminate as to how salesmen should or should not be treated. Remember that through them you can keep closely in touch with technical developments which will affect your business.

Usually you will find that the salesmen covering your trade follow certain schedules and count on making their calls at fairly definite and predictable intervals. Therefore, if at all possible, set a particular time during which salesmen can be sure of finding you in your office; then be there, even if it is not always convenient for you. And your allocated time should take into consideration the travel habits of the majority of salesmen who stop regularly at your factory.

Then, regardless of your worries or frame of mind, give each salesman an open-minded audience, and listen to his story courteously. Tell yourself there is no man on the road from whom you cannot learn something of value. Whether or not you intend to give him an order, treat him as respectfully as you would a customer. If you are alert and on the job he can be useful to you. He can tell you any amount of the personal bits of fact or gossip which it took years to collect and which are remembered only as they have some bearing on getting business. These salesmen can be of assistance before you set up your factory if they know you are contemplating such a venture. They can recommend favorable locations or possibly tell you about an established factory whose owner wants to get out of business and would sacrifice some of its value to do so quickly.

A salesman who is on friendly terms with you, even though he gets no business from you, may be the one who can tell you where a job-lot of miscellaneous bottles can be bought from a factory at less than quoted prices. When his own firm has had to repossess some equipment nearby, the salesman, it he wants to help you, can often divert it your way, at an attractive price, rather than have it go back to the warehouse. Or, if he knows a certain factory has gone out of business and its equipment and supplies are about to be auctioned off, he may call it to your attention, and you might buy up at bargain prices some needed supplies or piece of machinery.

If one of your best accounts is really shaky financially, some salesman's credit manager has probably discovered the fact and you may get a tip-off to stop your selling efforts and do some fast collecting instead. By the way, if that ever happens and you are not having much luck at collecting, ask the person with whom you have the account to sell you enough supplies from his stocks of raw materials to cancel the indebtedness.

Inventory Control

Perpetual inventory control arranges for practically automatic reordering. This system involves determining the maximum stock which should be on hand or on order at any one time and the minimum amount below which stocks cannot, with safety, be allowed to drop. Between these extremes is a reordering point.

For a small business an inventory card is the only record needed. On it show the maximum and minimum stock figures, the reordering point, the amount to reorder each time, and date of each order placed but not yet received. Carry running totals or orders received and of removals from the bin or rack to show, with reasonably accuracy, the stock on hand.

If you are extremely busy when the salesman calls, he may, to prevent delay, check the cards for you, give you the list of requirements, and facilitate making out of the order.

Perpetual Inventory Card

<table>
<tr><td colspan="2">Unit of issue:</td><td colspan="4">Description:</td></tr>
<tr><td colspan="2">Location:</td><td colspan="2">Unit value:</td><td colspan="2">Time required to obtain:</td></tr>
<tr><td>Max. stock:</td><td colspan="2">Min. stock:</td><td colspan="2">Reorder point:</td><td>Minimum ordering quantity:</td></tr>
<tr><td rowspan="2">Date</td><td colspan="4">Quantities in terms of units of issue:</td><td rowspan="2">Remarks</td></tr>
<tr><td>On Order</td><td>Rec'd.</td><td>Del'd.</td><td>Bal. on hand</td></tr>
<tr><td></td><td></td><td></td><td></td><td></td><td></td></tr>
<tr><td></td><td></td><td></td><td></td><td></td><td></td></tr>
<tr><td></td><td></td><td></td><td></td><td></td><td></td></tr>
<tr><td></td><td></td><td></td><td></td><td></td><td></td></tr>
<tr><td></td><td></td><td></td><td></td><td></td><td></td></tr>
<tr><td></td><td></td><td></td><td></td><td></td><td></td></tr>
</table>

Two or more times a year check the card total against the actual count for all items. Make occasional spot checks (a few scattered items at a time) in between regular inventory periods to make sure that employees are noting removals on the cards and that smaller items are not disappearing.

Make certain that you buy according to commonly understood specifications. On arrival of the merchandise, inspect it to see that you are getting just what you ordered.

Instructions on Use of Card

Unit of Issue. Release material from stock only in units as specified in this space. This may be dozens, pounds, feet, each, and so on.

Description. Describe the material as it should be entered on an order blank. This is for use in ordering and also enables proper replacement of cards.

Location. Make entries such as "Rack No. 1," "Office," "Cupboard No. 2."

Unit Value. Give cost per unit.

Time Required To Obtain. Show here how long it takes a shipment to arrive, starting from the day the order leaves the shop.

Maximum Stock: Indicate the largest amount of stock estimated to be on hand and on order at any time. Change the figure as business increases or to reflect seasonal variations.

Minimum Stock: Indicate the danger point below which stocks on hand must not drop without the risk of being out of stock.

Reorder Point: Considering time required to obtain, enter here the amount of stock on hand that is to serve automatically as the point (time) to place a minimum replacement order. If properly timed, the order will arrive before the stock on hand reaches the minimum figure but not so soon that it will increase stocks on hand above the maximum desired.

Minimum Ordering Guantity: Show in this space the size of order to be placed under average conditions. (Reread paragraphs on importance of prudent buying.)

Date. Put down the date of each entry on the card. Start a new line whenever necessary. Have cards printed with plenty of lines, fold in half if unhandy.

Quantity on Order. Make record of orders, when placed, in same terms as shown in *Unit of Issue.*

Quantity Received. Record shipments received and scratch out previous entry in *On Order* column.

Quantity Delivered. Remove material from stores to production only in exact quantities specified in *Unit of Issue.* Change these quantities as experience dictates.

Balance on Hand. This is the actual balance on hand. Make no entry in this column for materials on order. Keep this balance up to date after each receipt or withdrawal of material.

Remarks. If material has been allocated to a job scheduled for the near future, note it here so the reorder point can be stepped up, if advisable.

RECORDS AND BOOKKEEPING

Consider the number of bankruptcies and failures that are constantly in the courts and the relatively few small firms that survive their second year. Did these establishments know from day to day the status of their business?

The United States Department of Commerce has made a number of surveys in an attempt to find out why so many small businesses fail in the first, second, or third year. The purpose, of course, was to publicize the chief reasons so that small businessmen would know them and could avoid making mistakes.

Among the things brought to light was that a large percentage of such firms kept no records and an overwhelming majority kept inadequate ones. While the keeping of adequate records will not insure success, it will be a large factor in helping you avoid trouble.

Records are also necessary for other reasons. Banks and credit houses require financial statements before making loans. These firms employ men who are trained in interpreting and analyzing such statements and who can readily tell whether they are based on accurately kept records or are largely estimates.

Your government requires records to be kept so that taxes can be properly calculated. Tax returns must have definite supporting evidence Perhaps at some later date, you might wish to sell your business. If it can be shown in black and white just what the progress of the business has been and what it is worth, a better selling price can be obtained. It is human nature to want to pay only for obvious value; when there is a doubt, it will be assumed that the value is not present.

It may prove advisable to take in a partner when additional capital is needed to make profitable expansion. He is putting up cash, a known value. You are offering a share in your business, an unknown value, unless you have had the foresight to keep adequate records. If you have not, and this partnership arrangement is made notwithstanding, the agreement as to distribution of future profits may not be in your favor.

This section does not attempt to teach accounting. It can merely familiarize you with methods, forms, and terminology so that you can intelligently discuss your financial statements with your banker, tax collector, and others. It will also show you how to keep financial information, especially in the early period of your venture, so that it will be readily usable. Later you will find it necessary to have an accountant come in from time to time to check the work of your bookkeeper and prepare the financial statements.

Methods of Accounting

There are two basic methods of keeping accounts single-entry and double-entry.

The *single-entry* system is a simple system for accounting for the receipts and expenditures of cash. A checkbook, an accounts-receivable ledger, and a day book for entering memorandum type of information are all the equipment required. The check-book stubs give records of deposits and expenditures of cash. The accounts-receivable ledger keeps track of the money owed by customers. The day book, which need be nothing more than a notebook, is used to make daily notes as, "Received from John Doe, $10 on account." This amount can later be credited to the account of John Doe in the accounts-receivable ledger. This system records enough information to determine profits on a cash basis, and is used extensively by professional men.

The *double-entry* system is more complete and requires a knowledge of bookkeeping and accounting methods. Double entry means that the accounts are split into two sections, debit and credit. The debit side is always the left and the credit side the right. For each debit entry a credit entry is made also, and the books must always balance. This system provides the book-keeper with a means for checking his accuracy. Books required for the double-entry system are journals of original entry, the general ledger, and one or more subledgers. The general ledger and the various subledgers contain the actual accounts, which are divided into debit and credit by a vertical line down the center of each page. These books are of the loose-leaf or *post* variety so that new sheets can be inserted when needed. The subledger is a break-down showing several accounts, which are represented in total in the general ledger by a "controlling" account.

For example, we might discuss accounts receivable. The general ledger will contain one account headed *Accounts Receivable.* The accounts-receivable subledger will have an account for each customer. The total of the balances in all of the customer accounts should equal the balance in the general-ledger account. This makes it possible to balance the general ledger without having to balance all the customer accounts and vice versa. Various other subledgers are for accounts payable, notes payable, and other items. each being represented by a single account in the general ledger.

The journals of original entry are of many types, depending on their use and purposes. The journals are double entry, also, and are balanced before the totals are entered or posted in the general ledger and subledgers. In this way, the journals prove the balance of the items before they are entered in the books. If the books are in balance beforehand, and the items from the journals are accurately posted, the books will be in balance after the additional figures are added.

The journals also contain more detailed information and are cross-referenced with the ledgers. If you see an amount entered in one of the accounts in the general ledger that seems out of line, it will be possible, from the cross-reference symbol, or *folio,* to turn to the proper journal page and determine just what the figure represents.

In single-entry bookkeeping the records are kept on a cash basis. This means that all expenditures are considered expense for the year in which the check is issued. If you insured your equipment against loss from fire for a three-year period and the premium was $60, the entire $60 would be charged to the first year's operation although the business was being protected for three years. With double-entry bookkeeping, the *accrual* or *prepaid* method would be used. The $60, for example, would be set up as an asset, called *prepaid insurance,* and $20 would be charged to expense each year.

Thus, at the end of the first year, the books would show, as an asset, $40 for prepaid insurance. This is, of course, a very simple illustration. If you consider machinery or equipment, pay rolls, inventories of materials and supplies, buildings, and the numerous other types of expenditures, where the items purchased will not be used up in that year's operation, it is apparent that the only way to accurately determine profits or the financial worth of the business is to keep the books on an accrual basis.

When the business has developed to the point where such a record system becomes necessary, you will need the aid of an experienced bookkeeper or accountant.

It should be pointed out that the form of organization of your business will determine, from the outset, the type of bookkeeping set-up that will be required.

If you are incorporated, your liability for debts is limited to the amount of paid-in capital, no more, no less. The "no less" angle requires that an adequate set of double-entry books be kept on an accrual basis so that it can be proved that none of the paid-in capital has been withdrawn.

In the case of partnerships, the requirements for your books are not so important if the profits, by agreement, are evenly divided. However, if one partner is to receive, for example, the first $2,000, and the remaining profit is to be split evenly, the need for a complete system of books is apparent. It should be borne in mind that disagreements between partners are fairly common and often occasion termination of partnership. Where one buys out the interest of another, adequate records aid in determining the price; lack of them seriously complicates matters.

Good bookkeeping practice requires the recording of all financial transactions, a procedure which gives the partners a check on each other. This point alone removes many of the causes which lead to friction. If a double-entry set of books is not kept, accurate recording of each partner's withdrawals must be provided so that profits can be divided according to the partnership agreement.

Under a single-proprietorship form of business, the extent of the bookkeeping system is essentially at the discretion of the owner. Later in this chapter a simplified type of system adaptable for use by single proprietorships and simple partnerships is described.

Statements showing the financial condition and operation of the business are referred to as financial statements. The most important of these are the balance sheet and the profit-and-loss statement. The balance sheet indicates the financial condition and worth of a business as of a definite date. The profit-and-loss statement shows the results of the business operation over a definite period of time.

The Balance Sheet

The balance sheet deals with assets, liabilities, and net worth, and is usually presented in an account form. The assets are listed on the left-hand side, the liabilities and net worth on the right-hand. The totals of each side of the statement must be the same.

Here is a simple example. Assume you are the owner of a business, the assets of which total \$10,000, and the liabilities, or amount owed, \$4,000. It is easily seen that your equity, or investment, in the business is the difference between the two, or \$6,000. If your equity, or *net worth,* is totaled with the liabilities, then liabilities plus net worth will equal the total assets.

Let us start with an examination of the *Current Assets* part of the statement. It is customary to list current assets first in the order in which they will be liquidated in the normal operation of the business; in other words, in the order in which they will be turned into cash. Thus, cash will lead the list.

For purposes of illustration, we will assume the balance sheet is being prepared as of December 31. This statement will represent the assets, liabilities, and net worth as of the close of business on that date. In determining the amount of cash, would you check with your bank on the evening of December 31 to find our your balance? The answer is no. The cash figure from your books, or from your check stubs, represents the cash asset.

There may be a considerable number of checks still outstanding which have not been charged against your account at the bank but, so far as the business is concerned, the money is spent. Reconciliations, of course, must be made between your cash account and the bank's statements, which are sent out with your canceled checks at regular intervals. How this can best be done will be explained later.

After cash, accounts receivable will be listed. These should be divided into current and past due. Usually 60 days is the accepted dividing point. Reserve for bad debts is sometimes deducted from the figure for accounts receivable and sometimes shown in a *reserve* section on the liability side. Either method is acceptable, and both have supporting theories. Reserve for bad debts does not represent cash. It merely indicates the amount charged against profits for anticipated losses.

Next will come notes receivable, and then finished work on hand (which may be shown in accounts receivable if it is for a customer to whom credit has been extended), work in process, and inventory of materials. Your books will not show these items as such unless a cost system is introduced. They can be closely estimated, however, by a physical check of what you have on hand.

For inventory purposes, finished work should be valued at production cost rather than at the sales price. However, you will not have production-cost figures available. The amount to be collected, less a percentage comparable with normal profit, would be satisfactory.

Work in process, jobs on which you are working but which have not been completed, could be valued by totaling the labor and materials charged against the order sheets. Such a method would be conservative, however, because no charge would be included for overhead, since profits are included in overhead estimates and profits must not be anticipated in making financial statements. It would be best to show work in process at values thus computed.

Inventory of materials would be materials on hand which have not been charged to specific jobs. Care should be taken not to overvalue them.

Following *Current Assets* will appear *Fixed Assets*. These are assets that are not sold in the normal run of business. Items such as machinery and equipment, tools, and the building, if it is owned, are included. Reserves for depreciation may be shown as deduction from the assets or in the *reserve* section on the liability side of the statement. Reserves for depreciation are not cash but represent the amount that has been charged against profits as an expense.

The section below *Fixed Assets* is usually called *Prepaid Expenses*.

In the illustration of fire insurance it was indicated that at the end of the first year there would be an asset called *Prepaid Insurance* in the amount of $40. That item would appear in this section. Inventory of supplies on hand should appear also. Supplies may be defined as items that are used in the factory but which do not appear in the final product.

Unless items are of a sizable amount, however, it is not advocated that prepaid expenses be handled in this manner. It would be better to charge them as immediate expenses, which saves bookkeeping, and is, in addition, a conservative policy.

The liability side of the statement will start with *Current Liabilities.* First will be listed accounts payable and next, notes payable. If notes are to be paid monthly for a period of years on machinery purchased, it would be advisable to show the total due in the coming year as a current liability. The remainder would be listed in the next section with other liabilities, such as mortgages on property, if it is owned. Contingent liabilities, that is, liabilities for notes of others which you have discounted, or notes endorsed for others, should be shown as a memorandum entry, or as a footnote.

The form of the *Net Worth* section will depend on the organization of the business. A corporation shows the capital stock and surplus or deficit. A partnership lists the capital accounts of the various partners. The single proprietor shows the one figure, net worth.

The Profit-and-Loss Statement

As previously mentioned, the profit-and-loss statement shows the operation of the business for a definite period of time. Generally speaking, a profit-and-loss statement starts with total sales from which cost of goods sold is deducted, leaving gross profit.

In a manufacturing or processing company, cost of goods sold is ordinarily derived from the cost-accounting records. Since we are discussing a business too small for cost accounting, however, we will arrive at the same result by a slightly different method.

List all expenses, plus material cost and direct pay rolls, and then total. From this sum deduct estimated inventories as shown on the balance sheet for finished goods, goods in process, materials, and supplies. The result is actual cost of goods sold. Deduct the actual cost of goods sold from the total sales figure. The result is gross profit. Then list administrative and sales expense (explained in detail later), and total. Deduct this total from gross profit to obtain net operating profit.

Other income and other expense can then be added and deducted

Balance Sheet
December *31, 1968*

ASSETS		
Current Assets:		
Cash on hand and in bank		$ 5,000
Accounts receivable		
current	$ 4,000	
Over 60-day delinquent	200	
	4,200	
Less reserve for bad debts	100	4,100
Notes receivable		300
Inventories:		
Finished goods	150	
Goods in process	275	
Materials	640	1.065
Total current assets		$ 10,465
Fixed Assets:		
Machinery and equipment	5,000	
Less reserve for depreciation	300	4,700
Tools		600
Total fixed assets		5,300
Prepaid insurance		40
Total assets		$ 15,805

LIABILITIES	
Current Liabilities:	
Accounts payable	$ 2,000
Notes payable (Bank)	1.000
Notes payable (Finance Co. on machinery) due current year	2.000
Total current liabilities	$ 5,000
Notes payable (Finance Co.)	1,800
Total liabilities	$ 6,800
NET WORTH	$ 9,005
Total liabilities and net worth	$ 15,805

respectively to obtain net profit. "Other income" would be any income derived from other than regular sources. You might, for instance, rent your delivery truck to somebody. "Other expense" would be donations, contributions, and so on.

Profit-and-Loss Statement, January 1, 1968 to December 31, 1968

Sales			$ 27,000
Cost of Goods sold:			
Direct pay roll		$ 10,000	
Materials		5,000	
Total		15,000	
Overhead expenses:			
Power and light	$ 400		
Rent	1,800		
Heat	400		
Supplies	800		
Repairs to machinery	125		
Depreciation on machinery	300		
Tool expense	300	4,125	
		19,125	
Estimated inventories, Jan. 1, 1968			
Finished goods	$ 100		
Goods in process	200		
Materials	500	800	
		19,925	
Estimated inventories, Dec. 31, 1968:			
Finished goods	150		
Goods in process	275		
Materials	640	1.065	
Cost of goods sold			18,860
Gross profit			8,140
Administrative and selling expenses:			
Advertising		$ 500	
Entertainment		100	
Delivery expense		$ 1,500	
Bad-debt expense		100	
Office expense		500	
Office pay roll		1,400	
Telephone		250	
Miscellaneous		150	
Total administrative and selling expense			4,500
Net operating profit			3,640
Other income			100

	3,740
Other expense	400
Net profit	3,340

The two statements, balance sheet and profit-and-loss, have certain elements that tie together. For instance, inventories at the beginning of the period and inventories at the end of the period, as shown on the profit-and-loss statement, are the same as inventories on the balance sheet at the end of the preceding period and inventories on the balance sheet at the end of the present period.

Total net worth at the end of the preceding period, plus net profit as shown on the profit-and-loss statement for the present period, should equal net worth on the balance sheet at the end of this period. If there is a discrepancy, it should be traced down and an adjustment made.

The subject of supporting schedules should also be mentioned. As an example, let us again take accounts receivable. A supporting schedule would be a list of all accounts showing the amounts owed. These, of course, would total the amount shown on the balance sheet as accounts receivable.

Another example might be for the profit-and-loss statement. The cost of goods sold could be shown as one figure on the statement of profit and loss, with a supporting schedule showing the amounts of pay roll, materials, and shop expense with the inventory adjustments.

Another might be a list of machinery and equipment showing the types and valuation so that the figure for machinery and equipment on the balance sheet would be further explained. These additional schedules may be furnished where further classification of the financial statement is necessary.

A Simplified System of Accounting

A simplified method of bookkeeping that is not too difficult to understand and operate would probably be best for the early life of the business. The ease with which it could be converted to a double-entry system later when the business warrants is another important factor.

Probably the most difficult part of keeping records easily and handily is getting the habit. When money is expended, get a receipt and properly file it. When revenue is received, record it immediately in the *day book*. It soon becomes second nature to do those little things which lead to orderly business management. It actually takes less time than doing things in a helter-skelter manner.

A simplified system that will incorporate some of the features and advantages of double entry with the relatively elementary single-entry type of bookkeeping follows.

Of the books involved, the checkbook is the most important. On opening your checking account at the bank, you will probably be asked if a book showing your firm name on the checks would be desired. The answer definitely is yes. The small additional charge will be more than compensated for by the prestige and the appearance of permanence that your printed name will give.

The proper keeping of the checkbook has a direct bearing on the success of your business Every stub should show the date to whom the check is issued, the purpose for which issued, and the amount. Never fail to record these facts. The balance of your account should be shown in the margin provided. Deposits should be added and the total checks for the page deducted.

Great care should be taken to keep the checkbook absolutely accurately. Never use a check unless it is from your checkbook and the stub is properly filled out. Number each check. If it is necessary to handle a deal away from your place of business, take a check or two from the book, and fill out the stub as soon as you return. If these checks become soiled, or for some reason unusable, destroy them and write across the stub the word "void."

Never write a post-dated check and never write a check if your balance cannot cover it. Your cash balance should be brought forward as the checks are written, showing the deductions and adding in the deposits with the date of deposit indicated next to the amount. *If you are unable to keep your checkbook absolutely accurately, do not attempt to go into business for yourself.*

The other books, in order of their importance, will be the subledger, the sales book, and the expense book. These books are grouped together because they are supplementary and complementary to the checkbook. This subordination does not mean that they are insignificant or unimportant but, compared with the checkbook, they are definitely "small fry."

The subledger should be considered first because it contains, among other things, your accounts receivable. It is from these accounts that the monthly statements will be made, and they will control a considerable part of your sales volume, hence, your income. The subledger will also contain accounts payable, notes payable, and memoranda of various assets which are described below. Tabs might well be used for separating each of these sections.

This shows how accounts receivable are set up and how the entries are made.

Line 1. Sale for cash. On March 1, a customer has a job done for $25. The work is completed and he calls for it, paying you cash. Enter at once in your sales book $25 in the cash column, and $25 in the sales column.

Line 2. On March 1, Henry Brown, a customer, sends you a check for $100 on his account, which you receive that day. This pays his account in full for work done on February 10. Enter in the sales book $100 in the accounts-receivable credit column, and $100 in the cash debit column. Write the account name in the explanation column.

Line 3. Sale fo rcredit. On March 1, you deliver to John Doe a job which costs $50 to be charged to his account. Enter in the sales book $50 in the sales credit column and $50 in the accounts receivable debit column. Be sure it goes in the debit column. In the space for explanation, write John Doe's name.

The sales book should be totaled and balanced daily. The sum of the credit (Cr.) columns equals the sum of the debit (Dr.) columns. The total of the cash column represents checks and cash collected. See that this is deposited in the bank on the following day. The daily sales figures should be cumulated so that the total sales for the month are always quickly available.

The final book to be discussed is the expense book showing the following types of entries:

Line 1. Your supply house sends you $100 worth of material to be charged to your account.

Line 2. Your check stub shows $145 withdrawn for the pay roll.

Line 3. Your check stub shows $10 paid to the telephone company.

Line 4. Your check stub shows $20 paid to the electric company.

Most entries to the expense book will be made from the checkbook stubs. The other sources will be the petty cash analysis (to be explained later), and items purchased on credit. Entries to the expense book can be made at convenient intervals during the month. The book should be totaled and blanced at the end of each month. If the entries and additions are correct, the sum of the credit columns will equal the sum of the debit columns. The column for cash has no significance; it merely makes blancing the expense book possible so that you have a check on your work. Be sure all items in the accounts-payable column have been posted to the individual accounts payable, as shown in the lower part of the figure.

An explanation of what constitutes *shop expense* and what is *administrative expense* is in order. Shop expense is anything connected with the shop, such as rent, electricity, shop supplies. Administrative expense would be advertising, telephone expense, office expense, and so on. If you hired someone to handle the office work, his salary would be an administrative expense and should not be entered in the pay-roll column. Pay roll means the wages of the actual shop men you may hire.

Payments of accounts and notes payable should always be made by check. These items will be posted from the check stub directly to the individual subledger accounts. They will not be entered in the sales book or the expense book.

A definite procedure should be established for checking in all materials, supplies, and so forth, whether paid for in cash or charged. Each order will be accompanied by an invoice listing the various items. This list should be carefully checked against the merchandise purchased and the invoice should be O. K.'d by yourself or by someone in the shop appointed to do so. The invoice should then be filed. There should be two files for each company with whom a credit business is general practice, one for paid bills and one for pending bills. The pending-bill file should be posted periodically to the expense book. Make sure that each invoice has been O. K.'d and mark the invoice *posted* before restoring it to the pending file. In this way, you are sure that the goods are actually received before making payment.

Petty Cash

You cannot possibly write out a check every time some item is needed. Therefore, you will need to pay out cash from time to time. It is suggested that a reasonable supply of cash be kept in a locked drawer of your desk or in some other safe place to meet these needs. Always get a receipt when cash is used in making a purchase or, if this is not possible, make a notation on a slip of paper, showing the date, amount, and what the money was paid for. As more cash is needed, draw a check for the amount; do not take the money which should be deposited.

At the end of the month, separate the receipts and notations according to the purpose for which the cash was spent (material, pay roll, shop expense, or administrative expense) and list them. The compilation can be done on a scratch pad and the totals transferred to the expense book. The total for the entire amount should be entered in the cash column so that the balance will be maintained. Save the receipts so that they can be checked back if necessary.

You should determine a definite amount to be withdrawn weekly or at other intervals for personal use. Then a check should be issued and proper notation made on the check stub. It will be posted directly to your drawing account in the subledger in the same way that payments on notes and accounts are posted.

If payment is taken in cash, withdraw it from the petty-cash fund, and make out a slip marking it "drawing." This withdrawal, unlike totals of payments made from the fund, will not be entered in the expense book (cash column) but will be posted directly to the drawing account in the subledger. Remember the cash column has no bearing on your personal cash account but is included in the expense book so that a balance can be made to check your entries.

Your drawing account will show your personal withdrawals. You should also have a capital account headed by your name and marked *Capital.* An entry on the credit side should be made when the business is opened, showing the entire amount of cash or other assets you are contributing to the business, in other words, your investment.

If there are partners, a capital account and a drawing account should be set up in the subledger for each one. No entries should be made in the capital account until the end of the year, at which time profit, if any, should be credited or losses debited and the drawing account debited to the capital account. This will close the drawing account and show the new net worth balance in the capital account.

All entries in all books must be made in ink. Totaling should always be done in pencil so that corrections can be made and the figures will not be confused with the actual entries.

The figures for sales can be obtained from the sales book, which has been totaled and balanced daily. Sales will have been cumulated as previously mentioned. They must be carefully watched, for a falling-off of sales means you had better get out and hustle up some more business.

You now have sales for the month from the sales book, and you have monthly totals for materials, pay roll, and expenses from the expense book. Deduct them from your sales and you have a fair indication of profits for the month. The totals for shop expenses and administrative expenses can be further broken down by a work-sheet analysis, if so desired. The information for doing this appears in the explanation columns.

There are some other items to be considered in determining the profits. Foremost is the expense of depreciation on machinery. As mentioned before, you have listed all your machines purchased on one of the sheets in the subledger. This is a memorandum. It should show the type of each machine and the purchase price paid. The total of this account or memorandum will show the total paid for machinery.

Assuming that your equipment will wear out or become obsolete in about 15 or 16 years, depreciation could properly be charged at 6% per year, or 1/2% per month. Thus, if the total machinery cost is $5,000, the charge per month for depreciation expense is $25, or $300 per year.

Assuming your monthly sales were $2,000 and your expenses per the expense book were $1,500, your profit would be $500 less the $25 depreciation charge, or $475. As the depreciation charge is is relatively small and consistent, a special method for handling it is not considered necessary. In the profit-and-loss statement for the year, it should definitely be deducted as should interest expenses.

Interest expense can be shown by notations in the notes-payable accounts so that this amount can be determined for your end-of-the-year statements. If monthly notes are heavy, making sizable monthly interest costs, this expense should be handled accordingly.

Tools are set up as an inventory at the figure originally expended. In other words, for book purposes, it is assumed that these tools are the average amount needed. Additional tools purchased will be charged to shop expense in the expense book. Their cost will be posted from the checkbook stub if cash is paid for the tools. If the tools are purchased on open account, the cost will be entered from the invoice in the shop-expense column and in the accounts-payable column. The account to be credited will be noted so that it can be posted to the proper accounts payable. In the shop-expense column the notation would be tools expense.

The subject of bank reconciliations has a part in record keeping. You have been keeping your checkbook accurately and carefully noting all the necessary information. Eventually you will receive, through the mail, your bank statement with numerous canceled checks enclosed. Your checkbook will not agree with the bank statement because of outstanding checks, possible bank charges, and so on. Arrange the canceled checks in numerical order (you have been numbering your checks, of course), and list from your checkbook stubs the amounts of the checks which are not included with the bank statement. The following illustration will show the necessary details:

Balance per bank statement, March 31, 1968		$ 684.00
Outstanding checks:		
No. 6	$ 10.00	
No. 12	25.00	

No. 14	20.00	
No. 15	10.00	
No. 16	100.00	
No. 17	150.00	
Total outstanding checks		315.00
		369.00
Plus deposits (made after last one shown on statement):		
April 1	$ 100.00	
April 2	150.00	
Total deopsits since statement		250.00
		619.00
Plus bank charge		4.10
Balance as per checkbook, April 1, 1968		623.10

If the result does not agree with your checkbook, verify the additions and subtractions until the error is found. When in agreement, deduct the bank charge, if any, from your checkbook balance, and make a notation that the balance has been verified. File the checks with the statement and your reconciliation work sheet so they will be available for future use. An accurate reconciliation of the bank statement is a very important must. Be sure it is made every time the statement is received from the bank.

Other Necessary Records

It will be necessary for you to make statements for income-tax purposes. As a single proprietor, or under a partnership arrangement, you will pay taxes on the annual profit of the business regardless of whether this profit is withdrawn in the form of cash or left in the business. The amount credited to your capital account at the end of the year will be your net income. It is important to bear in mind that taxes may have to be paid to both federal and state governments and the responsibility rests upon you to see that this is done.

Attention is also called to your responsibility for maintaining adequate records of salary or wage paid each employee and for making returns as prescribed under the social security laws. The best system is to have a card kept on each employee, showing name, social security number, and the amount of salary paid weekly.

An accurate record of amounts deducted from the pay check must also be kept. It would be well to go to the local office of the Social Security Board and obtain direct information on the requirements. If no local office is listed in your telephone directory, ask your postmaster. The law requires that social security regulations be properly followed.

Familiarize yourself with record keeping because it is the basis for running a business successfully. If your business grows and prospers, you will soon need a bookkeeper. If the going proves tough, you may have to handle this end of the business yourself for some time.

The advantage of the bookkeeping system outlined above is in its comparative simplicity. The most important books are the checkbook, the sales book, and the accounts-receivable section of the subledger. These are necessary even in the simplest form of single-entry system. The instructions given for the others are valid even if the expense book is entirely ignored.

The only records you could omit would be accounts payable and notes payable, since the accounts payable are paid from the statements rendered and not from your record, and the notes payable would be evidenced by the notes and payment books. You would, however, be omitting the expense analysis, which is important for your personal information. You would still be getting, though in a sketchy fashion, an inkling of how things were going by comparing the labor and materials charges on the shop forms with your job estimates.

The important things for you to understand are the basic principles and purposes of a simple accounting system and to adapt that system to your business as a temporary means of keeping sound records. Above all, know and understand the profit-and-loss and the balance-sheet statements so that you can intelligently discuss your business problems with bank officials and other businessmen.

VIII Technical Help

IS THIS THE RIGHT TIME TO START YOUR BUSINESS?*

THINGS TO CONSIDER

Inflation changes consumer buying habits. Less discretionary income makes shoppers more selective in their spending.

Business failures are up. Weaker businesses didn't survive the recession; those that remain are sharper. A new business operator enters the marketplace against generally stronger competition.

Government red tape is growing, and it affects small businesses. This is a relatively less ominous factor, but definitely a disadvantage.

Going into business is always risky. Of all new firms started in a year, about one third are gone within the year; only one third live beyond five years, traditionally a measure of some stability for a small business. As important as general economic conditions are in figuring the odds for success, the personal characteristics and assets of the owner-operator count even more. Year after year, lack of managerial experience and aptitude is pegged as the cause of 90% of all business failures analyzed by Dun and Bradstreet, Inc.

It takes an astute manager to cope with the bad breaks and ill-winds that always seem to buffet small business.

You have a chance for success if you can answer these basic questions with a resounding yes:

1. Are you the type to run your own business? Do you have the inclination and the technical know-how to research the market potential for the product or service you want to sell, manage sets of financial books, direct people who work for you, buy and price goods and services, plan effective advertising and promotion stragegies to meet your competition?

*Changing Times, Washington, D. C., October 1975

Ask yourself whether you and your family are prepared for the financial and personal strain of operating a business. Government statistics reveal that men working for themselves as sole proprietors and in partnerships earn about as much as wage and salary workers but put in longer hours. Self-employed women have work-weeks about the same as wage and salary earning women but make less money than their counterparts. While traditionally women have been less likely to set up their own businesses, the majority of unpaid family workers in small businesses are women. Sad by-products of business failures often are broken family relationships and physical and mental health problems.

2. Are you going into the right kind of business? It should be a business you know something about from experience, special education or natural ability, and it should be one in which there is a real, not imagined potential for profit.

You will have to decide which way you want to go into business. You can start your own enterprise from scratch, acquire a going firm, take on a franchise operation or start out in your spare time with the aim of eventually going full time (more than a third of all wage and salary workers who moonlight are self-employed on their moonlight jobs.)

You'll have to decide, too, what legal structure your business should take: a sole proprietorship with you as sole owner; a partnership in which you share ownership and operating responsibility with one or several people, or a corporation, which is a legal entity subject to different laws and regulations than the other forms.

3. Do you have a well thought out plan? It should be a written blueprint for the organization and development of your business for four or five years. It should include completed financial worksheets, plus other information that will show the purpose of the business, what it will cost to operate, how much it should earn on sales and what goals you expect to achieve each year. A good business plan will make it a lot easier to get bank loans.

Naturally if your business doesn't exist yet, you can't predict exactly how it will run. But you can make educated guesses based on solid information that is widely available.

4. Do you have enough money? The average new proprietor uses his own cash and savings to supply the bulk of the funds needed to start his business. Another 15% usually comes from loans and about 6% in credit from suppliers. As a rule you need enough money in hand to cover your business operating costs and family household expenses for at least the first three or four months. It can take that long before your business produces the necessary cash flow.

WHERE TO GET INFORMATION

The Small Business Administration (SBA): This federal government agency with field offices in 90 cities, will help you in a variety of ways. For a nominal cost, or none at all, the SBA provides hundreds of booklets, worksheets and other printed materials to help you evaluate your business ideas. It runs periodic seminars and courses, many in conjunction with colleges and universities on various aspects of starting and operating a small business. It provides individual counseling by its own management and financial advisers, outside volunteer business men and women, graduate business students and professional management consultants. The SBA also offers a range of financial assistance services—primarily loan guarantees in cases where conventional financing is otherwise impossible to obtain, assistance in procuring government contracts and special help for socially and economically disadvantaged business operators.

LIBRARIES

Many libraries have how-to-do-it books on going into business, special reference materials, and research and other services of use to small business operators. Ask for the Bank of America's Small Business Reporter series of studies on individual small businesses and for current data on operating ratios for most kinds of business produced by firms like Robert Morris Associates, Accounting Corporation of America and the National Cash Register Co.

TRADE ASSOCIATION CHAMBER OF COMMERCE, GOVERNMENT OFFICES

Groups and agencies concerned with business conditions in specific industries or geographic areas usually have a wealth of statistical material available or can direct you to key sources. For example, local chambers of commerce and government departments of commerce should be able to fill you in on the local market potential for your product or service. At most Internal Revenue Service offices you can get a copy of Tax Guide for Small Business to size up your business tax liabilities.

SIMILAR BUSINESS OPERATORS

They may be reluctant to share too much information if you will be a direct competitor, but it's worth asking them and in some cases even working for the type of business you have in mind in order to pick up first hand knowledge and experience.

EQUIPMENT AND MATERIAL SUPPLIERS

Check library references, SBA office and other business operators for names of suppliers in your field. Because you will be a potential customer if you start a business, suppliers are happy to fill you in on what your business needs should be. They can also be a source of financial aid when you start out.

YOUR BANKER, LAWYER AND ACCOUNTANT

Each of them stands to benefit from an association with you if you go into business for yourself so they may be able to pass along helpful suggestions and data they've picked up in their dealings with other business operators in your community.

HOW MUCH CAPITAL DO YOU NEED TO START?

Starting costs	Sources of information	
Fixtures and Equipment	Visit other businesses and suppliers	________
Installation of features and equipment	Talk to suppliers and contractors	________
Decorating and remodelling	Consult landlord and contractors	________
Deposits with public utilities legal and other professional fees	Ask utilities Consult lawyers & accountants	________
Licenses and permits	Check with gov't. offices & lawyer	________
Advert. & promotion for opening	Ask media ad salesmen about rates & ad budget	________
Starting inventory	Get suppliers to help you estimate needs	________

Accounts receivable	Ask banker, other business operators about amount needed to buy more stock until credit customers pay ________
Cash	Ask banker about cash needed for unexpected expenses or losses, special purchase etc. ________
Miscellaneous	Itemize and total other costs you can think of ________

YOUR ESTIMATE BASED ON
PROJECTED ANNUAL SALE OF THE FORMULA TO USE

Ongoing monthly costs	$______	the formula to use	______
Salary of owner manager	$______	2 times figure at left	______
Other salaries & wages	______	3 times figure at left	______
Rent	______	3 times figure at left	______
Telephone & telegraph	______	3 times figure at left	______
Other items, utilities	______	3 times figure at left	______
Maintenance	______	3 times figure at left	______
Supplies	______	3 times figure at left	______
Delivery expenses	______	3 times figure at left	______
Advertising	______	3 times figure at left	______
Insurance	______	amount required by insurance Co.	______
Taxes, including social security	______	4 times figure at left	______
Interest	______	3 times figure at left	______

Legal and professional fees	______	3 times figure at left	______
Miscellaneous	______	3 times figure at left	______
		Total estimated cash you'll need to start with	$______

BOOKS

Good technical books give the results of years of experience of technical workers. They are especially invaluable to those who have not had the advantage of a technical education. The list that follows is not complete but has been selected carefully. For different chemical speciality manufacturing businesses, different books will be useful; therefore a varied list has been compiled.

Title	*Author*	*Publisher*
Handbook of Chemistry and Physics		Chemical Rubber Co., Cleveland, Ohio
Handbook of Chemical Engineering	Perry	McGraw-Hill Co. New York, N.Y.
Use and Applications of of Chemicals	Gregory	Reinhold Publishing Co. New York, N.Y.
Chemicals of Commerce	Snell	D. Van Nostrand Co. Princeton, N.J.
Chemical Formulary Vol. 1, 14	Bennett	Chemical Publishing Co., Inc., New York, N.Y.
Modern Cosmeticology	Harry	Chemical Publishing Co., Inc., New York, N.Y.
Modern Cosmetics	Chilson	Durg & Cosmetic Industry New York, N.Y.
Chemistry and Manufacture of Cosmetics	De Navarre	D. Van Nostrand Co. Princeton, N.J.
Modern Soap Making	Thomssen and Kemp	Amer. Perfumer, Pontiac, Mich.
Practical Emulsions	Bennett et al	Chemical Publishing Co., Inc., New York, N.Y.
Asphalts and Allied Substances	Abraham	D. Van Nostrand Co., Princeton, N.J.

Title	*Author*	*Publisher*
Sanitary Products	Schwartz	D. Van Nostrant Co. Princeton, N.J.
Industrial Waxes	Bennett	Chemical Publishing Co., Inc., New York, N.Y.
Oil, Detergents & Maintenance Specialties	Levitt	Chemical Publishing Co., Inc., New York, N.Y.
Plastic Molding Technique	Dearle	Chemical Publishing Co., Inc., New York, N.Y.
Organic Chemistry Simplified	Macy	Chemical Publishing Co., Inc., New York, N.Y.
Pyrotechnics	Weingart	Chemical Publishing Co., Inc., New York, N.Y.
Concise Chemical and Technical Dictionary	Bennett	Chemical Publishing Co., Inc., New York, N.Y.
Wines and Liquors, Chemistry and Technology	Hertstein	D. Van Nostrand Co. Princeton, N.J.
Flavors and Essences	Gazan	D. Van Nostrand Co. Princeton, N.J.
Essential Oils and Perfumes	Parry	D. Van Nostrand Co. Princeton, N.J.
Spices	Parry	Chemical Publishing Co., Inc., New York, N.Y.

CATALOGS

Catalogs of manufacturers of chemicals, equipment, containers, etc. are obtainable without cost. Many of them, in addition to descriptions, specifications and uses of their products, contain additional information that is useful. Such catalogs should be sent for and examined, then properly filed so that reference can be made to them in the future.

TRADE PAPERS

Trade papers are technical magazines that are published regularly. They are invaluable because they give the news of the industry promptly.

In them may be found announcements of new chemicals, equipment, containers, services, etc., price changes, new laws and regulations affecting the industry, changes in personnel, technical advances and new methods and many other valuable features.

The list of trade papers that follows is not complete but covers the most widely used papers in this field:

Adhesive Age.	New York City
Chemical Marketing Reporter	New York City
Chemical Week	New York City
American Perfumer	Pontiac, Ill.
Drug and Cosmetic Industry	New York City
American Paint Journal	St. Louis, Mo.
Rubber Age	New York City
India Rubber World	New York City
American Dyestuff Reporter	New York City
Food Industries	New York City
Soap	New York City
Detergent Age.	Denville, N.J.
Modern Plastics	New York City

CONSULTANTS

Consultants are expert chemists and engineers whose business it is to give practical technical advice and help. They are professional men who adhere to a high standard of ethics. Their fees are fair and they can save much time and money on problems where their expert knowledge can be applied. Do not hesitate to use their services when needed. The following consultants are well known.*

B. R. Laboratory	Miami Beach, Fla.
Harris Laboratories Inc.	Lincoln, Nebr.
Hudson Laboratories Inc.	New York, N.Y.
La Wall & Harrison Inc.	Philadelphia, Pa.
Truesdail Laboratories Inc.	Los Angeles, Cal.

*Since it is impossible to list all consultants, these have been selected without any discrimination. Other consultants may be found in the telephone directories of most large cities.

TRADEMARKS

A trademark is a distinctive word, emblem, symbol, or device, or a combination of any of these, used on goods actually sold in commerce to indicate or identify the manufacturer or seller of the goods. The mark must have been used in interstate or foreign commerce, before an application for registration can be filed in the Patent Office.

A trademark cannot be registered if it contains immoral or scandalous matter. No one can register a mark including the flag or coat of arms or other insignia of the United States of any simulation thereof, or any state or municipality or of any foreign nation, or of any design or picture that has been or may hereafter be adopted by any fraternal society as its emblem. Registration is prohibited of any name, distinguishing mark, character, emblem, colors, flag, or banner, adopted by any institution, organization, club, or society which was incorporated in any state of the United States prior to the date of adoption and use by the applicant provided use by the organization was prior to use by applicant. No portrait of a living individual may be registered as a trademark except by the written consent of the individual, nor may the portrait of any deceased President of the United States be registered during the life of his widow except by written consent of the widow. No mark which is identical with that used by another on the same class of goods, or which so nearly resembles it as to be likely to cause confusion in trade, can be registered.

Any mark which has been in actual and exclusive use as a trademark by the applicant during the 10 years preceding February 20, 1905, may be registered, and such a mark when once registered may be reregistered when used on other goods of the owner of the mark.

Registration under the act of March 19, 1920, does not give prima facie evidence of ownership, but does give the right to sue in the Federal courts.

On complying with regulations of the Treasury Department and filing there a certified copy of a registered mark the collectors of customs will prohibit importation of foreign goods copying or simulating the mark.

Many foreign countries will not register a mark for a citizen of the United States unless it is registered in the United States Patent Office.

The law makes no provision for the registration of a trademark by joint owners unless they apply as a firm.

The registration of a mark protects not only against the adverse use of that mark but against the use of any mark so like it as to cause confusion or mistake or to deceive purchasers.

A registrable mark is one used with merchandise. The law makes no provision for the registration of marks used only in connection with service, such as insurance, bonding, banks, collection agencies, laundry work, cab service, and the like.

The mere names of varieties of fowls, animals, fishes, and vegetation can not be registered as trademarks; e.g., anyone raising Jersey cattle or Fultz wheat has a right to sell the natural increase thereof under the same name.

The use of a registered mark on new articles of a kind similar to those named in the registration is generally protected, but a complete new application covering such goods may be filed. All the goods in a single official class on which the mark has been actually used may be included in one application.

Trademark Registration in Foreign Countries

Owners of trademarks having business or prospective business in foreign countries should ascertain the nature of the trademark laws in those foreign countries in order that they may take proper steps in time to protect their rights.

In many foreign countries a resident there may obtain a registration of a trademark without having actually used it in trade. And such a registration may be used to prevent the importation into that country of goods bearing the mark. Owners of marks in the United States might well seek registration of them in foreign countries. This may be done by direct private application before the appropriate departments of such countries. Where the owner may desire protection in countries maintaining its services, the Inter-American Trademark Bureau affords an alternative to the direct application.

To effect registration in foreign countries, whether the approach be direct or through the mutually supported Bureau, it is essential that registration be effective in the United States. Having secured registration domestically the owner frequently may secure registration in foreign countries before actually using the mark there. Duration of certificates of registration in foreign countries varies in accordance with the domestic laws of such countries. In some instances the duration is the same as in the United States.

Selecting a Trademark

It is far simpler to think of a trademark such as DUZ, REM, etc. than for the buyer to describe the product or its package. A good trademark for a good useful product is a great help in merchandising.

A trademark should not be adopted just because it appeals to its originator, his family or his friends. A symbol or a short coined word is the most commonly used trademark. A good trademark should subtly give an association with the product or its use.

The best method of selecting a good trademark is to see if it conforms with the following:

1. It is not descriptive.
2. It differs decidedly from similar trademarks.
3. It is simple in form.
4. It is attractive to the eye.
5. It is pleasant sounding.
6. It does not arouse dislike, mistrust or doubt.
7. It is easily read and pronounced.
8. It is easily remembered.
9. It creates a desire to possess or use it.

If it conforms to these standards, it is a good trademark.

Since a good trademark is usually of great importance in merchandising, its proper selection and registration are important. Advice from an expert in this field will be very helpful.

To Patent or Not To Patent

If you have made a novel product that you believe is patentable (see next section, Patent Laws) you must decide whether it is best to apply for a patent which, when granted, makes a complete public disclosure of the composition and or the process of manufacture—or if it is better to keep it secret so as to forestall or hinder competitive products.

PATENT LAWS ON CHEMICAL COMPOUNDS

While the Constitution of the United States makes no specific provision for the grant of a patent on a chemical composition or composition of matter, the patent statutes conferring such protection are based on Section 8 of Article I, which reads:

"The Congress shall have power . . . to promote the progress of science and useful arts, by securing for limited times to authors and inventors the exclusive right to their respective writings and discoveries."

It is required, of course, that the subject matter of a patent shall fall within the statutory definition of patentable subject matter before a grant of an authorized legal monopoly may be had for the limited term of seventeen years; under the patent laws "any new and useful art, machine, manufacture, or composition of matter, or any new and useful improvement thereof," may be the subject of a patent.

The particular phrase *composition of matter* is set forth in *Walker on Patents* to cover "all compositions of two or more substances. It includes, therefore, all composite articles, whether they be results of chemical union, or of mechanical mixture, or whether they be gases, fluids, powders or solids."

To be patentable, the composition must be not a mere aggregation of several ingredients without different or additional properties which the ingredients do not possess individually, but it must involve a true combination or be capable of a new use not obvious to those skilled in the art and remote from any known use.

Thus, an alloy has been held patentable for use as an electrical resistance where it had been known previously only as a non-magnetic material for making parts of a watch. And a solidified fuel has been held patentable which consisted of nitrocellulose, methyl alcohol and ethyl alcohol, all of these ingredients cooperating to produce the product and all contributing in the production of heat in the intended use of the composition.

It may be noted that a hair-treating or hair-food compound in which all except one of the ingredients had been used before in compounds for the same purpose, and this additional ingredient was equivalent to a substance used before in the same relation, was held to involve nothing patentable. There was no chemical combination of substances nor new function or effect, but merely the combined functions and effect of the separate ingredients. Also, a remedy for treatment of lung disease has been held unpatentable which involved merely an aggregation of ingredients, the medicinal value of which had long been known.

A variation in the proportions of the elements of a composition does not constitute patentable novelty or invention, unless a new material or substance, or an old material or substance with new characteristics or with substantially enhanced qualities of utility, has been created.

In connection with the infringement of claims for a composition of matter, it may be noted that changes in the proportions of the ingredients of a patented composition (except where such proportions are the essence of the invention), or additions thereto of other ingredients which do not change the principal characteristics do not avoid infringement; and addition to the patented composition of an ingredient of such minor consequence and inconsequential effect as not to change its essential character does not avoid infringement. But infringement may be avoided by the omission of an ingredient where no equivalent is substituted.

Infringement occurs only when the product has the same characteristics and is composed of the same or equivalent ingredients.

No *Serial Number* and/or date of filing is accorded an inventor, nor is an application for an invention complete, until all of the requirements provided for in the Rules of Practice of the United States Patent Office are met: a *petition* for the grant of a patent, duly signed by the inventor, or inventors; a similarly executed specification (and, if required for the better understanding, a drawing) describing the nature of the invention in such a manner that those skilled in the particular art can readily understand it, together with *claims* setting forth the features believed novel and patentable; and an oath, affirmed or sworn to before an authorized officer, setting forth among other things that the applicant, or applicants, is the original, first and sole inventor (or are joint inventors) of the subject matter described and claimed, and that the same has not been published, nor been in public use or on sale for more than two years prior to the date of application.

It is frequently desirable to make some record prior to the filing of an application, which may be useful as evidence of conception of the invention in the event of a subsequent interference proceeding in which the application may become involved. To this end, a statement describing the invention as fully as possible may be dated and signed by the inventor and witnessed by one or preferably several subscribing witnesses who fully understand the nature of the invention disclosed in the said statement. The unsupported statements of the inventors are of no use. Such statement is, also, preferably to be acknowledged before a notary public.

No provision is now in effect for filing Caveats in the United States Patent Office.

The grant of *Letters Patent* does not of itself confer the right of the patentee to make use of the invention covered thereby, but conveys merely the right, for a limited period, to exclude others from the manufacture, use and sale thereof. It may become necessary, therefore, to first secure a license under some controlling patent or to purchase an interest therein.

In the case of Letters Patent issued jointly to several patentees, as well as in the case of the owner of an undivided part of a patent, each party, in the absence of an agreement to the contrary, may proceed to dispose of his rights, grant licenses, and otherwise operate without first obtaining the consent of the other owner or owners, and without accounting to them for any returns.

In the case where patents are owned by corporations, or large manufacturers, difficulty may be encountered in securing a license to operate under the same, as the full rights are usually retained in such circumstances to secure a monopoly. When nothing has been done with respect to marketing the invention, it is generally possible to effect a satisfactory arrangement, as for a non-exclusive or an exclusive license, with royalty payments; negotiations may be initiated through the attorney of record of the particular patent, or directly with the patentee if his address is readily available.

The purchaser of a patent right should endeavor to buy the patent outright, if possible, a seller should usually retain title to a patent and grant only licenses thereunder.

Only the general principles are given here. There may be exceptions which it is not possible to consider without extending the length of this article; it is more satisfactory in matters of this nature to consult a compentent patent attorney.

IX Federal Laws and Regulations

Anyone who plans to market a food, drug, cosmetic or chemical specialty should be thoroughly familiar with federal laws and regulations pertaining to that particular product.

The Federal Food and Drug Law, passed in 1938, was comparatively lenient in its restrictions. But discovery of more potent drugs, new food additives, and powerful insecticides all posed safety problems. In some instances dangerous toxicity of a new compound was not suspected prior to marketing. Then when its use was widespread, serious poisonings and other deleterious actions were reported.

In 1962 very restrictive ammendments to the 1938 law were passed, demanding far more rigid proof of the safety of any drug or chemical used on or by humans, and also included products used for pet animals as well as food animals.

Responsibility is still divided among three government agencies. The Federal Trade Commission is charged with the responsibility of policing advertising, to check false, misleading or illegal statements. One exception is drug advertising to physicians, which comes under (FDA) jurisdiction. Food and Drug Administration.

The Department of Agriculture has jurisdiction over pesticides and their labeling and additives to foods of animals being raised for human consumption.

The FDA, a bureau of the Department of Health, Education and Welfare, has the broadest regulatory powers. It controls marketing of all foods, medications, and cosmetic products for human use and all veterinary medications.

Cosmetics

Cosmetics are not restricted as stringently as drugs, but the definition of a cosmetic is very narrow. Any preparation used *only* for cleansing or beautification of the skin, hair, or finger nails is considered a cosmetic. There are certain prohibited drugs or chemicals which are considered too dangerous for use in cosmetics, but on the whole it is up to the manufacturer to market products that are safe. Cosmetic claims can be quite flamboyant, as long as the claims are only for beautification. However, any claim of a medicinal nature, even if only implied, immediately places the product in the drug class, which means that official permission to market the product is mandatory.

The law does not forbid the use of poisonous substances in a cosmetic. The ban applies only when the quantity of a poisonous ingredient *may* be injurious to users, *under the directions of use as given by the manufacturer.* The product may be extremely hazardous if misused, yet still be acceptable under the law, *provided* safe directions are included, and proper cautionary statements made on the label.

While the Federal Trade Commission has charge of all cosmetic *advertising,* it is the FDA that controls *labeling.* The FDA has the authority to sieze any product in interstate commerce, which it considers misbranded. The term "labeling" is a rather all inclusive designation, and covers much more than just the label pasted on the container. Any written, printed, or graphic material accompanying the package is considered part of labeling, even if it is delivered by mail, separately from the actual product. A display in a store, along with the product is also "labeling"; so is any literature handed out by demonstrators.

The manufacturer of a cosmetic is walking a "verbal tight rope", as the following *partial* list of taboos illustrates. Inclusion of any of the following designations are sufficient to bring a charge of misbranding a cosmetic:

Circulating Cream, Contour Cream, Deep pore cleanser, enlarged pore paste, depilatories for *permanent* removal of hair, eyelash grower, eye wrinkle cream, hair color restorer, hair grower, hair restorer or revitalizing preparation, muscle oil, nail grower,nonallergic products, peroxide cream, scalp food, skin conditioner or tonic, skin food or nourishing cream, skin texture preparations, stimulating creams, tissue cream, wrinkle and double chin eradicator; any representation that a product is superior in cosmetic value beacuse it contains a vitamin.

The law does not require that a cosmetic must contain a list of its ingredients, but if the labeling puts it in the "drug" class it must list all ingredients with the quantities. Any statement that a cosmetic will cure or alleviate dandruff, will promote a "healthy skin or scalp," alleviate itching, or revitalize dry skin, cure acne, eczema, freckles, rash or any "skin trouble" places the product in the "drug" class.

The manufacturer or distributor of a cosmetic has the full responsibility for a truthfully labeled cosmetic, as defined by the FDA. The FDA is not authorized to approve the label or labeling of a cosmetic, but usually will give an informal opinion as to whether a proposed label is to be construed in fact as a drug label. But even such informal opinion does not relieve the manufacturer of his responsibility, because such opinion is not an official acceptance of the product nor its labeling.

HYPOLLERGENIC COSMETICS

Hypoallergenic cosmetics, as defined by the FDA (Food and Drug Administration) must cause significantly fewer adverse reactions than competing products of the same type. There are new rules for the testing and labeling of hypoallergenic cosmetics. If one wishes to establish the hypoallergenicity of his cosmetics, he must have his products tested in competition with others of the same type. All that is required is that these products represent combined a 10% share of the market. (A single product may be used as the competitor if it represents a 10% share of the market). Any dermatological testing that demonstrates that a cosmetic is hypoallergenic is acceptable. Human subjects must be used, but need not be extra sensitive people. The product in question must come out as clearly causing fewer adverse reactions. If the product tested is no more than 5% better than its competitors, it can't be called hypoallergenic. Once hypoallergenicity is established, the label must read "Less likely to cause adverse reactions than some competing products". The FDA sets a five-year limit on the results of the test data, once the tests are completed. The reason for this is that the market may have improved, and the term "hypoallergenic" may no longer apply.

For the testing, similar formulations may be combined to prevent having to test an entire line of shaded products, such as lipsticks or eye shadow. All that is necessary is that all the ingredients in the line be included, and that among the principal ingredients the amount of one ingredient does not exceed the greatest amount used in a single cosmetic. Products differing in fragrance are not considered similar.

If an ingredient or cosmetic formulation causes infrequent but severe allergic reaction, the ingredient or product may not be used, much less called hypoallergenic.

If the cosmetic product is in distribution now, claims of hypoallergenicity must be proven by June 6, 1977. If the cosmetic product is not yet in distribution, but hypoallergenic claims are to be made for it, they must be justified by the preceding rules before distribution.

COSMETIC LABELING

In a very recent ruling, the FDA (Food and Drug Administration) stated that all cosmetic ingredients, no matter how small the percentage, must be called by name and listed in descending order of predominance on the cosmetic label. (The label, of course is not necessarily the stick-on kind, but any sort of lettering by which the package is decorated. The size of the type is dictated by the FDA, and differs according to the size of the package). The exceptions to the rule are flavor, fragrance, and (for the time being), color; which may be listed simply as "flavor", "fragrance", or "color". These still must be listed in descending order of predominance. The reasons for flavor, fragrance, and color being excepted from the "name every ingredient" rule is obvious—they each may contain upwards of 20 different ingredients, making listing impossible. There are also trade secrets involved with these three. In the future, it may be necessary to list all the separate ingredients of the color used. There is to be a hearing, and if the FDA decides against the applicants, then all the color ingredients will have to be listed. As with color and perfume, there are other trade secrets involved in labeling. The person wishing to have a certain group of ingredients remain a trade secret must apply to the FDA. If the FDA agrees that this certain group of ingredients is indeed peculiar to his use, then these are retained as trade secrets, and may be listed as "and other ingredients" on the cosmetic label.

Ingredients present in quantities greater than 1% must be listed in descending order of predominance. Those present in quantities less than 1% may be listed in any order. All labeling ordered after March 3, 1976, and all packages labeled after September 3, 1976 must comply with this regulation. Packages labeled before September 3, 1976, but introduced into inter-state commerce after that date may be used up. However, any labels retained in stock after that date may not be used up. Petitions to extend the data for individuals will be considered by the FDA.

If the cosmetic is not manufactured by the person whose name appears on the package label, the label must say "Manufactured for __________" or "Distributed by __________". Formerly, the contract packager would simply list the person for whom the manufacturing was being done as the manufacturer. The FDA considers this false and misleading.

Cosmetic ingredients must be identified in the declaration of ingredients by the name established for that ingredient by the FDA. Reference sources include the CTAF (Cosmetic, Toiletry, and Fragrance Association, Inc.) Cosmetic Ingredient Dictionary, First Ed., 1973; The United States Pharmacopeia, 18th Ed., 1970; National Formulary, 13th Ed.,

1970; Food Chemicals Codex, Second Edition, 1972; United States Adopted Names and the USP Dictionary of Drug Names 1961–1971 Cumulative List, and 1973 Supplement. In the absence of such a listing, the name generally recognized by consumers is used.

Off-package labeling may be used for packages having a total surface area of less than 12 inches. Legislation regarding this should be completed within the next year. The FDA is exploring all sorts of possibilities–ingredient disclosure on the color chart accompanying shaded products too small to be labeled individually; ingredient disclosure in the form of a tablet of individual sheets attached (padded) to the display to be torn off by the consumer who wished to know what the product contains. This method requires that the words "Federal Law Requires Ingredients Lists to be Displayed Here" be visible when the last ingredient declaration has been taken. In all cases, ingredient lists must be available from the manufacturer or distributor. Any item enclosed in a carton or other outside container must have the declaration of ingredients displayed on it, or inserted into it, providing the carton can be opened at the point of purchase. Ingredients must be conspicuously listed on the front of color charts or displays. More complete information if available in the Federal Register, Volume 40, #40; Monday, March 3, 1975, Paragraph 701.3.

Pesticides

*A registration is a permit for use issued by the U.S. Environmental Protection Agency (EPA), once the agency is satisfied a pesticide is safe. EPA issues two types of registrations, experimental and commercial, in succession. Both types specify the use allowed (for example, on cotton crops), and the pests to be controlled.

†Indeed, *CE* plans to cover new pesticides per se in a separate article later this year.

Major provisions of the amended FIFRA are as follows:

1. All pesticides (intrastate and interstate) must be registerd (with a few exceptions).
    a. Data must be supplied in support of claims on the label.
    b. When used according to directions the pesticide will have no adverse effect on man, animals, or the environment.
    c. The use of a pesticide in a manner inconsistent with its labeling is in violation of the Act and may be subject to civil or criminal sanctions.
2. Pesticides will be classified for *general* use or *restricted* use.
    a. General use pesticides will ordinarily not cause unreasonable

adverse effects on the environment and will be available for use by the public.

b. Restricted use of pesticides because of their toxicity present a hazard to the applicator or other persons and may have an adverse effect on the environment and can be used only by *certified* applicators.

3. A certified applicator is an individual who is certified to use or supervise the use of any pesticide which is classified for restricted use. A certified applicator is classified as either a *private* applicator or a *commercial* applicator.

a. A private applicator is a certified applicator who uses or supervises the use of any pesticide which is classified for restricted use in agriculture on property owned or rented by him or his employer.

A person is also classified as a private applicator when a pesticide is applied without compensation other than trading of personal services between producers of agricultural commodities on the property of another person. Thus farmers, nurserymen, greenhouse growers and others who grow crops on their own or rented land must be certified before applying a restricted pesticide.

b. A commercial applicator is a certified applicator who uses or supervises the use of a restricted pesticide except as defined for a private applicator.

EPA has suggested the following categories for commercial applicators: 1) agriculture pest control–plant and animal; 2) forest pest control; 3) ornamental and turf pest control; 4) seed treatment; 5) aquatic pest control; 6) right-of-way pest control; 7) industrial, institutional, structural, and health-related control; 8) public health pest control; 9) regulatory pest control; and 10) demonstration pest control.

4. The EPA Administrator prescribes standards for the certification of pesticide applicators. To be certified, an individual must be determined to be competent with respect to the use and handling of pesticides, or to the use and handling of the pesticide or class of pesticides covered by such individual's certification. Although the States have the authority to submit a plan for the certification of pesticide applicators, the final approval, disapproval, or modification of any plan submitted by the States rests

solely with the administrator of EPA. Separate standards shall be established for commercial and private applicators. Standards for commercial applicators are more stringent than those for private applicators.

States assume *primary* responsibility for certification of applicators. State certification plans must have been submitted to the Administrator of EPA by October 21, 1975. States will charge fees for examination and certification of applicators to pay for the cost of this service since there will be no financial aid from the Federal Government.

5. Farmers and other private applicators can be fined up to $1,000 or given 30 days in prison or both upon criminal conviction for a knowing violation of the law. They are also liable to civil penalties of up to $1,000 on their second and subsequent offenses.

Any registrant, commercial applicator, wholesaler, dealer, retailer, or other distributor who knowingly violates the law is liable to a $5,000 civil or $25,000 criminal fine or one year in prison or both. These criminal penalties are now in effect.

6. The EPA Administrator may issue a "stop sale, use, and removal" order when it appears a pesticide violates the law or its registration has been suspended or finally cancelled. Pesticides may also be seized if they violate the law.

7. Pesticide manufacturing plants must be registered with the Federal government (EPA). Information on the types and amounts of pesticides produced, distributed, and sold must be submitted upon registration of the firm and annually thereafter. An EPA agent may enter and inspect such an establishment and take samples.

8. Federal assistance to the States to enforce provisions of the law and help develop and administer applicator certification programs is authorized.

The EPA administrator is required to develop procedures and regulations for storage or disposal of pesticide containers.

The Agency may issue experimental use permits, conduct research on pesticides and alternatives, and monitor pesticide use and presence in the environment.

9. Indemnification is authorized for certain owners of pesticides registrations which are suspended, then cancelled.

10. Applicants for registration of a pesticide retain proprietary

rights to their test data submitted in support of the application. However, the law authorizes a system whereby such data may be used by a second applicant upon payment of reasonable compensation to the original applicant.

11. States can be authorized to issue limited registrations for pesticides intended for special local needs.

The original purpose behind the amended FIFRA was to regulate the use of pesticides so that they do not endanger man or animals, or the environment, a worthwhile purpose by any standards. Unfortunately, rigid enforcement of the Act, some say with a minimum of common sense, has engendered such hostility from farm groups and other organizations that for a time it appeared the House Agricultural Committee would not recommend further appropriations for administration of the Act. Some members of Congress feel that EPA has far exceeded its authority and the intent of Congress with its rigid interpretation of the provisions of the Act.

We can sympathize with EPA in the administration of the Act when we realize how complex it is in all of its ramifications. Moreover, much of the personnel in the Agency is new with little or no practical experience with pesticides and their application. Another very heavy burden for EPA is that before October 1976 it must classify about 33,000 federally registered pesticides and 10,000 state registered pesticides into general-use or restricted-use materials. Nevertheless this does not make it easier for the farmer, pest control operator, pesticide manufacturers and others to live with the Act. Fortunately, it appears that EPA is becoming aware that intimidating users of pesticides is not the most effective way of protecting man, animals and the environment from harmful use of pesticides.

Drugs

Any product, whether used externally or internally, that purports to treat any abnormal human condition is classed as a drug; the definition is broad, and applies to products that can be considered preventive rather than curative or therapeutic.

Drugs are divided into two broad classes. The so-called "over the counter" drugs are considered reasonably safe (and effective) to be used by the general public without the recommendation and supervision of a physician, and can be purchased without a prescription.

The regulation of these drugs is very strict, specifying how directions should be stated, warnings and cautions to be prominently displayed on the label, and an accurate list of the ingredients, with quantities given for the active constituents. Also, curative claims are severly restricted. For instance, an aspirin containing tablet cannot claim to relieve pain of rheumatism; it must state "for the temporary relief of minor aches and pains of rheumatism", and in addition must carry the accepted warnings about the possibility of adverse reactions, and the precaution to keep out of reach of children!

All labels, claims and proposed advertising should be submitted to the FDA before any "over the counter" drug is marketed.

Prescription Drugs

Drugs that are considered potentially dangerous for self medication can be dispensed only by licensed pharmacists (or hospital dispensaries), on a physician's prescription, or by the physician to his own patients.

Prior to 1962, drugs were cleared for sale by proving safety when used according to directions. There was no legal basis for withholding permission to market a drug because of exaggerated or false statements of curative claims. The 1962 drug law ammendments change this. Now *all* drugs, whether prescription or nonprescription, must show acceptable proof that they are not only safe but will have the therapeutic effect claimed.

Drugs in both classes that had been approved for safety prior to the 1962 law must now have acceptable clinical proof that they are effective. If such proof cannot be presented, the drugs are ordered off the market, despite the fact that they may have been sold in great volume since 1938. Many an old time favorite has had to drastically revise claims and often the formula, or face economic extinction.

The FDA has the final decision as to whether a drug can be sold without a prescription or not. The manufacturer cannot designate the catagory.

New Drugs

All new drugs must of course comply with the 1962 law. This often presents a formidable obstacle. Before trying a new drug on humans, the FDA must be petitioned for permission. Unless there is incontrovertible evidence of safety, the FDA will not approve an Investigational Drug Application (IDA). This means, in the majority of cases, elaborate, carefully controlled experimentation on laboratory animals. Usually tests must be done with several species of animals, under supervision of qualified pharmacologists.

If the FDA is satisfied with the animal tests, it may consider the IDA favorably. Qualified physicians (usually specialists in their fields) with adequate hospital affiliations are permitted to administer the new drug to humans. This "clinical investigation" is hedged in by strict rules of reporting progress to FDA. The clinical testing may take three months or up to several years, and may require as few as 20 patients or may have to be tried on several thousand patients.

When all the clinical work is finished, the complete history of the project, along with the reports of the clinical investigators is then submitted to the FDA, as part of a New Drug Application (NDA). If government experts are satisfied with the report, the NDA is validated and the drug can be offered for sale.

But the FDA continues its surveillance and any violation of approved label claims can bring revocation of the NDA, thereby terminating the sale of the drug. Any change, no matter how slight, in the formula or method of manufacture requires sanction by the FDA; unauthorized changes invalidate the NDA.

An interesting side light is that all advertising material for a prescription type drug is considered part of labeling. An advertisement in a medical journal, a display at a medical convention, or a brochure or booklet mailed to a physician are all considered as much a part of a label as the piece of paper pasted on the drug container.

Another pitfall for the uninitiated is the definition of a "new drug." FDA contention is that combinations of drugs that are in proportions or identity different from those acknowledged as "safe and effective" are to be treated as "new drugs." This holds true even if every *single* drug in such a new combination has been used for years and has been approved. For instance, aspirin is a common drug, fully approved, as are Vitamin C, caffein. Yet if these well known constituents are combined and especially if offered in novel proportions and in a new vehicle or form, such as a liquid solution or an effervescent tablet, they are a "new drug" and must go through the formaility of obtaining an approved NDA.

Reading the provisions of the law as passed by Congress in 1962 is not too helpful to a manufacturer. There have been so many supplemental regulations and new rulings (they come with great regularity and rapidity) that any book purporting to give *all* interpretations of the law would be critically obsolete before it could be printed.

There is one other important provision that must be observed. Any person or company that manufactures drugs or cosmetics must permit inspectors of the FDA free access to the laboratories and manufacturing premises "during reasonable working hours". At least an annual inspection is supposed to be made to ascertain if facilities are adequate and if properly qualified personnel are in charge of the manufacturing and assay of raw materials and finished products. This also applies to any laboratory engaged in assay for other manufacturers, or any plant making products for distributors, even if the laboratory is not engaged in interstate commerce.

Faced with all these strict regulations it is only prudent for any manufacturer to check with the FDA on what will be required, before any serious committment is made on any new product, be it in the drug, food or cosmetic field. The Food and Drug Administration will give a reasonably prompt answer to any informal inquiry, and will in most cases indicate what will be needed for a formal application for a proposed drug or formula. All informal inquiries should be addressed to Food and Drug Administration, Arlington, Virginia. Household chemical specialties and insecticide inquiries should be addressed to the Department of Agriculture, Washington, D. C.

Chemical Specialties

Household cleaners, spot removers, polishes, floor and furniture waxes, bleaches and washing fluids as well as industrial chemicals are in a "twilight" zone, as far as regulation is concerned.

Many of these substances are toxic and potentially dangerous, yet there are not nontoxic substitutes for many of them. Farm chemicals are under control of the Department of Agriculture, but household chemical specialties come under surveillance of the FDA.

Special precautionary labeling is required, and in some cases an antidote must be printed on the label so that effective corrective measures can be taken promptly in case of accidental ingestion.

As in the case of drugs or cosmetics, any manufacturer contemplating marketing any household product should check with the FDA concerning restrictions and mandatory labeling. These restrictions are generally quite liberal and are designed only to protect the public. Accurate, official information is so readily obtained that there is no excuse for anyone to offer dangerous products for sale without the proper safeguards. To neglect this means not only official censure, but could weigh heavily against the manufacturer if he should become the defendant in a law suit for damages caused by his product.

Devices

This is a general term covering any device either mechanical, electrical, or hydraulic, that purports to cure or alleviate any physiological ailment. Massage vibrators, "whirlpool" baths, heat lamps, diagnostic instruments, ultra violet lamps are all subject to FDA regulations if the device is to be sold to the public and if some curative claim is made for it. Obviously "magnetic belts" that claim to cure all kinds of diseases would never get official permission to make such claims. But there are some devices, especially those used with specific medications (for instance, a special vaporizer for inhalation therapy), that have significant value and FDA will approve truthful and *proven* claims for these devices.

X Formulary

ADHESIVES AND SEALANTS

Greenhouse Glaze

The most important characteristics required of a Greenhouse Glaze are *a*) very slow drying, *b*) a high degree of internal softness, *c*) good adhesion to metal and glass, and *d*) the ability to be handled and formed for application without sticking to the hands. The following formula has been carefully balanced to achieve these properties.

"Coray 80"	75
Aged Linseed Oil	75
Soya Lecithin	2
"Atomite"	450
Putty Filler	323
7R Asbestos	75

Aquarium Sealant

Formula	No. 1	No. 2
"Enjay" Butyl LM 430	50	50
"Omya" BLH	60	–
"HiSil" 215	–	20
"Indopol" H-100	–	30
"Silane" A-187	4	–
GMF	3	–
PbO_2 (VFC 0.33μ)	–	8
Toluene	12	19

Window Putty

"Ameripol" 1009	50
"Ameripol" 1006	50

"Cumar" MH 2-½	95
"Calcene" TM	160
Heptane	200

White Linseed Oil Putty

Raw Linseed Oil	80
Aged Linseed Oil	20
Soya Fatty Acids	1
"Atomite"	300
Putty Filler	589
International Fiber	10
B. C. White Lead	–

Mastic, Trowelling

Water	180
SBR 2002 Latex (48% T.S.)	352
Whiting	336
Titanium Dioxide	5
"Carbopol" 934	5
Ammonium Hydroxide	to pH 9.5

Disperse the Carbopol 934 in the water with moderate mixing and neutralize to pH 9.5 with ammonium hydroxide. Slowly add the latex to this preneutralized Carbopol 934 mucilage with moderate mixing, such as with a Lightnin' propeller type mixer. Add the pigment, very slowly, with moderate agitation and stir until the product is smooth.

Masonry Grouts

(Crack Filling)

Rigid:

A	"ERL"-2774	100
	"Cellosolve"	4
	Whiting	50
	Talc	26
B	Diethylenetriamine	10.5
	Toluene	2.5

Whiting	50
Talc	11

A/B = 180 pbw/74 pbw

Flexible:

A As in Rigid

B	"Polysulfide LP"-3	50
	TETA	10
	Whiting	50
	Talc	10

A/B = 180 pbw/120 pbw

Cement Patching Compound

Portland Cement, Type I	100
Sand	300
"Dylex" Latex KCD-254B (51%)	40
Water	15

Nonsagging Masonry Sealant

A	Polysulfide Resin	100
	Calcium Carbonate (wet ground)	20
	Titanium Dioxide	10
	Tackifyer	5
	Sulfur	0.1
	Stearic Acid	1.0
	Thickener and Additives	To suit
B	Accelerator (lead peroxide in DBT)	15

A nonsagging polysulfide sealant containing a phenolic tackifyer can be prepared with 1.2 phr Cab-O-Sil M-5 and 0.3% ethylene glycol dispersed in the resin. Excellent shelf stability and nonsagging properties during pot life can be obtained with this formulation. Since phenolic tackifiers stain masonry, epoxy tackifiers or other substitutes are often used.

When an epoxy tackifier is used in the formulation, it is added to the accelerator portion. In this formulation 1.6 phr Cab-O-Sil M-5, 0.2% ethylene glycol, and 0.2% "Triton" X-100 is needed to produce a shelf stable sealant composition.

Masonry Surface Filler

"Pliolite" S-5A	62.00
Chlorinated Paraffin 40%	33.00
Mineral Spirits No. 10	65.00
"Solvesso" 100	65.00
"Nuosperse" 657	8.00

Mix until resin is completely dissolved and add slowly:

Asbestos Shorts 7R	50.00
Superwhite Silica	400.00
Calcium Carbonate	200.00
"Bentone" 38	5.00

Add sufficient solvent blend to prevent balling up of pigments, but do not over thin:

Methanol	1.50
Mineral Spirits No. 10	134.00
"Solvesso" 100	134.00

Portland Cement/-Latex Mix

Sand	300
Portland Cement	100
"Dow Latex" 560 (48% N.V.)	41.6
Water	13.4
Water/Cement	0.35
Sand/Cement	3/1
Latex/Cement	0.20
% Latex Solids (based on total solids)	5.0

The sand and cement are thoroughly mixed. The water and "Dow Latex" 560 are combined and added to the sand and cement. The entire composition is then mixed for 4–5 minutes. The use of slightly more or less water will adjust the material to the proper consistency for handling. For best results, the latex-modified cement should be placed as soon as possible.

Joint Cements and Spackling Compounds

	Prepared Joint Cement	*Prepared Spackle*
Triethanolamine	2	2
Oleic Acid	4	4
Water	226	135
"Methocel 65 HG" 4000 cps (3% in water)	33	125
Polyvinyl Acetate (55% NV)	136	136
"Duramite"	900	1100
"Snowbrite Clay"	150	–
1K Mica	50	–
"Troysan PMA-30"	0.3	0.3
Mineral Spirits	26	26

Mix in suitable equipment the triethanolamine and oleic acid until they are completely reacted. This is indicated by the formation of a clear gel. Water, methocel solution, and polyvinyl acetate emulsion are then added to the soap and thoroughly mixed. The pigments should be added slowly, followed by the preservative and the mineral spirits. The compound should be quite smooth and homogeneous throughout at the end of the mixing period. The material should be packaged in cans that have been lined to prevent rust and latex coagulation.

Ceramic Tiles Mastic

(Emulsion Type – White)

Triethanolamine	30
Oleic Acid	56
Aromatic Plasticizer #33	116
"Piccolyte S-40"	116
Rutile (Titanium Dioxide)	50
"Atomite"	250
"Snobrite Clay"	200
"Natrosol 250 H" (3% in water)	241
"Naugatex 2105" (62% N.V.)	80

Floor Tile Sealer

"Shanco" 110 Resin	6.8
Ethyl Cellulose N Type	1.2
Isopropanol – 99%	92.0

Break or crush "Shanco" 110 Resin into small pieces. Charge all materials and stir at room temperature until solution is complete. The batch may be allowed to soak for several hours after which it is stirred to uniformity.

Asphalt Mastic

A	Gilsonite selects pulverized	46
	Calcium carbonate or silica (100-mesh)	46
	Asbestos 7r	8
B	Asphalt – 100 penetration	38
	Gilsonite selects (lump or pulverized)	17
	Mineral spirts	23
	Xylol	22

Mix ratio:

100 parts A
70 parts B

This mixture will be of heavy consistency suitable for trowel application, such as for floor patching. It will have a workable pot life of about 30 min and will cure hard in 24 hr at ambient temperatures.

Each component may be varied to achieve specific properties. B can include such materials as rubber, oils, or resins for special applications. Similarly, A can include ground cork (insulation) or other fillers for specific uses.

Flexible Polyurethane Caulks

	Formula No. 1	No. 2	No. 3
	lb	lb	lb
A *"Multrathane" F-84	223	184	146
"Atomite"	223	184	146

*Tinting Part I very slightly with a blue will enable the mixing of the two halves to be readily checked visually.

B	"Niax" Triol LHT-34	362	325	275
	Tributyl Phosphate	8	16	32
	Dibutyl Tin di-laurate	0.5	0.5	0.5
	"Thixcin-R	35	25	15
	"Atomite"	377	616	854
	"TiPure" FF	50	50	50

While the polyurethane caulk formulations shown above exhibit excellent adhesion to a wide variety of surfaces the use of the following primer solution is suggested to be more assured of really good adhesion to new, smooth aluminum surfaces. This primer can be applied very easily with a paint brush and dries so rapidly that it can be caulked over in a matter of minutes.

Gun Grade Caulk

"Indopol H-100 Polybutene	8.50
Blown Soya Oil, Z_4 viscosity	16.69
Tall Oil Fatty Acids, 3% Rosin	0.51
"Atomite" Calcium Carbonate	48.69
Talc	23.59
Mineral Spirits	1.72
Cobalt Drier, 6%	0.30

Blend vehicle ingredients in a heavy-duty, sigma blade mixer. Add fillers in the order shown and mix for 30 min. Thin to desired viscosity with mineral spirits. Add drier and mix to smooth consistency.

Self-Curing Adhesive

Ribbed Smoked Sheets	10.0
Sulfur	0.1
Zinc Oxide	1.0
"Agerite" White	0.1
Naphtha	80.0

Nonflammable Pure Gum Cement

Milled Pale Crepe Rubber (RSS if darker color is permissible)	10
Trichloroethylene	90

Pressure Sensitive Adhesive

"Zonarez" B-85 Resin	50.0
Natural Crepe Rubber	50.0
SBR (28% Styrene)	50.0
Antioxidant	1.0
Heptane	600.0

Freeze-Resistant Pressure Adhesive

"Neoprene WRT"	100
Zinc Oxide	5
Magnesia	4
Antioxidant	2
"Shanco" 700	50

Tape Adhesive

FORMULA	NO. 1	NO. 2
"Ameripol" 4502 Crumb	25.00	–
"Ameripol" 1013 Crumb	75.00	100.00
"Pentalyn" H	57.00	57.00
"Staybelite" 10	57.00	57.00
Zinc Oxide	40.00	40.00
Titanium Dioxide	20.00	20.00
Toluene	315.00	315.00
Antioxidant 2246	5.75	–

NO. 3

"Zonarez" B-85 Resin	60.0
Butyl Rubber-2.5% Isoprene	70.0
Polybutene-MW 500	30.0
Polyisobutylene	30.0
Zinc Oxide	30.0
Polyparanitrosobenzene	0.8
Heptane	800.0

Hot Melt Adhesive

"Elvax" 240	24.75
"Wingtack" 95	24.75
Paraffin Wax (130 M.P.)	49.50
"A-C" Polyethylene 316	1.00

Paper to Vinyl Film Adhesive

Formula	No. 1	No. 2	No. 3
"Geon" 460X1, 51%	100	100	–
"Geon" 450X20, 55%	–	–	100
Tetrasodium Pyrophosphate, 100%	0.1	–	–
"Carbopol" 934 Resin, 5%	1.26	0.5	–
"Methocel" HG-65 (Methocel MC 4000), 5%	–	–	0.8
pH	7.2	5.5	–
Viscosity (cP)	28,000	2,900	3,000

Note: Ammonium hydroxide used to adjust pH for desired viscosity

Fabric Seaming Adhesive

	Wet
Butyl Latex 90-01	145
"Hi-Sil 233"	10
"Naugawhite NX 101A"	0.63
"Aerotex MW"	10
"Aerotex MX"	3.5

Blend ingredients slowly in order listed.

This adhesive, based on butyl latex, is a starting formulation developed to anchor chain stitching for automotive upholstery. This type of stitching is more economical than conventional stitching. The compound performs well and may have additional applications as a general textile-seaming adhesive. The compound should be diluted with water to 20% solids or less for longer shelf stability. The concentrated formulation, with approximately 65% solids, has a tendency to "set up" on long standing.

Easily Removable Adhesive

"Flexowax" C applied in a thin layer to paper, leather, wood, metal, and other materials when pressed onto a surface will hold indefinitely if the weight is not too great. It is removed by pulling and can be reattached by pressing on again.

COATINGS

Black-Sidewall Tire Paint

Asphalt (Softening Point 235°F)	30
Painters' Naphtha or Xylene	70

Gradually add the solvent to the melted asphalt under high-power agitation. The product is a concentrate which can be diluted further with five to eight times its volume of thinner.

Gun Stock Varnish

Bodied Linseed Oil Bodies @ 590°F for Z_4 viscosity	144
Mineral Spirits	256
Lead Naphthenate	4
Cobalt Naphthenate	2

Moisten cloth and rub on.

Allow 24 hr between applications. Apply 10–12 thin coats or until you achieve the desired effect.

Wood Filler

Cellulose Acetate-Butyrate (½-sec)	100.0
"Aroclor" 5460	214.0
Magnesium Silicate	642.0
Toluene	85.7
Butanol	71.5
Butyl Acetate	314.0

Finger Paint

Masterbatch:

"Carbopol" 934	200
Sodium Tetraborate (anhydrous)	156
Color	200

Paint Formulation:

Water	94.7
Masterbatch (Above)	5.3

Ball mill the masterbatch until desired pigment fineness is achieved. Carefully disperse the ball-milled masterbatch in the water, in the proportions noted.

A test formulation has a final viscosity of 28,600 cps (Brookfield, 20 rpm) and a pH of 2.6. The paint has excellent stability and is easily worked with the fingers. An unusual aspect of the Carbopol formulation is that dried "creations" can be rewet and then be reworked or corrected.

It is important that anhydrous sodium tetraborate (borax) be used in the formulation above if the masterbatch is to be stored more than a few days. If hydrated borax is used, the associated water causes the masterbatch to ball up in a tough, rubbery mass.

Anti-Tarnish Coating for Metal

(Aerosol)

PVP/VA 1-335	10.0
"Santicizer" B-16	0.5
Isopropanol	14.5
Propellent 11/12 (50/50 or 40/60)	75.0

For nonaerosol application, this mixture may be modified by the use of methanol instead of propellent.

Sound Deadening Acoustic Coating

Petroleum Asphalt (85-100 penetration vacuum reduced)	10.0
Gilsonite BV	2.5
Petroleum Naphtha	12.5

Asbestos 7R	2.1
Kaolin Clay	6.0
Wetting Agent	0.2
Ground Limestone	66.7

Apply by airless spray and dry by heating up to 450°F.

Driveway Sealer

Asphalt (85 to 100 penetration steam-vacuum)	24.0
Gilsonite selects	16.0
Mineral Spirits (35 KB)	60.0

Basement Wall Sealer

Into a Cowles dissolver premix:

"Pliolite" AC-3	45
"Varsol" 18	270

After mixing for approximately 45 min add:

"Pliolite" VT	45

After the mixture is dissolved, add:

"Titanox" RCHT	172
"Cold Bond" R	247
White Portland Cement	319
"Varsol" 18	135

RCM, $0.90/gal; viscosity, 115 KU; coverage, 75 to 125 ft^2/gal.

2% Cal Ink 6600 series colorant can be added for coloring based on the total weight of the paint.

Aluminum Paint

"Pliolite" AC or "Pliolite" S-5	13.5
Aroclor 5460	13.5
Xylol	62.5
Aluminum Paste	10.5

Aerosol Heat Resistant Aluminum Paints

Concentrate:

	FORMULA NO. 1	NO. 2
"Aroplaz" 7323 (60% solids)	55.0	55.0
G.E. SR-82 Resin (60% solids)	18.0	18.0
Catalyst*	2.0	2.0
Aluminum Paste #1571	25.0	–
Aluminum Paste #1593		25.0

Blend all ingredients of the concentrate together. Be sure to use sufficient agitation to thoroughly disperse the pigment to avoid aggregates that might cause valve plugging.

Aerosol Loading:

	NO. 1	NO. 2
Concentrate	15.0	15.0
Toluene	40.0	18.0
Methyl Chloroform	–	22.0
Propellent 12	45.0	45.0

The can should be held 8–12 in. from the object being sprayed. Use oscillating motion while spraying to insure a uniform, smooth coating. Overspraying in one place causes sags and runs. The No. 1 formulation, because it utilizes a leafing pigment, has a silver colored coating but will show some slight pigment rub-off unless cured by heat. Curing during use is sufficient and at temperatures from 300–400°F, this occurs quite rapidly. The No. 2, because it has a nonleafing pigment, provides a much duller surface than No. 1, but when tackfree shows no pigment rub-off and a tougher film. However, its properties are also much improved duirng exposure to heat.

*6% Cobalt Octoate	21.7%
6% Manganese Naphthanate	21.7%
Xylene	23.9%
n-Butanol	32.7%

Baking and Frying Pan Coating

(Nonstick Surface)

GE SR-323 Resin (50% Solids)	4.5
GE SR-418 Resin (20% Solids)	1.0
Odorless Mineral Spirits	1.0
1, 1, 1 Trichloroethane (inhibited)	24.5
Propellent 12/11 (50/50)	69.0

Mix the silicone resins and solvents together. Moderate agitation will insure complete dispersion of the resins. Charge the solution of solvent and silicone resin to an aerosol can and load propellent.

Moisture Proof Paper Coating

"Pliolite" NR, P-1230	85
Paraffin Wax, m.p. 134°F	15
"Tenox" BHA	0.85
Toluene	400

Shoe Glaze, Aerosol

"Dow Corning" Resin R-5581	2.00
"Dow Corning" 510 Fluid	.05
"Chlorothene"	97.95
Nitrous Oxide (Propellent)	4.63

Latex Paints

	Light-Tone Formulations		*Masonry Paint Formulations*	
	(White)	*(Tint Base)*	*(White)*	*(Tint Base)*
FORMULA	NO. 1	NO. 2	NO. 3	NO. 4
Water	293.0	302.0	370.0	365.0
"Tamol" 731	8.0	10.0	8.6	9.0
"Victawet" 35-B	–	–	–	–
"Advawet" 33	–	–	–	–
Potassium Tripolyphosphate	0.3	–	0.3	–
"Tergitol" NPX	1.5	1.5	1.5	1.5

Phenyl Mercuric Acetate (30%)	5.0	5.0	5.0	5.0
Ethylene Glycol	20.0	20.0	20.0	20.0
"Carbitol" Solvent	15.0	15.0	20.0	20.0
Butyl Cellosolve Acetate	–	–	–	–
2-Ethylhexyl Acetate	5.0	5.0	5.0	5.0
Defoamer	5.0	5.0	5.0	5.0
"Cellosize" QP-4400	5.0	6.0	6.0	5.0
Rutile Titanium Dioxide	175.0	175.0	175.0	185.0
Anatase Titanium Dioxide	50.0	–	50.0	–
Red Iron Oxide	–	–	–	–
Chromium Oxide	–	–	–	–
Black Iron Oxide	–	–	–	–
Brown Iron Oxide	–	–	–	–
ASP-600	–	–	–	–
Basic Silicate of White Lead	–	–	–	–
Zinc Oxide	–	–	–	–
Mica, Water Ground	25.0	25.0	–	–
"Lorite"	–	125.0	–	–
Talc	125.0	–	–	–
Calcium Carbonate	–	–	150.0	180.0
"Ucar" Latex 180	400.0	398.0	315.0	319.0

Flameproof Coating

"Cereclor" 70	7.5
"Cereclor" 48	3.0
Antimony Oxide	5.0
Zinc Borate	1.5
Toluene	15.0
"Synthrapol" GP	1.0
Sodium Polacrylate	2.0
"Cellofas" A	0.5
Sodium Hexametaphosphate	0.1
Water	64.4

Fire Retardant Paint

	FORMULA	No. 1	No. 2
A	Monoammonium Phosphate	56	56
	Dicyandiamide	10	10

	Pentaerythritol	22	22
	Titanium Dioxide	12	12
	Diphenyl Phthalate	5	–
	Anionic Wetting Agent (acid-stable)	1.25	–
	Water	47–34	46–61
B	"Dow" PVC Latex 744-B	40	42.8
	"Santicizer" 141 or Dioctyl Phthalate	–	10
	Suspension Agent, Acid-Stable	–	0.6
	Anionic Emulsifier, Acid-Stable	–	1.2

Mix the intumescent agents, pigment (and plasticizer and wetting agent for Formula No. 1) and water. Disperse with two passes through a high-speed stone mill.

Add the latex (plasticized for Formula 2) with medium stirring.

Red Lead Ship Paint

Red Lead (97% Grade)	220.0
Vinyl Resin	145.0
Tricresyl Phosphate	15.0
Methyl Isobutyl Ketone	295.0
Toluene	295.0

Ship Paint

	(White)	*(Orange)*
Titanium Dioxide	50	–
Molybdate Orange	–	24
Vinylidene Chloride/Acrylonitrile Copolymer		
Methyl Ethyl Ketone	625	625

Both white and orange resin solutions should be manufactured in a porcelain or stone line pebble mill or any other explosion-proof totally enclosed equipment which will provide good pigment dispersion. A steel ball mill is not satisfactory since contact with steel may result in gelation of the resin and/or discoloration of the finished products.

Marine Antifouling Coating

Pigmentation:

	PVC%
Cuprous Oxide	8.0

Red Iron Oxide	8.0 or nil
Zinc Oxide	3.0

Binder:

	% Volume on Total Binder
Chlorinated Polyethylene 20 cp or Chlorinated Rubber 20 cp	7.5 or 15
Natural Wood Rosin	60.0 or 67.5
Chlorinated Paraffin 42 or Methyl Ester of Hydrogenated Rosin	to 100

Suspending Agent:

Modified Hydrogenated Castor Oil 1.3% w/w on Binder

Solvent:

Mixed Trimethyl Benzenes/Mineral Spirits 3:1

Marine Corrosionproofing Paint

	(Red Lead)	*(Lead Sulfate)*	*(Aluminum)*
Chlorinated Rubber 10 cps	12.1	12.3	12.0
Chlorinated Paraffin 42	4.0	4.1	4.0
Chlorinated Paraffin 70	7.1	7.2	7.0
Read Lead	18.2	–	–
Red Iron Oxide	18.2	6.7	0.3
Basic Lead Sulfate	–	13.3	–
Barytes	–	6.7	18.0
Nonleafing Aluminum Paste 65%	–	9.2	12.0
Titanium Dioxide	–	–	6.0
Modified Hydrogenated Castor Oil	1.6	1.6	1.6
Xylene	31.0	31.1	30.5
Mixed Trimethyl Benzenes	7.8	7.8	7.6
PVC	33.5	31.0	34.2

Paint and Varnish Remover

Methylene Chloride	72.0
Methanol	6.0
Ethylene Glycol Monoethyl Ether	4.0

Water (Distilled)	3.0
Paraffin Wax (Type I, Grade A)	3.0

Flush-Off & Scrape-Off Paint Remover

	(Flush-Off Remover)	*(Scrape-Off Remover)*
Paraffin	1.7	1.5
EHEC-75H	0.7	0.5
Methylene Chloride	73.4	92.0
Toluene	2.1	–
Methanol	7.9	6.0
Water	1.3	–
"Nopco" Soap C	5.0	–
Di-Triisopropanolamine	7.8	–
Triethylammonium Phosphate (40% in ethanol)	0.1	–

Dissolve the melted paraffin in the methylene chloride or methylene chloride-toluene solvent. Add the EHEC-75 and stir until wetted. Add the menthanol slowly with agitation until the EHEC is dissolved. Add the Nopco Soap C, di-triisopropanolamine, triethylammonium phosphate, and water in that order.

COSMETICS AND DRUGS

Bath Salts

"Calsoft" F-90	10
Sodium Sesquicarbonate	85
Magnesium Carbonate	5
Perfume and Dye	q.s.

Bubble Bath

"Miranol" CS Conc	20
Sodium Lauryl Sulfate 28% Active	10
Triethanolamine Lauryl Sulfate 40% Active	10
Lauric Diethanol-Amide-High Active	3
Water	57

Hand Lotion

Oil Phase:

Isopropyl Isostearate	4.0
Stearic Acid	5.0
Glycerol Monoisostearate	4.0
Sorbitan Monostearate	1.0
Cetyl Alcohol	1.0

Water Phase:

Propylene Glycol	2.0
Triethanolamine	0.9
Distilled Water	82.1

Heat each phase to 80°C and mix together thoroughly. Cool with slow stirring and add perfume at room temperature.

Astringent Lotion

Menthol	0.05
Perfume	0.50
Aluminum Chlorhydroxy Allantoinate	0.20
Propylene Glycol	3.00
S.D. Alcohol #40	50.00
"Chlorhydrol" (50%)	5.00
Water Distilled	41.25

Dissolve menthol and perfume in S.D. Alcohol #40. Dissolve rest of ingredients in water and add to the alcohol solution.

Cleansing Cream

	Cream	*Lotion*
"Provol" 10	1.0	2.0
"Novol"	1.5	2.5
"Volpo" 10	3.0	3.0
"Polychol" 15	4.0	2.0
"Polawax" A31	4.0	–
"Super Corona"	3.0	2.0
P E G 400 Monostearate	–	2.0

Mineral Oil	42.0	–
Mineral Oil	–	45.0
Stearic Acid XXX	1.0	1.0
Triethanolamine	1.4	0.7
Propylparaben USP	0.1	0.1
Methylparaben USP	0.1	0.1
"Carbopol" 940 (1% aq sol'n)	14.0	–
"Carbopol" 941 (1% aq sol'n)	–	7.0
"Carbowax" 200	2.0	2.0
Propylene Glycol	2.0	2.0
Water	21.0	28.6
Perfume and Color	q.s.	q.s.

Combine the Carbopol solution and water; dissolve the preservatives in the TEA and glycols and add to the water phase. Heat the oil to 75°C and add to the water phase at the same temperature with mechanical agitation. When the emulsion is uniform, cool to 50°C and add the perfume, q.s. the lost water and continue cooling to room temperature.

Skin Freshener for Dry Skin

"Volpo" 10	1.0
"Volpo" 20	1.0
Witch Hazel	20.0
Perfume	up to 0.3
"Crotein" SPN	2.0
Water	75.7
Color and Preservatives	qs

Premix Volpos and perfume at 45–50°C. When clear, add Witch Hazel at room temperature with stirring until clear and then add water. To achieve maximum clarity, the ratio of Volpo 10 to Volpo 20 should be adjusted for optimum solubilization of the perfume. A 1:1 ratio is a reasonable compromise. A cool color can be achieved by using a mixture of D&C Green #5 and a trace of FD&C Yellow #5.

Almond Lotion

Cellulose Gum (CMC-7HP)	1.3
Almond Oil	9.9

Ethanol (90%)	7.5
Oil of Bitter Almonds	0.2
Oil of Geranium	0.8
Water	80.3

Moisturizing Lotion

A	Cetyl Alcohol NF	0.35
	Ethoxylated Cetyl-Stearyl Alcohol	0.15
	Decyl Oleate	3.00
	Stearic Acid, Triple Pressed	5.00
	Mineral Oil, NF	15.00
B	Deionized Water	76.50
C	Perfume, Dyes and Preservatives	q.s.

Heat A to 80°C. Heat B to 80°C. Add A to B with vigorous stirring. Allow to cool to 40°C. Add C at 40°C. Continue stirring until room temperature is reached. Homogenize and fill off.

Moisturizing Cream Cologne

FORMULA	NO. 1	NO. 2	NO. 3
"Volpo" 3	1.0	–	–
"Volpo" 10	–	3.0	1.5
"Polychol" 5	–	–	1.5
"Crodafos" N10 Neutral	1.0	–	–
"Crodamol" IPM	–	5.0	5.0
Perfume	2.0	1.0	1.0
Ethanol SD 40 (95%)	40.0	32.0	32.0
Water (Distilled)	36.0	34.0	32.0
2% Carbopol 941 Stock Sol'n.	20.0	–	–
1.2% Carbopol Stock Sol'n.	–	25.0	27.0
Triethanolamine	qs to pH 6.5	–	–
Diethanolamine	–	qs to pH 6.4	qs to pH 6.4

Dissolve the emulsifiers in the oils and add the water to the oils at room temperature with efficient mechanical agitation. Then add the

Carbopol stock solution followed by the required amount of amine dissolved in the ethanol.

Moisturizing Lotion with Protein

"Novol"	5.0
"Super Corona"	1.0
"Polawax"	2.0
"Skliro"	1.0
"Polychol" 15	0.5
"Crodamol" IPM or Apricot Kernel Oil NF	3.0
Stearic Acid (T.P.)	1.0
Oleic Acid	1.0
Glycerine	10.0
"Crotein" SPO	2.0
Triethanolamine	1.0
Water	72.5
Preservative System, Perfume, Color	q.s.

Mix all oil phase ingredients by melting the solids in the oils and bring to a maximum of 85°C. Dissolve the water phase ingredients and bring to 75–80°C. Add the water to the oils with mechanical agitation. Continue stirring and cool to 25–30°C after perfuming at about 50°C.

Cationic Hand or Body Lotion with Protein

"Emcol" E607S	1.0
"Skliro"	2.0
"Polawax"	1.0
"Super Hartolan"	0.5
"Crodamol IPM"	2.0
Glycerine	10.0
"Crotein SPO"	2.5
Water	79.2
"Crotein BTA" – Acid (add after emulsion is made and cooled to 40°C)	1.5
Citric Acid	q.s. to pH 5
Preservative System, Perfume, Color	q.s.

Skin Protective Cream

"Crodafos" N.10 Neutral	2.0
"Polawax"	5.0
"Crodacol" S	5.0
"Novol"	5.0
Apricot Kernel Oil NF	10.0
Mineral Oil (70 visc.)	19.9
Petrolatum	15.0
Myristic Acid (90%)	2.6
Triethanolamine	1.2
Water (dist.)	14.2
"Carbopol" 940 (1.2% aq. sol'n)	20.0
Butylated Hydroxy Anisole	0.1
Perfume, Color, Preservation System	q.s.

Combine all oils and emulsifiers and warm to 70–75°C. Combine TEA and water and warm to 70–75°C and add to oils with agitation. Cool and continue slow agitation. Add the Carbopol stock solution (acid form) at about 65–70°C and continue cooling till nearly solid. Hot fill into jars.

Germicidal Skin Cleansers

FORMULA	NO. 1	NO. 2
"Polawax"	5.0	–
"Miranol" C2M	40.0	–
Sodium Lauryl Sulfate (WA Grade)	17.0	–
Water	27.0	19.5
"Triton" X 100 (28% active)	–	70.0
"Polychol" 5	–	1.0
"Volpo" 3	1.5	–
"Super Hartolan"	0.5	0.5
"Cetrimide"	3.0	3.0
Propylene Glycol	5.0	5.0
Petrolatum White	1.0	1.0

Formula No. 1: Blend the water phase and warm to 60°C. Warm the oil phase until the Super Hartolan has melted with the hexachlorophene and then add to the water phase slowly with stirring, cool to room temperature with slow agitation. No. 2: warm both phases separately to 55°C and disperse the Triton well. Add the water phase to the oil phase with stirring and continue to cool to room temperature with slow agitation.

Cosmetic Cleaner

FORMULA	NO. 1	NO. 2
Sodium Lauroyl Lactylate	10.0	5.0
Stearyl Alcohol	5.0	5.0
Isopropyl Palmitate	–	3.0
Hexachlorophene	–	3.0
Water	85.0	84.0

The pH is adjusted to 7.0 with sodium hydroxide. Formula No. 1 is a white pearlescent lotion and it is used in the laboratory as an all purpose liquid skin cleanser. It is used for washing hands with satisfactory results. The skin feels supple and well lubricated after frequent daily washings. Formula No. 2 is a modification of A to impart increased emolliency as well as antibacterial action.

Leg Make-Up

Water Phase:	
"Carbose" 514	16.8
Water	40.0
Oil Phase:	
Amerlate LFA	9.0
"Isopar" (cosmetic grade)	13.0
Pigment Blend:	
TiO_2	4.7
Talc	14.1
Iron Oxides	1.2
Ammonium Hydroxide (28%)	1.2
Preservatives	q.s.

Heat oil phase to 65°C; add pigment blend with agitation; add water phase at 65°C to oil phase at 65°C; add ammonium hydroxide, cool with agitation to RT; keep mixture tightly covered during processing.

Face Mask

(Acid)

A	"Veegum" K	6.00
	Water	82.76
B	Ethanol	4.00
C	Glycerin	4.00
	Sulfated Caster Oil	3.00
D	"Vancide" TBS	0.12
	"Vancide" FP	0.12
	Color	q.s.
	Buffered to pH 5.5	

Buffer Solution: 60 parts – 1M citric acid, 40 parts – saturated sodium citrate solution.

Add the "Veegum" K to the water slowly, agitating continually until smooth. Add B to A. Blend C and add to A and B. Add D to a small portion of the mixture, blending well. Add the concentrate to the remainder. Add the buffer solution with mixing.

Prickly Heat Lotion

Menthol	1
Glycerin	1
Salicylic Acid	4
Alcohol (95%)	to make 100

Prickly Heat Cream

A	Stearic Acid	50 g
	Mineral Oil (light)	250 cc
B	Triethanolamine	7 g
	Water	700 g
C	Bismuth Subgallate Powder	150 g

Form cream with A and B as in the Insect Bite Formula. When cool, incorporate C, mixing thoroughly till cream is smooth and free from lumps.

Diaper Rash Cream

A	White Bees Wax	115 g
	Spermaceti	85 g
	"Sterolan"	150 g
	Mineral Oil (heavy)	470 g
B	Water	175 g
	Borax	2.5 g
	1-Methionine	30 g

A – Melt all ingredients at 150°F. B – Dissolve the two powders in the water, heat to 150°F. Pour hot B into hot A, stirring constantly till cool.

Acne Cream

A	"Veegum";1.75	1.75
	CMC (7 MSP)	0.40
	Water	34.60
	Glycerin	5.00
	Allantoin	0.25
B	Resorcinol	3.00
	"Triton" X-100	0.20
	Water	29.70
	"Nytal" 300	16.00
	Titanium Dioxide	2.90
C	Iron Oxides	1.10
	Sulfur	5.00
D	"Vancide" BN	0.10

Dry blend Veegum and CMC and add to the water slowly, agitating continually until smooth.

Add B to A.

Pulverize C and add to A-B. Incorporate D by making up concentrate in portion of finished cream, triturating thoroughly.

Add this to product, milling or homogenizing finished cream.

Psoriasis Cream

FORMULA NO. 1 w/o olly		NO. 2 o/w vanishing	
"Super Hartolan"	4.0	Stearic Acid TP	16.0
Petrolatum	5.0	"Novol"	6.0
Paraffin Wax (high M.P.	12.0	"Super Corona"	2.0
Min. Oil	30.0	Coal Tar	2.0
Coal Tar	2.0	Triethanolamine	0.6
Allantoin	0.25–0.5	Allantoin	0.25–0.5
Water (dist.)	qs to 100	Water (dist.)	qs to 100

Formula No. 1: heat both phases to 70°C, add the water slowly to the oil phase with H.S. stirring, allowing time for each addition tu incorporate. When all in, cool with agitation to just above melting point and fill off. A homogenizer may be used to improve stability.

Formula No. 2: heat both phases to 75–80°C, add water to oils with H.S. stirring. Cool to 40°C with stirring and fill off.

Hand Protective Cream

FORMULA	NO. 1	NO. 2	NO. 3
Triple-Pressed Stearic Acid	50.0	8.5	4.0
Polyethylene Glycol 1450	60.0	–	4.5
Polyethylene Glycol Monostearate	–	4.3	–
Anhydrous Lanolin	20.0	1.0	1.5
Terpineol	0.3	0.1	0.1
Triethanolamine	4.0	–	–
Potassium Hydroxide	2.0	0.5	0.2
Propylene Glycol	36.0	5.0	7.4
Hydroxyethyl Cellulose	45.0	15.0	7.0
Perfume	0.3	0.4	0.4
Zinc Stearate	21.0	–	1.6
Water	200.0	65.2	73.0

A paddle-type stirrer is preferred for No. 1 to prevent aeration, but a higher-speed, mechanical mixer can be used to make the liquid creams (No. 2 and 3).

Melt the stearic acid, PEG-1450, monostearate, and lanolin and bring the temperatue to 70°C. Add the terpineol. Dissolve the potassium hydroxide in about an equal weight of water, add this solution and the triethanolamine (where used) to the melt, and stir until most of the soap is in solution. Bring the temperature to 70°C. Heat the water to 70°C and stir into the mixture or stir the mixture into the water, if more convenient. Continue stirring until a homogeneous mixture is obtained. Add the propylene glycol and hydroxyethyl cellulose and continue stirring slowly until the cream has cooled to about 40°C. Stir in the zinc stearate and the perfume.

The cream can be packaged at 40°C as it has a better consistency than when allowed to cool to room temperature before packaging. The liquid creams (Formulas No. 2 and 3) should be stirred slowly, but continuously, until cooled to room temperature to prevent increased viscosity on standing. The stirring of any of the creams should be fast enough to maintain thorough mixing without aerating.

Night Cream

"Polawax"	6.5
"Solan"	4.0
"Super Corona" Lanolin (USP)	1.0
"Crodacol" C	1.0
Apricot Kernel Oil	6.0
Mineral Oil	16.0
Hydrogenated Vegetable Oil	0.5
Water	65.0

Heat the oils and the aqueous phase separately to 75°C and add the water to the oils with stirring down to near room temperature; hot fill and allow to set.

Cold Cream

A	"Lanocerin"	5.0
	Beeswax	8.0
	Mineral Oil (80–90 vis.)	52.0
B	Borax	0.8
	Water	34.2

Heat ingredients listed under A at 145°F until a clear melt is obtained. Dissolve borax in water at 145°F and emulsify A with B solution. Pour into jars at 110°F.

Rolling Massage Cream

(Fluid)

"Veegum"	2.0
Water	78.8
Triethanolamine	1.0
Stearic Acid	3.0
Paraffin	15.0
Perfume	0.2
Preservative	q.s.

Add the "Veegum" to the water slowly, agitating continually until smooth. Add *b* and heat to 65–70°C. Heat *c* to 65–70°C, add *a* to *b* and *c*, and mix until the temperature drops to 50°C. Then add *d*. Allow mixture to stand overnight. Stir well before packaging.

Apply to hands in the same manner as a hand lotion. Continue rubbing hands together. The cream will roll off and carry with it dirt, sweat, and loose dry skin.

Facial Astringent

Rose Water	6 fl. oz.
Witch Hazel	10 fl. oz.
Tannic Acid	10 g

Cream Makeup

Stearic Acid TP	1.60
Isopropyl Myristate	4.00
"Cetinate"	1.00
"Amerlate" P	0.25
"Robane®"	4.00
Carboxymethylcellulose	0.10
"Veegum" Regular	1.50
Propylene Glycol	5.00

Lecithin	0.40
Triethanolamine	0.80
Sorbitan Monolaurate	0.10
Titanium Dioxide, Talc, Kaolin and Pigments	30.00
Deionized Water	
Perfume & Preservatives	
q.s.	100.00

Beauty Mask, Enzyme

A Base:

Chymotrypsin	15/Cht.U./TEE
Perfume	0.045 g
Calcium Carbonate up to	15.00

B Diluent:

Glycerol	9.60 g
Buffer Solution	7.27 g
"Tylose"	0.1152 g
Color	0.0192 g

Parts A and B to be mixed before use.

Facial Clay Mask

Bentonite	12.0
TiO_2 3328 Type	6.0
"Cellosize" (low visc.)	0.5–1.0
Ethanol SD 40	10.0
Water	q.s. to 100
Certified Colors	q.s.
Methylparaben USP	0.2
Perfume Solubilized with "Volpo" 10 (1:1)	q.s.

Whitening Clay Mask, Face

"Albagel"	17.0
Ethanol SD 40	10.0
TiO_2 (3328 Type)	5.0
"Methocel" 60 HG 50	0.5–1.0

Water	q.s. to 100
Perfume Solubilized with "Volpo" 10 (1:1)	q.s.
Methyl Paraben USP	0.2

Micropulverize the TiO_2 and mineral gum. Dissolve perfume in Volpo 10 and alcohol. Disperse the organic gum in the water and sprinkle in the Albagel/TiO_2 blend under high speed kitchen-aid type mixing. When smooth add alcohol phase and continue mixing until alcoholic phase is well blended.

Sunscreen

FORMULA NO. 1

"Solulan" PB-20	2.0
Mineral Oil (70 visc.)	26.5
Isopropyl Palmitate	70.0
"Escalol" 506	1.5

NO. 2

"Polychol" 5	1.0
"Volpo" 3	3.0
Mineral Oil (65–75 visc.)	4.0
"Robane"	8.0
"Filtrosol" A-1000	2.0
Glycerin	4.0
Water	78.0

Warm the Polychol 5 in the Volpo 3 and other oil phase ingredients. Cool to room temperature and add the water to the oils with vigorous mechanical agitation. No other heat or homogenization is required.

Since the oil phase is external, these emulsions may easily be converted to aerosols by addition of suitable propellants such as the hydrocarbons or P 12/11 blends to the emulsified concentrate.

Sunscreen Lotion

(Wash-Off Resistant)

A Oil Phase

"Adol" 158	3.00
"Hydrofol" Acids 1655-CG	0.30

	"Adol" 66 (Isostearyl Alcohol)	3.00
	"Adol" 52-NF (Cetyl Alcohol)	0.04
B	Water Phase	
	Borax	0.08
	Methyl "Paraben" U.S.P.	0.10
	Water (Demineralized)	91.28
C	Final Additions	
	"Filtrosol" A-1000	2.00
	Perfume	0.20

Heat phases separately to 75°C. Add oil to water slowly with agitation. Mix 5 min; add Filtrosol slowly; mix 5 min. Cool to room temperature with moderate agitation, adding perfume at 40°C.

Aerosol Suntan Lotion

A	"Veegum"	1.5
	Water	88.0
	Propylene Glycol	3.0
B	Diglycol Stearate SE	3.0
	Silicone 556 Fluid	1.0
	Cetyl Alcohol	0.5
	"Acetulan"	2.0
	"Giv-Tan" F	1.0
	Preservative	q.s.

Add the "Veegum" to the water slowly, agitating continually until smooth. Add the propylene glycol and heat to 75–85°C. Heat B to 70–75°C. Add A to B, mix until cool.

Aerosol package: Concentrate 90%; Propellant 12/114, 40/60 10%.

Skin Browning, Aerosol

Erthrylose	3.0
Propyleneglycol	0.2
Alcohol (Anhydrous)	96.8

Mix the above. To 33 parts add:

"Freon" 11		33.0
"Freon" 12		33.0

Suntan Cream

A	G-E SF-96 (350) Silicone Fluid	2.0
	Isopropyl Myristate	18.5
	Cetyl Alcohol	0.2
	Stearic Acid	4.0
	"Giv Tan" F	1.5
B	Water (Sterile)	67.5
	Triethanolamine	1.2
	1% "Carbopol" 934 Sol'n	5.0

Indoor Tanning Lotion

A	"Veegum"	4.0
	Sodium Carboxymethylcellulose (low visc.)	0.5
	Water	77.5
	Propylene Glycol	2.0
	"Acetol"	2.0
	Buffer Solution	q.s.
B	Dihydroxyacetone	4.0
C	Synthetic Iron Oxide	3.0
	Titanium Dioxide	3.0
D	Ethanol	4.0
	Preservative	q.s.

Dry blend the "Veegum" and CMC and add to the water slowly, agitating continually until smooth. Add the balance of A and mix. Buffer to pH 5.0 with saturated citric acid. Add B to A gradually with mixing. Blend C and then add to A and B, mixing until uniform. Add D and mix until uniform.

Directions for use: Spread a thin film evenly over the area to be tanned.

Wave Set Texturizing Conditioner

"Resyn" 28-2930	3.00
Ammonia (18%)	0.21

"WSP"-A200 Alco-Pro	3.00
"Ammonyx" 4002	0.50
Ethanol	68.14
Deionized Water	25.00
Perfume	q.s.

Permanent Wave

Potassium Hydroxide	10
Sodium Sulfite	20
Monoethanolamine	50
Sulfonated Castor Oil	15
Water	900

Add the potassium hydroxide to 100 ml of the water, and stir until *completely* dissolved. Then dissolve the sodium sulfite in another 100 ml of the water. Mix the sulfonated castor oil and monoethanolamine. Then mix the three solutions and add the remaining water.

This is a strong solution. Be sure ingredients are dissolved, and instruct user to wash off carefully after hair is waved.

Translucent Vegetable Pomade

FORMULA	NO. 1 *(Soft)*	NO. 2 *(Hard)*
Castor Oil AA (refined)	88.1	82.4
Japan Wax	8.8	13.5
Castor Wax (refined M-80)	1.0	1.0
"Polychol" 5	1.0	1.0
"Novol"	–	1.0
BHT	0.1	0.1
Perfume	1.0	1.0
Color	q.s.	q.s.

Melt and blend together all raw materials, after complete mixing, cool rapidly to 5–10°C in bulk or fill into jars at 60–65°C and cool immediately to 5–10°C. Quick cooling produces translucent appearance.

Pomade

	Formula No. 1	No. 2
"Lantrol"	5.00	–
"Lanfrax"	–	30.00
Mineral Oil (355 visc.)	55.00	55.00
Isopropyl Myristate	15.00	10.00
"Multiwax" 445	20.00	–
Polyethylene Glycol 400 Dioleate	5.00	5.00

Heat all ingredients to 78°C. Stir down to 50°C and perfume. Stir down to 5° above solidifying point and fill.

Kinky Hair Straightener

Formula No. 1

(Pressing Oil)

Mineral Oil (heavy)	15.0
"Modulan"	5.0
Petrolatum	80.0
Perfume	q.s.
Color	To suit

No. 2

(Pressing Oil)

Mineral Oil (heavy)	20.0
"Modulan"	5.0
"Amerlate" P	5.0
PEG 400 Monostearate	5.0
Silicone 556 Fluid	3.0
Petrolatum	62.0
Perfume	q.s.
Color	To suit

No. 3

(Caustic Type)

Gum Tragacanth	2.0
Boric Acid	1.0
Water	40.0

Make a uniform paste and stir in previously dissolved:

Sodium Carbonate	1.0
Potassium Hydroxide	1.0
Glycerin	2.0
Water	8.0

Apply the paste to hair with combing. Allow to remain for ½ hr. Then wash well with water to remove all paste from the hair.

NO. 4

(Thio Type)

Water	2400.0	
Sodium Hydroxide	24.0	
Ethylene Glycol	15.0	slurry
CMC 70 (high visc.)	7.5	slurry
Ammonia 26°B	74.0	
Ammonium Thioglycolate 60%	324.0	
Perfume	26.0	

NO. 5

Water	1000.0
"Veegum"	40.0
Lauryl Monoethanolamide	20.0
Ammonium Thioglycolate	205.0
Ammonia 20°B	52.0
Sodium Hydroxide	12.0

Cream Rinse Concentrate

"Fluilanol"	3.0
"Ammonyx" 4	10.0
"Crodacol" S	2.0
PEG 400 Distearate	3.0
Water	81.2
Sodium Chloride	0.8
Color and Perfume	q.s.

Melt the oils together and bring to 65–75°C. Heat most of the water to the same temperature, retaining the remainder in which the salt should be dissolved. Add the water to the oils and cool to 40°C with stirring; add the salt solution, color and perfume. Continue agitation and cool to about 32°C before filling.

Balsam Type, Protein Enriched Conditioner Rinse

"Ammonyx" 4	2.0
"Crotein" BTA	2.0
"Fluilanol"	2.0
"Crodacol" S	1.0
Polyethylene Glycol 400 Distearate	1.5
NaCl	0.15
Water (distilled)	88.85
Ethyl or Isopropyl Alcohol	2.5
Perfume and Preservative System	q.s.

Heat oil phase ingredients to about 60°C and add the water at about the same temperature, reserving a small amount to dissolve NaCl solution. Then allow to body for about 12 hr and add alcohol and perfume. Alternatively in place of using alcohol, the system can be thickened by adjusting the pH to 2 with citric acid q.s.

Proteinized Clear 'Creme' Rinse

"Crotein" BTA	0.5
"Lanexol" AWS	1.0
JR Polymer*	0.7
Water (distilled)	q.s. to 100
Perfume (Lemon/Lime type water-soluble) and Preservative System	q.s.

Dissolve JR Polymer in water with heating to 50°C. When clear add Lanexol and Crotein BTA. Perfume and preservatives if needed may be dissolved in the Lanexol prior to addition. For additional body Methocel 90 HG4000 (4) is compatible (up to 0.5%).

---

*A new, water-soluble, cationic cellulosic, substantive to hair and other substrates.

Amphoteric Protein Shampoo

1	"Tegobetaine" C	25.00
2	Water	38.15
	"Sipon" LT 6	12.20
	"Sipon" ES	9.40
	"Trepoline" M3	5.00
	WSPX-250	9.00
3	Citric Acid Monohydrate	0.25
	Water	1.00

Mix each phase separately, then add 3 to 2; then add 1 quickly and with good agitation.

Baby Shampoo

"Richodet" X-6311	20.0
"Natrosol" 250 HR	0.6
Formalin	0.2
Methyl Parasept	0.1
"WSP"-X250	2.0
Triethanolamine to pH 6.5	7.0
Water, Dye, Perfume to make	100.0

Using Lightning agitation, disperse the "Natrosol" in cold water and mix for one hour (it must be mixed long enough to thoroughly hydrate the "Natrosol"). Then heat to 55°C., add the "Richodet" X-6311, protein, formalin, and methyl parasept. Cool to 40°C. and add dye and perfume. A small amount of Triethanolamine may be necessary to adjust the pH to 6.5-7.0.

Quick Breaking Hair Styling Foam

Ethanol SD 40	29.0
"Ucon" 50HB 5100	0.5
"Volpo" 10	1.0
"Polychol" 15	0.5
"Resyn" 28-1310	4.0
Amino Methyl Propane Diol	0.45
Water	64.0

Fill above concentrate 95%, propellant 152a 4% and propellant 142b 2%.

Clear Hair Groom Stick

"Croda" S42 Polyamide Resin	18.0
Refined Castor Oil	72.3
Cocoanut Diethanolamide	2.0
"Volpo" 3	1.0
"Satulan"	0.7
"Crodamol" IPM	3.0
Perfume	3.0
Antioxident	q.s.

Heat all except the IPM, perfume and antioxident to 110–120°C with stirring; vacuum conditions should be used on large scale production. Add the other ingredients, heated to 80–90°C and add to the original solution. Pour the mixture into molds and set at room temperature for one hour. Flaming the stick will add to the clarity and minimize the tendency to 'sweat'.

Hair Spray Aerosols

Many of the resins used in aerosol hair sprays are quite hygroscopic, a beneficial property if the humidity is not too high, since most of the resins are also quite brittle and the moisture can act as a plasticizer. But few areas if any possess just the proper humidity, so small inclusions of additives are employed to remedy the inherent resin deficiencies. The category includes most of the Croda line. The surfactants aid in washability, since a good hair spray must be easily rinsed from the hair. In extremely dry areas, humectants such as Lanexol AWS, Solan, Polychol 20 and Volpo 20 are added to the formula. In extremely humid environments, where the film may become too tacky, most hair sprays contain additives such as Cetyl or OLEYL Alchol, Procetyl AWS, or Skliro lanolic acids to help make the film less hygroscopic. Other additives such as Lanexol AWS and Super Hartolan are primarily used to make the resin film more flexible, while still others like Liquid Base, Fluilan and Fluilanol are principally utilized to increase sheen. Invariably, the additive chosen will have a multiple functionality.

Aerosol Hair Spray for Men

PVP/VA E-735 (50% resin sol'n in alcohol)	3.0
"Solan"	0.5

"Fluilan"	0.3
"Ucon" 50HB660	1.5
Ethanol (anhydrous)	24.5
Perfume	0.2
Propellant 11	45.5
Propellant 12	24.5

Dissolve the Solan, Fluilan and Ucon in the alcohol and then add the PVP sol'n and the perfume. Fill into cans, chill, cold fill and crimp valves – or pressure fill.

Wig Luster Spray

FORMULA	NO. 1	NO. 2
PVP/VA E-635 (50% solids)	4.80	–
PVP/VA E-535 (50% solids)	–	3.5
Dimethyl Phthalate	0.25	0.1
Silicone	0.10	0.1
"Ethylan" 35%	–	0.5
Ethanol (SDA-40, anhydrous)	24.70	25.6
Propellant 11/12 (65/35)	70.00	70.0
Perfume	0.15	0.2

Coloring Gray Hair Black

FORMULA NO. 1

Precipitated Sulfur	20
Glycerin	40
Lead Acetate	15
Perfume Oil	5
Water q.s.	1000

Dissolve the perfume oil in glycerin. Dissolve lead acetate in water; add sulfur and perfume oil-glycerin. Mixture: The mixture must be well shaken before bottling and use. After application, clean hands.

NO. 2

Sulfur (precipitated)	3.0
Lead Acetate	3.0

Glycerin	25.0
"Solulan" 5	2.0
"Isopar" M	5.0
Cetyl Alcohol	2.0
Mineral Oil (light)	5.0
Petrolatum	55.0

Hydrogen Peroxide Gel

(Hair Bleach)

"Pluronic" F127	25
Hydrogen Peroxide* (30% sol'n)	10
Water (deionized or distilled)	65

Cool the deionized water to 40–50°F. and place it in a mixing container. Add the Pluronic F127 slowly with good agitation and continue mixing until solution is complete, maintaining the temperature below 50°F. Add cool hydrogen peroxide solution (40–50°F.) slowly, with gentle mixing. Immediately transfer to containers and allow to slowly warm to room temperature.

*The peroxide content may be varied by increasing the amount of hydrogen peroxide used, up to 20% of the 30% H_2O_2 solution (or the equivalent percent of a more concentrated sol'n) while decreasing the amount of water in the formulation.

Preshave Talc Stick

A	"Veegum"	1.90
	Water	q.s.
	Zinc Stearate	4.70
B	Light Magnesium Carbonate	1.90
C	"Nytal" 300	91.50

I. Add the Veegum to about 30 parts of water slowly, agitating continually until smooth.

II. Blend B and add to C.

III. Add I to II. Work into a smooth paste (add more water if necessary.

IV. Pack in stick mold and allow to dry.

These talc sticks do not break easily and have excellent rub-off qualities. For a softer stock, the Veegum content should be reduced.

Pre-Electric Shave Lotions

	Formula No. 1	No. 2
"Crodamol" DA or Crodamol IPM	15.0	–
"Cetrimide"	0.1	0.1
"Procetyl" 30	–	15.0
Perfume	q.s.	q.s.
Ethanol SD 39C	q.s. to 100	100

Blend, filter and fill. Aging may be required depending upon the perfume content and level of alcohol insoluble resinoids in the perfume.

No. 3

Alcohol SD 40	89.5
Perfume	0.5
"Neobee" M 20	10.0

No. 4

Phase I	
"Lantrol" AWS	2.50
Centrimide	0.10
Oleyl Alcohol	2.50
Isopropyl Myristate	5.00
Dipropylene Glycol	4.00
Alcohol SDA 40 (anhydrous)	70.80
Phase II	
Allantoin	0.10
Water	15.00

Add ingredients of Phase I to mixing vessel along with perfume. Agitate till solution forms. Do same with Phase II, and very slowly add Phase II to I with agitation, color and fill.

After Shave Foam Lotions

Formula No. 1

"Polychol" 40	1.5
"Crodacol" C	1.5

Ethanol SD 40, 95%	59.0
"Hyamine" 1622	0.1
Menthol	0.1
Perfume	1.0
Water	37.8

Fill concentrate 92%, 3% propellant 11 and 5% propellant 12/114 (40:60).

NO. 2

S.D. Alcohol #4	100.00
Perfume	0.63
Hexadecyl Alcohol	
Glycerin	
Deionized Water	

Color with certified green or blue dye.

Protein After Shave Lotion

SDA-40 Alcohol	69.50
"Tegester" 504-D	2.00
"WSP"-A200	2.50
FD&C Blue #1	q.s.
Fragrance	1.00
Water	25.00

Shaving Cream

(Self-Heating)

Stearic Acid TP	4.12
"Emery" 621	1.37
"Cetinate"	0.55
"Ethoxylan" 100	0.22
Glycerine	7.33
Triethanolamine	3.21
"PVP" K30	0.25
Deionized Water	
Perfume & Preservative q.s.	100.00
Potassium Sulfite	9.00
Fill into co-dispensing	

Package:

Above Concentrate	80.00
Hydrogen Peroxide (7 X Sol'n)	20.00
Pressurize with Isobutane/Propane (84:16) Mix	4.50

Aerosol Shaving Cream

A	Stearic Acid (triple pressed)	4.5
	Lauric Acid	1.5
	Dow Corning F-157 Wax	1.1
B	Triethanolamine	3.6
	Glycerin	10.0
	Water	79.3
C	Perfume	q.s.
	Concentrate	95
	Isobutane	5

Mix and heat A and B separately to 158°F (70°C), then add A to B with gentle stirring. Cool to 122°F (50°C) and add C. Package in lacquer lined can with foam valve.

Aerosol Shaving Foam

"Fluilanol"	5.0
Lauroyl Sarcosine	3.0
Oleoyl Sarcosine	4.0
Mineral Oil	1.0
Propylene Glycol	5.0
Triethanolamine	1.0
Water	81.0
Perfume	q.s.

Fill: above concentrate 95%
propellants isobutane/propane 46 lb mix 5%

Women's Leg Shave Cream

(Tube Dispensed)

A	GE SF-96 (350) Silicone Fluid	3.5
	Stearic Acid	3.5
	Cetyl Alcohol	1.0

B	"Carbopol" 934	0.2
	Triethanolamine	1.0
	Propylene Glycol	7.5
	5% "Veegum" Sol'n	20.0
	"Cinaryl" 200	0.2
	Water (Sterile)	63.1
C.	Perfume	q.s.
	Color	q.s.

Heat A and B to 80°C (176°F). Add A slowly to B with high speed agitation. Continue slow mixing until the temperature reaches 30°C (86°F). Then blend in C.

Application: Apply a thin uniform film to pre-wetted skin and shave. Pre-wetting the skin improves greatly the lubricity of the film.

Brushless Shaving Cream

A	Stearic Acid	18.0
	Mineral Oil	5.0
	"Tween" 60	5.0
	"Sorbo"	5.0
	Borax	2.0
B	Triethanolamine	1.0
	Water	64.0
	Preservative	q.s.
C	Perfume	q.s.

Heat A to 90°C., B to 95°C. Add B to A with agitation. After the cream reaches the set point, stop mechanical agitation, perfume, and stir occasionally by hand until the preparation reaches room temperature.

Depilatory

Formula No. 1

A	Calcium Thioglycolate	5.0
	Thioglycolic Acid	5.0
	De-ionized Water	25.0
B	Sodium Hydroxide	2.2
	De-ionized Water	22.0

C	Calcium Hydroxide	1.0
D	"Polawax"	5.0
	"Cabosil" M-7	1.0
	De-ionized Water	34.8

Add A to B slowly, and cool. Add C to AB and, finally, add D.

NO. 2

Strontium Sulfide	35.0
Titanium Dioxide	21.0
Corn Starch	43.0
Benzocaine	0.2
Perfume	0.8

Lipstick Base

Ethoxylated Lanolin	7.0
Beeswax	8.5
Lanolin	3.0
Carnauba Wax	2.0
Candelilla Wax	5.0
Paraffin Wax	2.7
Castor Oil	57.0
Isopropyl Myristate	5.5
Bromo Acid	2.5
Lipstick Lake Colors	6.8
Perfume	q.s.

Melt waxes separately to 80–90°C. Grind Bromo Acid in oils. Next day add waxes to the oil in melted form, bring temperature to 80°C and add color and perfume. Grind finally, cool, and mold as required.

Lipstick

FORMULA NO. 1

"Robeyl"	25.
"Spermwax"	7.5
Castor Oil	30.
Beeswax	5.

Petrolatum	7.
Carnauba	7.
Lanolin (anhydrous)	10.
Eosin Acid	
q.s. Pigment	8.

NO. 2

Candelilla Wax	5.0
Carnauba Wax	2.0
Ceresin Wax	1.5
"Emerwax" 4266-0	1.5
Cetiol V	4.0
Mineral Oil (65/75 Saybolt) q.s.	100.00

Deep-Staining Lipstick

"Volpo" 3	10–20
Bromo Acid Dyes D&C Red #21 or Orange #4 eg	1–3
"Novol"	20–10
Castor Oil	40
Candelilla Wax	10
Carnauba Wax	10
Perfume	q.s.
Insoluble Pigments and Lakes	q.s.

Dissolve the bromo in the Volpo or Polychol with as little heat as possible. To the resulting clear solution add the Novol – note that the solution remains clear. Then add the castor oil and the waxes all at elevated temperatures. No precipitation of any kind should occur; the pigments are added subsequently and the milling is carried out in any way adapted to your equipment, or the pigments may be milled in the Novol and castor oil before combining with the waxes. The resulting sticks will possess superior gloss (without flaming) plasticity and mold release properties.

Lip Pomade

"Spermwax"	7.
Petrolatum	40.

"Cetina"	26.
"Robane"	26.
Perfume, Preservative	q.s.

Eye Shadow Powder

Micronized Talc	q.s. to 100.0%
Kaolin	8.0
Zinc Stearate	4.0
"Croda" Liquid Base	3.0
"Polychol" 20	4.0
"Fluilan"	4.0
TiO_2, Inorganic Oxides and Bismuth Oxychloride (% depends on desired coverage and pearl)	q.s.

The dry powders are blended in a twin shell blender, the hydrophillic binder is sprayed on and blended thoroughly before pressing into pans.

Eye Shadow Stick

"Crodamol" IPM	46.02
"Fluilan"	6.66
Carnauba Wax	8.00
White Beeswax	3.33
Mineral Oil	6.66
"Crodacol" C	2.66
Pigments, Inorganic Oxides, TiO_2 or Bismuth Oxychloride	26.67

Prepare as you would a lipstick. Melt waxes and oils together, add colors and mix well. Run entire mass hot thru a heated roller mill, then repeat at a narrower setting. Remelt mass, vacuum to remove air, perfume if desired and pour into molds.

Mascara Block

"Polychol" 15	15.0
"Polychol" 5	10.0
Triethanolamine	3.3
Stearic Acid TP	6.7
Paraffin Wax 45–50°C	12.0

Carnauba Wax	10.0
Talc	q.s. to 20.0
Carbon Black	q.s. 2–4
Brown Iron Oxide	4.0

Melt waxes, add color and mix well. Run entire mass through a heated roller mill. Remelt material and pour into molds.

Cream Mascara

FORMULA NO. 1

A	"Marcol" 52	4.10
	"Parmo" 10	5.50
	"Emersol" 132	9.10
	Triethanolamine	2.75
	Silicone Fluid "SF-96"	5.00
	"Tegosept" P	0.05
B	Titanium Dioxide 3328	9.10
C	Deionized Water	64.30
	"Tegosept" M	0.10

Blend B. Heat A to 70°C. Heat C to 60°C. Stir B slowly into C. Continue stirring and heat the mixture to 65°C. Stir in A. Cool to 30°C. with stirring.

NO. 2

A	Carnauba Wax	1.00
	Yellow Beeswax	8.00
	Ozokerite, 170°F	6.00
	"Emersol"-132	2.00
	"Emersol"-233LL	2.00
	"Tegosept"-P	0.10
	"Trisolan" LP	2.00
B	Deionized Water	q.s.

NO. 3

Glycerol Monostearate	10
Triethanolamine	3

Stearic Acid	15
Petrolatum	20
Gelatin	2
Water	17
Beeswax	25
Pigment (Lampblack, Umber, Ochre)	8

Eye Liner

A	"Veegum"	2.5
	Water	75.5
B	Polyvinylpyrrolidone	2.0
	Water	10.0
C	Pigment	10.0
	Preservative	q.s.

Add the "Veegum" to the water slowly, agitating continually until smooth. Dissolve the polyvinylpyrrolidone in water, using a little heat. Add B to A and then add C. Mix well.

For a product with a soft, cream like viscosity that can be applied with a brush, increase the "Veegum" content to 3.5%.

Nail Polish Remover

FORMULA NO. 1

(Gel)

Butyrolactone	10.0
Ethylene Glycol Monobutyl Ether	20.0
"Laneto" 50	4.0
Acetone	45.0
"Carbopol" 940	0.7
Water	20.0
"Ethomeen" C/25	q.s. to gel

Disperse the Carbopol 940, with rapid agitation, in the mixture of Butyrolactone, Gafcol EB, Laneto 50, acetone and water. In small increments, add the Ethomeen C/25 to produce the desired gel, mixing vigorously upon each addition. The amount of Ethomeen will run between 0.5–1.0%. The final pH of the gel should be 7.0, or slightly on the acid side. Color and perfume.

NO. 2

(Liquid)

Wax Emulsion	1.5
Alcohol SD #40	22.0
"Synotol" LM 60	1.5
Acetone	35.0
Ethyl Acetate	5.0
Water (distilled)	35.0

Dissolve the Wax emulsion in warm alcohol and then add the rest of the ingredients. Fill concentrate while still warm.

Cuticle Remover

A	"Veegum"	1.00
	Water	74.75
B	"Vancide" TBS	0.25
	Triethanolamine	11.00
	Glycerin	10.00
C	"Brij" 30	1.20
	"Brij" 35	1.80

Add the "Veegum" to the water slowly, agitating continually until smooth. Add B to A and heat to 70°C. Heat C to 75°C. Add to A and B and mix to 40°C.

Antiperspirant Lotion

(For Bottle or Roll-On Dispenser)

Veegum HV	1.0
"Starfol" GMS-480	8.0
"Adol" 66 (Isostearyl Alcohol)	2.0
Water (demineralized)	49.0
"Chlorhydrol" 50%	40.0
Perfume and Color	q.s.
	100.0

Add "Veegum" slowly to water (room temperature), mix until smooth, heat to 70°C. Heat GMS-480 and "Adol" 66 to 70°C, add slowly to "Veegum" dispersion (70°C). Cool with moderate agitation to 50°C. Add "Chlorhydrol" (heated to 50°C). Continue cooling to below 30°C, adding perfume and color at about 40°C.

Roll-On Antiperspirant

A	"Lantrol" AWS	2.00
	Propylene Glycol	2.50
	S.D.A. 40 (anhydrous)	20.00
	"Methocel" HG 0.65–4000 cps.	0.75
B	Water	29.75
C	"Chlorhydrol" (50% sol'n)	45.00

Slurry A with agitation and perfume. Slowly add B. Agitate slowly and add C.

Emollient Antiperspirant Cream

"Polychol" 10	2.0
"Super Hartolan"	0.5
"Crodacol" C	1.5
Spermaceti Wax	2.0
Perfecta Petrolatum	4.0
"Arlacel" 165	4.0
Glycerin	2.0
Chlorhydrol (50% sol'n)	40.0
Veegum	1.5
Water	42.5

Underarm Deodorant

"Irgasan" CH 3565	0.3–0.5
Cetyl Alcohol	1.5
Lanolin	1.0
Mineral Oil	2.0
Isopropyl Myristate	1.0
Stearic Acid	15.0
Propyleneglycol	5.0
Triethanolamine	0.9
Potassium Hydroxide	0.1
Water	73.5

Deodorant Stick

A	"Vanseal" CS	30.0
	Sodium Hydroxide	2.0
	Cetyl Alcohol	21.4
	Glycerin	4.0
B	"Carbowax" 6000	16.8
	Stearic Acid	20.0
	Propylene Glycol	5.3
C	"Vancide" TBS	0.5

I. Blend A over low heat until melted.
II. Add B to I and stir over low heat until melted.
III. Add C to II stirring until "Vancide" TBS goes into solution.
IV. Add III to stick mold while hot and refrigerate for 3 hour minimum.

Odor and Moisture Removal Block

Activated Carbon (powder)	75.0
Silica Gel (powder)	75.0
Polyvinyl Acetate Latex (55%)	27.3
Water	77.7

Mix well, press into forms, and dry at 200°C.

Air Freshener

Water	47.5
Odorless Kerosene	10.0
"Span" 85	1.5
"Tween" 81	1.0
Butane	40.0

The mixture of emulsifying agents is dispersed in the odorless kerosene. The water is added slowly with constant stirring until a thick w/o emulsion is formed. This is then filled into the containers, and after the valve has been sealed on, the propellant is injected under pressure. The emulsion stability depends to a large extent on the choice of surface active agents, the most successful being mixtures of polyoxyethylene sorbitan esters, sorbitan esters, and esters of polyethylene glycols. In the example quoted above, separation occurs within 2 or 3 minutes but the phases are easily redispersed.

Aerosol Deodorant

(Concentrate)

Heaxachlorophene	0.5
Zinc Sulfocarbolate	1.0
"Acetol"	1.0
"Brij" 30	0.5
SDA 40 Anhydrous	52.0

Mix all ingredients with agitation until in solution. Fill with either of the propellants:

	A	*B*
Concentrate	55.00	55.00
"Freon" 114	19.35	27.00
"Freon" 12	26.65	18.00
	(colder)	(warmer)

Space Deodorant

"Neutrolair" D7	1.0
Alcohol, Anhydrous	2.5
Propylenegycol	6.0
Triethyleneglycol	8.0

Toilet Drip Deodorizer

Kerosene	85
Mineral Oil (100 visc.)	95
Eucalyptus Oil	10
Nitrobenzol	10

Feminine Hygiene Deodorant Spray

"Cetrimite"	0.10
Propylene Glycol	0.50

Isopropyl Myristate	0.70
Anhydrous Denatured Alcohol (SDA 40-1)	3.68
Perfume	0.02
Freon 12 Propellant	95.00

Dissolve all ingredients in alcohol with mechanical agitation. Make up any weight loss with alcohol before loading the concentrate. Pressure fill the propellant.

Men's Cologne

(Boat Type)

"Givaudan's" Perfume W-127-2	2
Specially Denatured Alcohol #40	80
Distilled or Deionized Water	18

Solid Cologne

A	Stearic Acid	5.00
	"Sorbo" (70%)	2.00
	Alcohol	87.36
B	Sodium Hydroxide	0.64
	Water	5.00
C	Perfume	q.s.

Heat (A) to 65°C., (B) to 70°C. Add (B) to (A) with agitation. Add perfume at 55–60°C. Pour into heated molds. Cool slowly to prevent formation of air pockets. Wrap in foil and package in airtight containers.

Lilac Perfume

Heliotropine	2.5
Musk Ketone	5.0
Methyl Ionone	6.5
Lily of the Valley Synthetic F. B.	15.0
Synthetic Red Rose F. B.	19.5
3-methylnonan-3-ol	51.5

Frictional Cologne

Hexadecyl Alcohol	5.5
Ethyl Alcohol SD-40 (anhydrous)	22.1

Distilled Water	68.0
"Carbopol" 940	0.2
Diisopropanolamine (10%)	2.2
Perfume	2.0

The "Carbopol" is dissolved in water with agitation and then the diisopropanolamine is added to the stirring solution. The remaining ingredients are combined and slowly added.

Perfume for Lotion

Bergamont	29
Benzaldehyde	41
Lemon Oil	10
Clove Oil	10
Mace Oil	10

Hindu Incense Powder

Sandalwood (powdered)	80.00
Orris Root (powdered)	4.00
Licorice Root (powdered)	12.00
Patchouly Leaves (powdered)	2.50
Potassium Nitrate (pulverized)	0.50
Oriental Rose Compound	1.00

Place all of the ingredients in a ribbon mixer and mix at low rate of speed for 30 min.

Foot Spray Deodorant

"Polawax" A 31	2.0
Stearyl Alcohol	0.6
Cetyl Alcohol	0.6
Hexachlorophene	0.3
Dichlorophene	0.6
Perfume Oil	0.9
Alcohol Anhydrous	61.0
1,2-Propylenglycol	2.8
Distilled Water	31.2

90% solution is filled with 10% propellant 12/114 (40:60).

Foot Sweat Control

Salicylic Acid	1.12
Boric Acid	14.5
Zinc Oxide	4.5
Formaldehyde	1.8
Ointment Base	q.s. 100.0

Athlete's Foot Remedy

(Ringworm)

Equal parts of:

Camphor, Gum (U.S.P.)
Phenol, Crystals (U.S.P.)

On a water-bath melt together the camphor and phenol.

The repeated application, with caution, to the infected foot has been known to help.

The warning is again emphasized; the foot (infected area) must be dry when the camphor-phenol mixture is applied. And the foot must be kept dry for some hours after applying the mixture. Suggested that the treatment be upon a dry foot just before retiring.

Store the preparation in a well-stoppered bottle, as the mixture tends to be hydroscopic.

Athlete's Foot Control Cream

Undecylenic Acid	5.00
Triethanolamine – enough to buffer to pH	6.50
Zinc Undecylenate	20.00
Hydrophyllic Ointment Base USP	75.00

Rub up Zinc Undecylenate with ointment base until smooth and add the buffered acid. Mix slowly.

Foot (Ease) Bath

Epsom Salts	3 lb
Sodium Bicarbonate	½ lb

Methyl Salicylate	½ cc
"Aerosol" OT (75%)	1 cc

Dissole in gal. water.

Mouthwash

FORMULA NO. 1

Cetylpyridinium Chloride	0.100
Citric Acid Monohydrate	0.100
Cinnamon Oil (U.S.P.)	0.050
Peppermint Oil (U.S.P.)	0.100
"Polysorbate" 60	0.300
Ethyl Alcohol (U.S.P.)	10.000
Sorbitol Solution (U.S.P.)	20.000
Color	q.s.
Distilled Water q.s. ad.	100.00

NO. 2

"Hyamine" 10-X	0.03
"Hamposyl" L-30	0.06
"Pluronic" F-68	0.50
Glycerin	10.0
Ethanol	5.0
Water, Flavor, Color	q.s. 100

Toothpaste

FORMULA NO. 1

"Sorbo"	1000
Sodium Saccharine	6
Sodium Lauryl Sulfate	70
"Carbopol" 934 Dispersion (6% in Water)	600
Water	484
Oil of Spearmint	40
Sodium Hydroxide (50% in Water)	31
Dibasic Calcium Phosphate Dihydrate	1800

Dissolve sodium lauryl sulfate in one-third of the water. Mix the "Sorbo," sodium saccharine, and 6% "Carbopol" solution with the rest of the water. This will be a thin solution. Blend the two solutions carefully to avoid foaming. Slowly add the sodium hydroxide to neutralize the "Carbopol 934" while gently stirring. The solution will become thick as the neutralization progresses. Slowly mix in the calcium phosphate. Stir until the particles are wetted out and all the lumps are broken up. Mill and deaerate the batch on conventional equipment. Add the flavoring during the last stages of milling to avoid evaporation loss. Fill into tubes.

NO. 2

Calcium Carbonate	20.0
Calcium Phosphate Dibasic (hydrous)	32.0
"Quso" G 32	4.0
Sodium Lauryl Sulfate	1.2
Sodium Carboxymethylcellulose Sol'n (0.5%)	3.2
Glycerin	6.0
Mineral Oil	1.0
Sorbitol Solution (70%)	32.0
Saccharin Sodium	0.6
Peppermint Oil	as needed

Dissolve the saccharine sodium in the sodium carboxymethylcellulose solution and mix this solution together with the glycerin and sorbitol. Slowly add the "Quso" with thorough mixing. Blend the dry powders, calcium carbonate, calcium phosphate and sodium lauryl sulfate together, and incorporate into the binding solution. Finally, add the mineral oil and peppermint oil.

NO. 3

A	"Veegum"	1.00
	Water	18.50
B	CMC (7HP)	0.50
	Glycerin	30.00
C	Dicalcium Phosphate	47.00
D	Flavor	1.00
E	Sodium Lauryl Sulfate	2.00
	Preservative	q.s.

Add the Veegum to the water slowly, agitating continually until smooth. Wet the CMC with glycerin and add to I slowly with agitation.

Add C slowly with agitation.

Add D and mix.

Add E slowly with just enough mixing to obtain a smooth batch.

Denture Cleaner Tablet

A	"Veegum" WG	5
	Sodium Perborate	13
	Tetrasodium Pyrophosphate Anhydrous	25
	Sodium Chloride	13
	Tartaric Acid	9
	Sodium Phosphate Dibasic	12
	Citric Acid	7
	Sodium Bicarbonate	16
B	Isopropyl Alcohol	25
C	"Veegum" WG	2

I. Granulate A with B. Pass through a No. 10 mesh screen. II. Dry the granulation one hour at 105°C. and pass through a No. 16 mesh screen. III. Dry blend with C and compress.

Face Powder

A	Magnesium Carbonate	2.00
	Calcium Carbonate	2.00
	Perfume Oil	2.00
	Italian Talcum (best grade)	66.00
	Magnesium Stearate	4.00
	Titanium Dioxide	6.00
B	Kaolin	16.00
	Arrowroot Starch	2.00
	Zinc Oxide	2.00
	Prepared Color Pigments	2.00

Dusting Powder, Body

FORMULA NO. 1

Talc	50
Precipitated Chalk	25

"Thixcin" R	4
Perfume	1
Boric Acid	5
Colloidal Kaolin	15

NO. 2

Talc	96.00
Magnesium Carbonate	2.00
Zinc Stearate	1.00
Perfume	1.00

Talcum Powder Spray

Talc (325 mesh or finer)	9.50
Pyrogenic Silica	0.50
Perfume	0.50
Propellant 12/11 (30:70)	89.50

Bed Sore Dusting Powder

Tannic Acid	10
Zinc Stearate	90

Fungicidal Dusting Powder

FORMULA NO. 1

Zinc Undecenoate	100.0 g
Undecenoic Acid	20.8 g
Pumilio Pine Oil	4.7 ml
Starch	500.0 g
Light Kaolin, to make	1000.0 g

Triturate the pumilio pine oil and undecenoic acid with the light kaolin, incorporate the starch and the zinc undecenoate. Sift and mix.

NO. 2

Talc USP	95
Isopropyl Myristate	1
Polyethylene Glycol 1500	2
"Vancide" R.E.	2

Crab Louse Ointment

γ-Benzenehexachloride	1
Stearyl Alcohol	3
Beeswax	8
Petrolatum, White	82

Warm and mix until uniform.

Scabicide

Benzyl Benzoate	50
Castor Oil	50
Pot. Carbonate	30
Water	70

Itch Reliever

Talc	90.0
Undecylenic Dialkylolamide	5.0
Menthol	0.5
Benzocaine	0.1
Perfume	to suit

Add water and triturate to make a smooth paste.

Burn Dressing

Petrolatum	500
Sorbitol Monoöleate	20
Sorbitol Monoöleate	20
Sulfathiazole Powder	50
Distilled Water	500

Analgesic Back Rub

Methyl Salicylate	40
Menthol	5
Oil Turpentine	20
Oil Clove	5
Propylene Glycol	5

Oleic Acid	3
Triethanolamine	5
Water	17

Mix the first six ingredients together. Mix the last two ingredients together. Then blend all with stirring.

Analgesic Balm

"Polawax"	15
Methyl Salicylate	10
Menthol	5
Mineral Oil U.S.P.	10
Distilled Water	60

Hemorrhoidal Suppositories

Aluminum Dihydroxy Allantoinate	0.2
"Metycaine"	1.0
Zinc Oxide	10.0
"MYRJ" 52	2.0
Glyceryl Monostearate S.E.	5.0
"Wecobee" SS q.s. ad	100.0

Melt the "Wecobee", "MYRJ" 52 and glyceryl monostearate S.E. Triturate the powders, add part of the melted base and levigate. Pour the mixture into the melted base and mix thoroughly.

Stomach Antacid

Magnesium Hydroxide	8.00
"QUSO" G 32	5.00
Sodium Benzoate	0.20
Peppermint Oil	0.05
Purified Water	q.s.

Dissolve the sodium benzoate in eight percent of the final volume of water. Slowly add the "QUSO" G 32 while stirring, then the magnesium hydroxide. Make the suspension up to final volume, add the peppermint oil and pass through a colloid mill.

Corn and Callous Remover

1.	Salicylic Acid	160 g.
2.	Lactic Acid	32 g.
3.	Acetone	60 cc.
4.	Chlorophyl (oil soluble)	10 g.
5.	Flexible collodion	880 cc.

Dissolve 1, 2, and 3 in 5; add 4 to the solution.
Caution: Flammable and highly volatile.

Anti-Microbial Spray

Ethanol, SDA #40	587 g
Triethylene Glycol	214 g
Propylene Glycol	128 g
2-Chloro-*m*-5-xylenol	14 g
Perfume	57 g

Insect Repellent Foam

"Polawax"	3.0
N,N Diethyl Toluimide	15.0
Ethanol SD 40 (Anhydrous)	39.6
Water	32.4
Propellant 12/114 (40:60)	10.0

Anti-Tarnish Solution for Silver

Morpholine	1
Water	99

After silver is cleaned, wipe the objects containing the above solution. As water evaporates a residual alkaline film remains that prevents tarnishing in jewelry case, storage, etc.

Silver Dip

(Silver Tarnish Remover)

Thiourea	300
Sulfamic Acid	100

"Triton" X-100	50
Water	1000 cc

Aluminum Brightener

Citric Acid	4
Phosphoric Acid (85%)	3
Nonionic Detergent	2
Methyl Ethyl Ketone	3
Water (distilled)	88

Aluminum Cleaner

"Renex" 690 or	
"Renex" 698	2
Phosphoric Acid (85%)	3
Citric Acid, C.P.	4
Methyl Ethyl Ketone	3
Water	88

Apply with sponge, soft brush or rag. Leave on surface no longer than 15 min. Rinse with water. Caution: Wear waterproof gloves when handling. Do not use on magnesium. It may be necessary to add a corrosion inhibitor, e.g., 1% sodium chromate.

Rust Remover

FORMULA NO. 1

For vertical and overhead surfaces brush on to 1/8″ thick, wipe or wash off.

"Santocel" C	21.00
Water	178.00
Nonionic Surfactant	1.25
H_3PO_4 (85%)	60.00

NO. 2

Immersion Type

Sodium Hydroxide	54.0% max
Sodium Gluconate	25.0% min

Trisodium Salt EDTA	13.0 % min
Foamers & Surfactants	7.0% max

Use 5 lb/gal, temp. up to boiling. Wash with water, dry at once. Coat surface with light oil unless finish is applied at once.

Will clean copper, brass, magnesium. Will strip phosphate coatings.

NO. 3

(Gel)

This brush-on, thickened product removes rust from ferrous surfaces, as well as oxides and scale from other metals. Its consistency may be varied by adjusting the percentage of "Methocel" in the formula.

Water	33.00
Diethylene Glycol	5.00
"Carbitol"	5.00
"Surfonic" N-95	0.75
Phosphoric Acid (75%)	45.00
Hydroxyacetic Acid (70%)	10.00
"Methocel" 65HG4000	1.25

In a polyethylene tank, stir the first three ingredients together. While stirring vigorously, slowly sift in the "Methocel". Agitate till solution has thickened and is free of lumps. Now add the remaining ingredients in order, and stir until uniform. Pack or store in glass or polyethylene.

NO. 4

Abrasive

Olein	10
Petroleum	40
Aqueous Ammonia, 0.910%	3
Water	10
Denatured Alcohol	2
Emery Powder with a little Alumina	40

NO. 5

Triethanolamine	21.5
Oxalic Acid	18.5

Useful on porcelain, china, tile and chrome plate.

Add the acid crystals to the amine and stir until dissolved.

Apply the concentrate and allow it to remain in contact with the stain for a few minutes. Rub briskly and rinse thoroughly with water. Repeat if necessary.

This liquid may also be used on fabrics that will subsequently be washed.

Rust Remover and Anti-Corrosive

FORMULA	NO. 1	NO. 2
Phosphoric Acid	40	30
Potassium Permanganate (10%)	10	10
Zinc Monophosphate	10	10
Aluminum Monophosphate	5	5
Methanol	20	20
Butanol	5	5
CrO_3 (in H_3PO_4)	10	10
Hydrochloric Acid	–	10

Immerse article for 1–5 min in either solution. Rust is removed; steel is not attacked and a gray corrosion-proof layer is formed that acts as a good base for paint.

Metal Cutting Oil

FORMULA	NO. 1	NO. 2	NO. 3
"Surfactol" 318	–	18.00	12.50
"Surfactol" 340	7.50	18.00	12.50
"Surfactol" 365	42.50	–	–
Tertiary Nonyl Polysulfide	–	5.00	–
Petroleum Sulphonate	–	5.00	5.00
Dipropylene Glycol	50.00	54.00	44.83
Dye	–	0.02	–
Masking Agent	–	0.15	–
Water	150.00	150.00	600.00
Hexylene Glycol	–	–	20.00
Dibenzyl Disulfide	–	–	2.00
Polyoxyethylene Octadecyl Amine	–	–	2.00
Phenyl Mercuric Acetate (18% sol'n)	–	–	1.00

The base grinding fluid is prepared by mixing the two "Surfactols" with the dipropylene glycol and then forming an emulsion of this non-aqueous mixture in water.

Finger Mark Remover (for Metals)

FORMULA	NO. 1	NO. 2	NO. 3
Paraffin Oil	15.1	–	–
Triethanolamine Oleate	14.5	15.0	8.0
"Pentasol"	21.7	40.0	26.5
Lanolin	–	4.0	5.3
Stoddard Solvent	9.0	13.5	–
Water	39.7	50.0	33.8
Butyl "Carbitol"	–	25.0	13.2

Rusting of polished steel surfaces is often caused by fingerprint residues. The following materials will remove such residues. They should then be followed by an application of a thin rust inhibiting film of any desired type.

Methyl Alcohol (min 90%)
Ethyl Alcohol
"Carbitol" (min 80%)
"Alkaterge" 4% in Methyl Isobutyl Ketone

POLISH

Wash n' Wax, Auto

"Calsuds" CD-6	50
Water	47.5
"GE" Silicone SF 96	2.5

Mix the silicone into the CD-6 and add the water last. Dilute 1–2 oz./gal.

Spray Auto Polish

"A-C" Polyethylene 629 Wax	5.45
Silicone 200 (500 cs)	3.27
"VM&P" Naphtha	91.28

Melt the wax, add the Silicone 200, then add preheated "VM&P" naphtha with constant stirring. The wax is soluble in warm naphtha and settles out as solution cools. If it is necessary to fill the solution while cold, agitation will keep the waxes in suspension.

Auto Polish, Liquid

Oil Phase:	
"Epolene" E-10 Wax	1.6
"DC-200" Silicone Fluid (350 cs)	2.0
Tall Oil Fatty Acid	3.0
"Stoddard" Solvent	34.5
Morpholine	2.0
Abrasive Cleaner:	
"Snow Floss"	12.0
Water Phase:	
Gelwhite GP	2.0
Water	42.9

Combine the silicone fluid, "Epolene" E-10 wax, tall oil fatty acid and "Stoddard" solvent. Heat the mixture to 180–190°F (82–88°C) and maintain this temperature until a clear solution is formed. Add "Snow Floss" or "Kaopolite" SF-O abrasive cleaner with good agitation. Throughly disperse the "Gelwhite" GP in water with strong agitation and heat the dispersion to 170°F (77°C). Add morpholine to the oil phase. Then add the "Gelwhite" GP dispersion to the oil phase with good agitation. Cool rapidly.

Vinyl Auto Top Polish

FORMULA	No. 1	No. 2
"Efton" D Super	50.0	50.0
"AC"-540	50.0	50.0
Oleic Acid	3.0	3.0
KOH at 45%	5.0	5.0
Diethyl Amino Ethanol	14.0	14.0
Methyl "Carbitol"	–	6.0
Water	299.0	293.0

Furniture Polish

FORMULA NO. 1

"A-C" Polyethylene 629 Wax	10.0
Silicone 200 (500 cs)	6.0
"VM&P" Naphtha	84.0

Melt the wax, add the Silicone 200, then add preheated "VM&P" naphtha with constant stirring. The wax is soluble in warm naphtha and settles out as solution cools. If it is necessary to fill the solution while cold, agitation will keep the wax in suspension.

Aerosol:

Concentrate (above)	30.0
"Genetron" 12/11 (50:50)	70.0

NO. 2

"Shanco" W-1120 Wax (60%)	3
Silicone 350 CTS (40%)	2
Mineral Spirits	95

Heat the "Shanco" W-1120 Wax, silicone and a small portion of the solvent to a clear solution. Add the balance of solvent and heat to a clear solution at approximately 160° or higher. Cool rapidly with vigorous agitation.

Aerosol Shoe Polish

FORMULA NO. 1

A	"Viscasil" (10,000) fluid	1.0
	Wax OM	1.5
	Wax V	0.6
	Lanolin	0.5
B	1, 1, 1 Trichloroethane (inhibited)	96.4
	Dye	q.s.

The procedure for manufacturing this polish calls for heating A to 100°C. (212°F.) to melt the waxes, using moderate agitation. Next, heat B to 50°C. (122°F.) to prevent precipitation of the waxes. Finally, add B slowly to A with constant agitation. Continue agitation until temperature has reached 30°C. (86°F.) The concentrate (40 parts by weight) is then loaded into the aerosol container and the propellant blend of 12/11 (70/30) (60 parts by weight).

NO. 2

A	"GE" SR-82 Resin (60% Solids)	5.2
	99% Isopropanol	42.0
	Butyl Acetate	2.4
B	"Acryloid" B-72 Resin	8.8
	Toluene	41.6

1. Dissolve "Acryloid" B-72 resin in toluene. Add remaining material in A to B solution.
2. Aerosol Loading

Concentrate	25.0
Propellent 12/11 (50/50)	75.0

Pipe Bowl Polish

Mineral Spirits	95.0
Carnauba Wax	3.0
Silicone (10,000 cp)	2.0

Warm with care mineral spirits to 90°F and dissolve the wax in small pieces till in solution. Add silicone. Note – any hard wax may be used.

Dip a cloth into solution and spread onto outside of bowl and stem. Allow to dry.

Metal Polish

"Monamine" ARA-100	4.0
Cottonseed Fatty Acids	0.2
Abrasive	18.0
Water	77.8

Silver Polish

FORMULA NO. 1

A	"Veegum"	2.0
	Water	68.9

B	"Triton" X-102	5.0
	Silicone Oil (350 cps)	2.0
	Ammonia	2.0
	Abrasive	20.0
	"Rotax"	0.1

Add the "Veegum" to the water slowly, agitating continually until smooth. Then add the components in B in the order listed with mixing after each addition.

NO. 2

Soap Flakes	12
"NSAE" Powder	30
TSPP	5
Sodium Thiosulfate	30
Water	175
Kieselguhr to make cream	

Brass Polish

FORMULA NO. 1

"Xynomine" Powder 20	20.0
"Neutronyx" 600	5.0
Coconut Oil	1.0
Chalk	25.0
Infusorial Earth	10.0
Propylene Glycol	25.0
Glycerol	14.0

	NO. 2	NO. 3
"Carbowax" 1500	35	50
"Tergitol" Nonionic NPX	–	3
"Tergitol" Anionic 7	3	–
Citric Acid	5	–
Sodium Chloride	5	–
Bentonite	8	9
"Multi Cel" 000	19	21
Water	25	29

Stir "Carbowax" 1500, water, and surfactant until solution is clear. If Polish No. 1 is being formulated, add the citric acid and sodium chloride, and stir until dissolved. Then, for both polishes, add the "Multi Cel" 000 and the bentonite, and stir until a smooth paste is obtained.

DETERGENTS & SANITIZERS

Aerosol Laundry Starch

"LE-463" 35%	1.00
Vegetable starch	4.00
Methyl "Parasept"	0.17
Propyl "Parasept"	0.03
"Tergitol" NPX surfactant	0.10
Distilled water	94.70
Perfume	q.s.

The "Parasept" preservatives are added to prevent starch spoilage and "Tergitol" NPX serves to wet the fabric. If foaming occurs during preparation or filling of the starch, 0.5 parts of "SAG 470" silicone antifoam may be added.

Slurry starch and preservatives with one half of the water and heat slowly with moderate agitation to 85–90°C. Hold at this temperature for 15 min. Force-cool by adding remaining water and (if possible) use water jacket. After cooling, add "LE-463," "Tergitol" NPX, and perfume, with agitation.

Laundry Bluing

Prussian Blue	5 lb
Water	25 gal
Glacial Acetic Acid	0.5 gal

Household Bleach

FORMULA NO. 1

(Dry)

Anhydrous Sodium Sulfate	80
Sodium Tripolyphosphate	15
Alkylaryl Sulfonate	5

This is a commercial spray-dried mixture to which 5–10% "Halane" is added.

NO. 2

(Dry)

"ACL"-59	13.4
Sodium Tripolyphosphate	30.3
"Santomerse" 85	1.9
Sodium Sulfate	53.5
Mositure	0.9

Normal usage of a dry chlorine bleach by the housewife requires a concentration of 50–100 ppm available chlorine in the wash water for bleaching and removal of stains. To remove stubborn stains, soaking in solutions of higher available chlorine content may be necessary.

In packaging household bleaches based on chlorinated cyanuric acid, a foil laminated or foil overwrap carton is suggested.

NO. 3

(Concentrated)

"Halane"	30
Sodium Tripolyphosphate	40
Alkylaryl Sulfonate (85%)	15
Sodium Toluenesulfonate (85%)	15

Clear Household Ammonia

Water, Soft	71.00
Ammonia (26° Bé)	11.00
"Versene"	0.25

Cloudy Household Ammonia

Water	71.00
Ammonia (26° Bé)	10.00
Dedecylbenzene-Sulfonic Acid	0.44
"Rhoplax" AC	18.00
"Cheelox" BF 13	0.25

Cold Water Woolens Wash

"Calsuds" CD-6	20
"Calsoft" T-60	20

Optical Brightener	0.1
Preservative	0.1
Water	59.8

Fine Fabric Detergent

FORMULA NO. 1

"Igepon" T-73	30.0
Sodium Hexametaphosphate	35.0
Sodium Bicarbonate	34.2
"Blancophor" FFG	0.3
Polyvinylpyrrolidone PVP K-30	0.5

NO. 2

"Sandopan" DTC	3.00
Nonylphenol Ethylene Oxide Nonionic (9 moles Ethylene Oxide)	30.00
"Leucophor" WS Liquid	5.00
"Sandoz" TH-40	1.00
Water	61.00

Carpet Shampoo Base

"Petro BAF" Liquid	15
Sodium Lauryl Sulfate (30 percent actives)	75
Water	10

Rug and Upholstery Shampoos

FORMULA NO. 1

"Avirol" 103	15.00
Coco Diethanolamide (1:2)	2.00
Water, Color, Perfume, Preservative	83.00

NO. 2

(Dry)

Wood Flour	20.0
Sodium Perborate	0.5

Water	50
Stoddard Solvent	20.0
Methylene Chloride	0.5
"Triton" X-100	0.5
Tetrasodium Pyrophosphate	0.5
"Sequestrene" NA4	0.5
Terpineol	0.5

Wall Cleaner

(Aerosol)

"Neutronyx" 600	3.04
"Nacconal" NRSF	1.42
TSPP	0.95
"Carbitol"	3.80
Distilled Water	85.79
"Genetron" 12/114A (70:30)	5.00

The container should be shaken before use and held so that the orifice of the valve is 6 to 8 inches from the surface to be cleaned. Allow foam to remain in contact with surface for a few minutes and then wipe it clean with a damp cloth or sponge.

White-Wall Tire Cleaner

"Plurafac" A-38	5
Na_2SiO_3	5
Na_2CO_3	5
Na_2SO_4	85

In using, scrub the powder on the wet tirewall with a damp brush. A liquid solution (4–8 oz/gal) can also be brushed or sponged onto the tire. After application, flush the tire thoroughly with water.

Toilet Bowl Cleaner

	FORMULA NO. 1	NO. 2
"Sandopan" DTC	10.00	10.00
Hydrochloric Acid (38%)	60.50	60.50

Pine Oil	–	1.00
Water	29.50	28.50

Wall and Floor Tile Cleaner

(Spray and Wipe)

FORMULA	NO. 1	NO. 2
Water	85.0%	85.0%
TKPP	5.0	5.0
Soda Ash	–	2.0
Sodium Metasilicate (anhydrous)	2.0	–
"Tergitol" 08	3.0	3.0
Butyl "Cellosolve"	5.0	5.0
pH of formulated solution, as is	12.85	11.0

Combine the ingredients in the listed order with continuous low speed mixing until homogeneous.

Masonry Cleaner

Formula No. 1: for use on rough porous surfaces.
Formula No. 2: for use on smooth polished surfaces.

FORMULA	NO. 1	NO. 2
Phosphoric Acid 85%	23.8	9.5
Ammonium Acid Fluoride	35.2	4.7
Nonionic Surfactant	0.3	15.6
Cellulose Gum	1.4	–
Water to 100% in each case		

Dissolve the gum in the water and when in solution and clear add nonionic, phosphoric acid and ammonium acid fluoride.

Dishwashing Detergent

(Nonphosphate)

Trisodium Citrate $2H_2O$	20.0
Hydrous Sodium Silicate L-20	2.5
Hydrous Sodium Silicate H-20	27.5

"Plurafac RA-43"	3.0
"CDB"-63	1.7
Sodium Sulfate	43.8
"Micro-Cel C"	1.5

A 0.2 percent solution of above has a pH of 10.5.1. Blend sodium citrate, sodium sulfate and sodium silicates for 15 minutes in a Patterson-Kelley Co. blender. 2. Add the surfactant and blend for 15 min. 3. Add the "Micro-Cel C" and sodium diisocyanurate and blend the total for 2–2½ hr.

Lens Cleaner

(Aerosol)

Isopropanol	98.98
Silicone Oil-"DC"-200	0.82
Masking Agent	0.04
Perfume	0.16
Propellent:	
Nitrous Oxide	4.43

Dissolve silicone oil in alcohol. Add masking agent and/or perfume as necessary.

Window Cleaner

Formula No. 1

(Aersol)

"Veegum"	2.0
Water	84.0
"Emcol" 5138	0.5
Odorless Mineral Spirits	8.0
"Triton" X-100	0.5
"Super Floss"	5.0
Propellent:	
Nitrous Oxide	1.48

NO. 2

Water	1000 g
Ethylene Glycol	20 g
Amyl Acetate	1 drop
Dye Solution	To suit

Windshield Cleaner

"Petro" BAF	20
"Alfonic" 1012-60	1
Water	79

Combine in proportions shown.

Dilute above concentrate with water at a 1:50 ratio. Finished product dries well and leaves no films or streaks.

Removing Haze from Weathered Glass

In dilute ammonium bifluoride solutions, it will polish glass, and can be used for removing from windows the haze that results from weathering. It should be rinsed completely from the glass to avoid frosting.

Chrome Cleaner

Citric Acid	25.0
"Solar" 90	25.0
Water	50.0

Add the citric acid to the water in a vessel in which they can be heated and agitated. After the citric acid is in complete solution, follow with Solar 90. Heat batch, with agitation, at 150°F for 15 to 30 min, or until the mixture becomes optically clear. Cool to room temperature.

Aluminum Cleaner

Citric Acid	4
Phosphoric Acid 85%	3
"Neutronyx" 600	2
Methyl Ethyl Ketone	3
Water	89

Heavy-Duty Metal Cleaner

Water	85.3
Caustic Soda	0.3
"Triton" X-102	2.0
"Pilot" ABS-99	2.9
Trisodium Phosphate	5.0
"Tamol" N	2.0
Sodium Tripolyphosphate	2.5

Dilution rate, 4 oz or more to gal.

Rifle Bore Cleaner

1. T.E.A. Oleate	12
2. Butyl "Carbitol"	10
3. Isoamyl Alcohol	15
4. Tetralin	15
5. Water	28
6. Paraffin Oil	20

Boat Bilge Cleaner

FORMULA NO. 1

Naphtha Aromatic	1
"Petromix" 7	1

This gives a clear emulsion which is diluted with water for use.

NO. 2

Perchlorethylene	$8^1/_2$ lb
Orthodichlorobenzene	$8^1/_2$ lb
"Tergitol" 15-S-9	16.8 lb
"Calsoft" 60L.	16.8 lb
Water, soft	420 lb
Butyl "Cellosolve"	42 lb
Water	Up to
Dye: Green "Pylaklor" LX6391A	1 oz

Marine All-Purpose Cleaner

	Materials	Weight	Volume
1	Water, soft		50 gal
2	Isopropyl Alcohol		$2^1/_2$ gal
3	Pine Oil		$2^1/_2$ gal
4	"Igepal" CO T10	35 lb	$4^1/_4$ gal
5	"Alipal" CO 432	35 lb	
6	Water, soft		100 gal
7	Dye Urainine Conc.		

Charge vessel with items 1, 3, 3, 4, 5, 6. Mix until a clear solution develops. Then add the dye solution to the desired shade.

Waterless Hand Cleaner

FORMULA NO. 1

A	Oil Phase	
	Deodorized Kerosene	39.0
	Pale Oleic Acid	7.5
	"Neodol" 25-3	2.0
B	Water Phase	
	Monoethanolamine	0.8
	Triethanolamine	2.6
	Propylene Glycol	2.5
	Glycerin	1.0
	Lanolin	0.5
	Butyl Oxitol Glycol Ether	0.4
	Water	43.7
	Perfume, Color	As desired

Heat both phases A and B separately to 70–75°C. Pour A into B while stirring continuously. Continue until a homogeneous smooth gel is formed, pour into jars. Let cool to room temperature.

NO. 2

(Germicidal)

A	Maypon 4C	4.6
	Stearic Acid T.P.	11.8

	Cetyl Alcohol	0.8
	Triethanolamine	5.6
	Propylene Glycol	5.4
B	"Deo Base" (Deodorized Kerosene)	42.8
	Fragrance	0.3
C	Distilled Water	25.5
	Formaline (40%)	0.2
	Polypeptide #37-S	3.0
	Color, as desired	

Heat A gently until all parts are dissolved. Mix B and add to A with agitation. Warm C and add the A/B mixture slowly to it with good agitation. Continue the agitation until the product "firms up."

Nicotine Finger-Stain Remover

Hydrochloric Acid (pure)	4
Glycerin	100
Rose Water (triple)	900

To use the lotion, saturate a piece of cotton with it and rub gently over the stained area. Several applications may be necessary before desired results are obtained.

Stain Remover, Stick

"Carbowax" Polyethylene Glycol 4000	60.0
Triethanolamine	21.5
Oxalic Acid	18.5

Stick No. 2

"Carbowax" Polyethylene Glycol 4000	70
Sodium Bisulfite	25
"Carbitol" Solvent	6

Melt Carbowax polyethylene glycol 4000, add the triethanolamine and oxalic acid, and stir until the acid crystals disappear.

Blood Stain Remover

"Solar" NP	17.9
Hexyleneglycol	10.7
Ammonia, Aqua	17.9
Water	53.5

Allow to act on stain for 5–10 min and wash off.

Spot Remover, Aerosol

"Syloid" 255	7.0
Isopropanol	3.9
"Shell Sol"-36	6.5
Chlorothene	29.1
"Freon" 11	28.5
"Freon" 12	25.0

The concentrate is prepared by mixing the isopropanol, Shell Sol-36 and Chlorothene together, then stirring in the Syloid 255. The slurry is passed through a fine screen (30 mesh) to remove any foreign matter. The screened slurry is milled by passing it twice through a colloid mill or high-speed dispenser to obtain a fine grind.

Plastic-Leather Cleaner

Water	48.5
"Renex" 20	3.0
Morpholine	1.0
Stoddard Solvent	25.0
Oleic Acid	2.0
L-45, Silicon Fluid (350 cs)	3.5
Water	19.8
"Carbopol" 934	0.2

Saddle Soap

1. Palm Oil Fatty Acids	80
2. Microcrystalline Wax	30
3. Caustic Soda (50% Sol'n)	23
4. Water	322

Mix 1, 3, 4 warm and mix until uniform. Add 2, warm and mix until emulsified. Color with pink dye.

Removing Chewing Gum

Spray or wet with "Freon" 12. This will harden and crack gum which can be brushed off.

Transparent Soap

FORMULA	NO. 1	NO. 2
Tallow	100	80
Coconut Oil	100	100
Castor Oil	80	80
Caustic Soda (38° Bé)	161	133
Alcohol	50	30
Glycerin	25	–
Sugar	80	90
Water in which above are dissolved	80	80

The first four items are warmed together and mixed slowly until foaming. Then the stirrer is switched off and the kettle left covered for some time.

The sugar solution is prepared in a separate container or kettle by dissolving the sugar in water heated to 80°C. Any froth formed on the surface is removed.

The next step is to add glycerin and then the warm sugar solution to the soap mass. This is carried out under constant stirring. At this stage, the free alkali content is adjusted. If it is found to be too high, coconut fatty acids are added. Roughly, every percent of free NaOH corresponds to 5.3% coconut fatty acids. The free alkali content should not be higher than 0.15% as NaOH. The kettle is then covered again and left standing until the temperature drops to 60°C., at which point perfume and color are added. The color solution is prepared beforehand by dissolving water soluble color in hot water and filtering it through a cloth. After the color and perfume are added, the soap mass is run into frames.

NO. 3

	oz.
Tallow	75
Coconut Oil	75

Rosin WW	50
Caustic soda (37° Bé)	100
Alcohol	80
Sugar	40
Water	40

The soap proper may be made either by the semi-boiled or fitted process. If made according to the semi-boiled principle, the temperature during processing must be kept as low as possible, with ample time allowed for saponification. Upon completion of saponification, the sugar solution is added at the same temperature as the soap. When all of the ingredients have been added and the soap is thin and free from lumps, the alcohol is added, together with any coloring matter. The perfume is incorporated just before the soap is run into frames.

Tobacco Pipe Sweetener and Cleaner

Methanol	75.0
Isoproponal	22.5
Glycerin or Propylene Glycol	2.0
Ethyl Vanillin	0.5

Mix alcohols and dissolve the vanillin in same. When in solution add glycerin or glycol.

Pipe Bit Cleaner

"Micro-Cel"	10.0
Water	90.0
"Surfonal"	0.5

Dry Cleaner Soap

FORMULA NO. 1

Use at a level of 1¼ oz./Gallon chlorinated solvent. Add 1% water to the bath.

"Calimulse"	50
Butyl "Cellosolve"	25
Mineral Oil	24
Optical Brightener	1

NO. 2

"Dry Cleaners" Solvent	33.5
Water	33.5
"Petronate" L	32

Pre-Spotting Agent for Drycleaning

Equal parts of

"Gafac" RS-610
Perchloroethylene
Water

Drain Opener

Caustic Soda	89.7
Sodium Peroxide	10.0
"Micro-Cel"	0.3

Paint-Brush Cleaner

Pine Oil	61.0
Castor Oil	15.8
Caustic Soda (48° Bé)	3.3
Water	19.9

Mix the water, castor oil, and caustic soda, and heat to boiling.

Wallpaper Remover

"Veegum" T	2.0
Water	73.0
Butyl "Carbitol"	12.5
"Tergitol" Anionic 08	12.5

Add the "Veegum" T to the water slowly, agitating continually until smooth; add the Butyl "Carbitol"; then add the "Tergitol" Anionic 08.

For Use: Dilute one part of concentrate (thick gel) with one to two parts water.

The advantage of using a viscous product for this purpose is that it does not drip or run. It removes wall paper with considerably less effort than thin watery preparations.

Detergent-Sanitizer Deodorant

"Hyamine" 2389	10
"Triton" X-100	10
STPP	2
"Versene"	1
Water	77

Perfume and dye as desired

Use 1 oz per 2 gal of water.

Hospital Cleaner Disinfectant

BTC® 2125M (50%)	25.0
Sodium Sesquicarbonate	3.0
"Perma Kleer" 100	9.0
"Neutronyx" 656	7.0
Water	56.0

Bactericidal activity at 1/2 oz per gal against Staphylococcus aureus ATCC #6538, Salmonella Choleraesuis ATCC #10708, Pseudomonas aeruginosa ATCC #15442.

Fungicidal activity at 1/2 oz per gal against Trichophyton interdigitale, Strain #640.

Industrial Disinfectant Cleaner

Sodium or Potassium Dichloroisocyanurate	3.5
Sodium Tripolyphosphate	30
Sodium Metasilicate Pentahydrate	5–10
Anionic Surfactant	5
Soda Ash	Balance

Soluble Pine Oil Disinfectant

Steam-Distilled Pine Oil	80.0
Potash Soap (Anhydrous)	10.0
Isopropanol (99%)	3.0
Water	7.0

Sanitizing Scouring Powder

"Santomerse" 85-B	4.0
Sodium Tripolyphosphate	5.0
Sodium Silicate	1.0
"ACL"-66 (granular)	0.6
Silica	89.4

Room Deodorant

"Barquat" MB-80	1
Triethylene Glycol	4
Propylene Glycol	4
Ethanol (anhydrous)	5
Perfume	1
Freon 12/11 50/50	85

Food Plant Deodorant, Sanitizer

"Arquad" S (50% active)	2 lb
"Ethofat" 242/25	1 lb
Soda Ash	2 lb
Water	100 gal

Garbage Can Deodorant

Ortho-Dichlorobenzene	40.0
Pine Oil, NF	25.0
Methylene Chloride	35.0

Combine ingredients and stir. The aerosol formulation should consist of 75 wt/% of concentrate, 25 wt/% of propellent "Genetron" 12.

Swimming Pool Chlorine Stabilizer

Sulfamic acid added in an amount sufficient to acidify chlorinated water allows slow release of chlorine, thus extending life of latter. Ten times as much chlorine is used but consumption is reduced 60%. Maintain about 5 parts per million when recirculating (about 4 oz of above to 50,000 gal of water).

Swimming Pool Algicide

"Hyamine" 1622	98.8
Water	1.2

In freshly filled pools add 1 lb of above to 10,000 gal of water.

Sanitizing Kennels

After thorough cleaning use 1 oz of "Hyamine" 3500 in 2 gal water.

Bubble Blowing Solution

"Monamine" LM-100	5.0
Isopropopanol	4.0
"Monaquest" CA-100	0.5
Water	90.5

Soap Bubbles, Long Lasting

Formula	No. 1	No. 2	No. 3
Sodium dibromostearate (4.49%)	2	1	1
Glycerin	5	2	1
Polyvinylalcohol (5%)	1	–	–
"Plasdone" C (5%)	–	1	–
"FC-128" (270)	–	–	4

Bubbles from above have lasted from 300–700 days.

Mop-Treating Oil

Paraffin Oil	92.0
"Solar" F-342	8.0

MISCELLANEOUS

Pet Shampoo

(Insecticidal)

"Miranol" C2M (conc.)	25.0
Lauric Diethanolamide	4.0

Hexyleneglycol	2.0
"Pyrocide" 5192	0.5
Water	68.5

Insect Spray Repellent

FORMULA NO. 1

Diethyltoluamide	15.0
Isopropanol	15.0
Propellant 11	35.0
Propellant 12	35.0

NO. 2

Diethyltoluamide	7.0
"MGK" 264	2.0
"MGK" Repellent 11	0.5
"MGK" Repellent 326	0.5
Isopropanol	30.0
Propellant 11	30.0
Propellant 12	30.0

Rodent Repellent

*"Bio Me T" 12	1-5
Toluene	To make 100

*Hazardous to eyes & skin.

Rat Poison

Corn Meal	95
"Prolin" Conc.	5

Rodenticide Bait

FORMULA	NO. 1	NO. 2
Corn Meal	60	–
White Corn Meal	–	60
Oat Meal	10	–

Fish Meal	5	20
Wheat Germ	10	–
Vegetable (or Whale) Oil	5	–
Hog Protein Supplement		5
Corn Oil		5
Sugar	9.5	9.5
5-p-Chlorophenyl Silatrane	0.5	0.5

Pigeon Repellent, Sticky

Polybutene #32	800
"Cab-O-Sil"	16
Anthraquinone	16
"Thompson"-1 Clay	300

Warm and mix.

Dog Repellent for Shrubbery

Oil Lemon Grass	5.0
Denatured Alcohol	95.0

It is also recommended for use in gardens to keep cats and rabbits out.

Insecticide

"Aldrin" (95%)	25.6
Xylene Type Solvent	70.0
"Emcol" N-300B	2.2
"Emcol" N-500B	2.2

Termite Control

Dig a trench 6–30 in deep and 8 in wide, walls long, but do not go below foundations. Apply 1 gal trichlorobenzene at bottom of trench; 1 gal three in. below surface. On infected wood the best method is to force trichlorobenzene in under pressure.

Insect Bite Cream

A	Methyl "Pyrelene"	20 g
	Mineral Oil (light)	30 cc.

B	Stearic Acid	45 g
	White Bees Wax	10 g
	Mineral Oil (light)	240 cc.
C	Triethanolamine	7 g
	Water	675 cc.

Heat B to 140°F till melted. Heat C to 140°F. Form cream by pouring hot C into hot B; stir constantly till cool. Form smooth paste by rubbing the mineral oil with the methapyraline A. When cream B+C is cool, incorporate paste A, mixing thoroughly.

Pearly Insect Repellent Lotion

A	"Amerchol L-101"	7.00
	Diethyl Toluamide	12.00
	"Escalol"	1.20
	"Arlacel" 165	5.00
	Antimicrobial (oil soluble)	q.s.
B	Purified water	to 100.00
	"Carbopol" 940	0.14
	Antimicrobial (water soluble)	q.s.
C	Triethanolamine	0.20
D	Perfume	q.s.
E	"Timica Pearlwhite"	1.00

Mix A, heat at 70°C. until dissolved. Sprinkle "Carbopol" into stirred water, add antimicrobial, heat to 70°C. Stir B into A. Add C with stirring. Disperse pearl pigment in small increments with stirring. Add perfume with stirring.

To Preserve Flowers in Plastics

Select flowers with thick petals and without blemishes, as defects are exaggerated during the preserving process. Best results are achieved when flowers are immersed in the preserving solution immediately after cutting.

As a result of experiments with more than fifty types of flowers, eight preserving solutions have been developed. Generally, colors ranging from red to pink are preserved in solutions having an acid reaction; those ranging from green to blue require a solution having an alkaline reaction.

FORMULA NO. 1

tert-Butanol	1000
Thiourea	10
Citric Acid	20

NO. 2

tert-Butanol	1000
Thiourea	10
Sodium Citrate	20

NO. 3

tert-Butanol	1000
Thiourea	10
Sodium Bicarbonate	20

NO. 4

tert-Butanol	1000
Thiourea	10
Citric Acid	20
40% Formaldehyde	50

NO. 5

tert-Butanol	1000
Thiourea	10

NO. 6

tert-Butanol	1000
40% Peracetic Acid	100

NO. 7

tert-Butanol	1000
Citric Acid	20

No. 8

tert-Butanol	1000
Thiourea	10
Citric Acid	20
Sodium Sulfite	20

Herbicide

Formula No. 1

Monosodium Acid Methane Arsonate	4.00 lb
"Atplus" 401	1.75 lb
Water, to make	1.00 gal

No. 2

Butyl 2,4-D	40.0
Heavy Aromatic Solvent	43.0
Kerosene	14.0
"Emcol" N-712	2.7
"Emcol" N-714	0.3

Weedkiller, Broad-Leaf

"Monamine" ACO-100	29
2.4-D	21
Water	50

Usage: 5 to 10 oz in 1 gal of water.

Hydroponic Nutrient Mix

Potassium Nitrate	16.49
Ammonium Sulfate	4.80
Potassium Sulfate	42.22
Calcium Dihydrogen Phosphate	14.44
Calcium Sulfate	16.06
Magnesium Sulfate	5.98

Micronutrient Concentrate

"Hamp-OL" 9% Iron Chelate	55.000
"Hamp-OL" 9% Manganese Chelate	27.500

"Hamp-OL" 9% Copper Chelate	14.000
"Hamp-ene" Zinc Chelate	9.000
"Solubor"	2.500
Sodium Molybdate	0.025

Dry-blend ingredients and package in containers that prevent moisture pickup and can be resealed if contents will not be rapidly used up. The micronutrient mix can be applied either bimonthly by diluting 1/3 oz (1 level tablespoon) in a gal of water or mixed with home and garden fertilizers at a rate of 50–200 lb per ton.

Orchid Culture Fertilizer

Epsom Salts	2
Potassium Phosphate (monobasic)	24
Ammonium Nitrate	28
Calcium Nitrate	16
Ferrous Sulfate	2
Ammonium Phosphate (dibasic)	6
Molybdic Acid	1/2
Zinc Sulfate	1/2
Copper Sulfate	1/4
Manganese Sulfate	1/4
Boric Acid	1/4

Plant Leaf-Shine Spray, Aerosol

PVP/VA 1-535	3.00– 5.00
"Carbowax" 1500	0.20– 0.35
Isopropanol	31.80–29.65
Propellents 12/11 (40/60)	65.00

Dissolve the copolymer in isopropanol. Add "Carbowax." Pressure fill.

Waterproofing Leather

"Oronite" Polybutene No. 128	20
"Polypale" Resin	30
Mineral Spirits	50

This mixture will penetrate into the interior of leather almost as well as the solvent when used by itself. The resulting leather exhibits longer wear, greater resistance to moisutre, nonskidding.

Protective Leather Dressing

Dichlorophene	0.3
Neatsfoot Oil	17.7
Cyclohexanone	12.0
"Genetron" 12/11, 50/50	70.0

Waterproofing Spray, Aerosol

No. 1

Silicone	5–10
Perchlorethylene	30
Methylene Chloride	30
Propellant	35–30

No. 2

SS-4098 Water Repellent (G.E.) (50% Solids)	3.0
Mineral Spirits	29.0
Propellant 12/11 (50/50)	68.0

1. Mix SS-4098 water repellent and mineral spirits together. Moderate agitation will insure complete dissolving of the water repellent in the solvent.

2. Charge the solution of the solvent and silicone water repellent into the aerosol can and load propellant.

To Check Leaks

(Air or Other Gases)

A 5% solution of "Antaron" FC-34 is applied to joints and other areas of equipment that are suspected of being faulty; where there is pressure of an escaping gas,the foam produced by the "Antaron" reveals the presence of a leak.

Liquid Radiator Anti-Leak

"Veegum"	3.0
Water	76.0
Methanol	15.0
Asbestos Powder	6.0

Add the "Veegum" to the water slowly, agitating continually until smooth. Add methanol and mix until smooth. Add asbestos and mix until smooth.

Anti-Fog Cloth for Glasses, Windshields Etc.

	(25% Sol'n)		*(50% Sol'n)*	
FORMULA	NO. 1	NO. 2	NO. 3	NO. 4
"Aerosol" OT-75% (lb.)	113	113	113	66
"Varsol" No. 2 (gal)	15	6	–	2
Isopropyl Alcohol (gal)	3	6	8	–
Water (gal)	19	25	29	3

Impregnate cotton flannel with above.

The solvents and water are mixed at 80–100°F. The Aerosol OT-75% is added slowly with continued mixing. After the Aerosol OT has dissolved, all solutions will remain clear and liquid at temperatures as low as 50°F. No scumming of the surface occurs because of evaporation during a normal run of the cloth through the bath. The solutions that are allowed to stand should be covered to prevent loss of solvent.

The optimum amount of Aerosol OT to be left on the cloth is 40–50%, based on the weight of the cloth. However, a range of 25–50% will give satisfactory results.

Flame Retardant

(Cotton Duck)

"Geon" 652 (52%)	193
"Firemaster" T-23P Emulsion (65%)	62
Antimony Oxide Dispersion (50%)	32
Calcium Carbonate Dispersion (50%)	70

Fire Extinguisher, Powder

Ammonium Phosphate	50–90
Ammonium Sulfate	15–30
Barium Sulfate	8–20

Plastic Wood Filler

Camphor	1/4 lb.
Xylol	1 qt.
Acetone	1 1/4 pt.
Ethyl Acetate	1 pt.
Castor Oil	3/4 oz.
Pyroxylin	1 1/4 lb.
Wood Flour	q.s.

Make paste

Wood Hardener

Wood is impregnated by vacuum-pressure method with the following mixture:

"Aroclor" 4465	70
Microcrystalline Wax	20
Sulfur	10

Wood Preservatives

Formula No. 1

For heavy decay or rot resistance, surfaces not to be painted:

"Nuodex" Copper, 8%	25
Kerosine, Mineral Spirits, or Similar Solvent	75

This mixture contains 2% copper expressed as metal.

No. 2

For heavy decay or rot resistance, surfaces to be painted:

"Nuodex" Copper, 8%	25
Linseed Oil	35
"Nuodex" DMR Cobalt 6%	0.5
Petroleum Naphtha	39.5

Vinyl Floor Tile

Formula	No. 1	No. 2	No. 3
PVC Polymer	100	100	100
DOP	35	–	–
"Cereclor" S.52	–	44	51
Ground Limestone	350	350	362
Asbestos	100	100	108
Barium Zinc Stabilizer	3	3	3
Epoxy Stabilizer	3	3	3
Titanium Dioxide	5	5	5
Colorant	2	2	2

Vinyl Asbestos Tile

"Vinylite" VYHH	85
"Pliovic S-50"	85
"Santicizer 160"	18
DIDP	37
"Paraplex" G-62	15
"Temex 3"	10
7R Asbestos	125
6N Fiber	125
Marble Dust	500

Investment Casting Wax

Formula	No. 1	No. 2	No. 3	No. 4	No. 5
Candelilla Wax	27	–	–	–	31
"Staybelite" Ester 10	6	–	–	–	–
"Piccovar" 420	20	–	–	–	–
"Be Square" 170/175	15	25	–	25	17
"Castorwax"	20	15	65	30	22
"Alpco" 16	–	10	–	–	–
"Piccopole" 100-SF	–	30	–	30	–
Burgundy Pitch	–	15	10	10	7
"Arochlor" 4465	–	–	25	–	–
"Arochlor" 5460	–	–	–	–	23

Aerosol Lubricant

Molybdenum Disulphide Lubricant	50.0
Naphtha	50.0
Methylene Chloride	40.0

Aerosol Formulation

Concentrate:	50.0
"Genetron" 12/11 (30:70)	50.0

The ingredients in the concentrate are mixed at room temperature. The formulation can be packaged in an unlined, sideseam container fitted with a standard valve and spray actuator. Shake well before using.

Graphite Lubricant

(Suspensions)

Graphite	12
"Baymal"	6
Water	82

In a one-quart blender is placed 410 ml of water, and with slow stirring, 30 g of "Baymal" is added. The mixture is blended at high speed for two minutes; then 60 g of graphite powder is added and blending at high speed continued for five minutes. The mixture is usually not foamy and develops sufficient thixotropy to keep fine graphite from settling.

Lubricant, Silicone

A	G-E SF-1023 Silicone Fluid	10.0
	"Ucon" LB-1145	80.0
B	"Cab-O-Sil" HS-5	10.0

Blend components of A thoroughly. Add B slowly and agitate until a uniform mixture is obtained. A high shear agitator with sufficient power to mix a grease-like material is required. Give mixture A-B two passes through a three-roll paint mill.

Penetrating Oil

(For Rusted Bolts)

Used to loosen rusted-in-bolts, or other combinations where rust and dirt have made separation difficult.

Methyl Oleate	10
Light Mineral Oil	10
Kerosene	10

Rustproofing Lubricant

Soya Lethicin	0.8
Isopropyl Myristate	5.0
Soft Paraffin	3.0
Anhydrous Lanolin	2.0
Mineral Oil	15.2
White Spirit	40.0
Propellant 12/11 (50:50)	34.0

Marking Crayon

"Castorwax"	18
Ceresin	40
Carnauba Wax	17
Paraffin Wax	20
Beeswax	5
Talc	50
Pigment	15

Silk Screen Ink

(Oil Based)

Precipitated Chalk	14.9
Ti O_2 Anatase	26.4
Blanc Fixe	12.4
"Beckosol" 9610 (50% NV)	45.9
"Thixcin R"	0.4

Driers are 0.5% lead, 0.025% cobalt and 0.025% manganese; solvent is mineral spirits.

Water Printing Ink

Zein	100
Sodium Resinate	20
Diethylene Glycol	230
Hexylene Glycol	50
Water	30
Pigment	50–150

Heat Set Printing Ink

Formula	No. 1	No. 2	No. 3	No. 4
50% Resin Varnish	71	71	71	71
Ink Oil	10	5	5	5
Carbon Black	19	19	19	19
Decyl Alcohol	–	5	–	–
Tridecyl Alcohol	–	–	5	–
Lauryl Alcohol	–	–	–	5

Fast Drying Printing Ink

Carbon Black	14
Prussian Blue	6
Ester Gum	47
"Carbitol" Monobutylacetate	31
White Petrolatum	2

Laundry Marking Ink

Formula No. 1

(Blue)

Bronze Blue	2.0
Titanium Dioxide	8.0
Chlorinated Rubber	6.5
Paraffin Wax	4.5
Xylol	13.5
Tetrahydrofurfuryl Alcohol	65.5

No. 2

(Black)

Channel Black	10.0
Bronze Blue	6.0
Bleached Shellac	30.0
"Cellosolve"	54.0

Stamp Pad Ink

Glycerin	70
Water	30
Dye	2–5

Magnetic Ink

Metallic Iron, Carbonyl	68
Coconut Oil Fatty Acids	5
Oleic Acid	25
Tall Oil, Refined	1

Screen Silver Ink for Ceramics

Gold Powder	10.0
Mercurous Chloride	10.0
Lead Borosilicate	3.5
Anthracene	2.5
Amyl Acetate	24.6
Ethyl Lactate	8.8
Nitrocellulose	3.5
Methyl Abietate	2.2
Castor Oil	2.2

Invisible Ink

Formula No. 1

Cobaltous Chloride	2
Dextrin	2
Glycerin	10

Warm together and mix. This writes pale pink and turns blue on heating. It is best used on pink or colored paper.

NO. 2

Phenolphthalein	1
Denatured Alcohol	9

This writes colorless and turns to red on application of ammonium hydroxide or ammonia gas. The writing disappears again when the ammonia evaporates, in a few minutes.

NO. 3

Ferrous Sulfate	1
Water	9

This writes practically colorless and turns blue on treating with a solution of potassium ferricyanide, in water.

NO. 4

Starch, Soluble	1
Water	9

This writes colorless and turns blue on treating with a water solution of potassium iodide.

Trick (Toy) Ink

(Vanishing Ink)

Thymolphthalein	0.1 g
Denatured Alcohol	10.0 cc
Water	90.0 cc
Ammonium Hydroxide or Propylenediamine	2.0 cc

When this is thrown on light clothes a blue stain appears. This disappears in a few minutes.

Ball Point Ink

FORMULA NO. 1

(Black)

Oleic Acid	44
Estergum	33

Nigrosine NB Base	23

Mix dye and oleic; add resin; heat to 90°C for 1 hr; filter and roller mill.

NO. 2

Spirit Fast Blue MBSN	1
"Carbitol"	1

NO. 3

Victoria Green WB	1
1-3 Butyleneglycol	1
Octylenglycol	1

NO. 4

Victoria Blue Oleate	56
Victoria Blue Phosphotungstic Toner	27
"Crill" 5.6	2
"Carbowax" 1500	8
Polyethyleneglycol 400	7

Carbon Paper Ink

(One-Time)

"Petrolite" Wb-10	4
"Ceraweld" Brown	20
Paraffin 143/145 AMP	36
Iron Blue #400	15
"Sunpar" 110	25

Typewriter Ribbon Ink

"Petrolite" WB-5	15
Carnauba N.C. #3	15
Methyl Violet Base	4
Iron Blue	14
Furnace Black	12
100 SUS Mineral Oil	40

Fuel Oil Additives

To reduce clogging of filters, rusting of tanks, etc. Will dissolve tank sludges. Particularly useful if oil remains in tank for longer time periods.

Application—addition of concentrate to tank when it is filled.

Materials: calcium, barium or ammonium sulfonates.

Use the sulfonate as a 20% concentrate in light oil or kerosene. Add to furnace or fuel oil

Charcoal Lighter

Odorless Mineral Spirits	99
Tall Oil Fatty Acids	1

The odorless mineral spirits should be high boiling with an initial drop of about 164°C (327°F) to final drop of 205°C (400°F). The tall oil is to give it a wood smoke-like odor.

Mildewproofing for Textiles, Cordage, Hair Felt

FORMULA NO. 1

1	Dichlorophene	3
2	Isopropanol	26
3	Mineral Spirits	271

Dissolve 1 in 2 and stir into 3. Apply by padding at 120–125°F.

To Soften Casual Leather

In the processing of casual shoe leather or other types of soft leather, the incorporation into the fatliquors of 20 to 40% "Antaron" FC-34 on the weight of the sulfated oils produces mellowing and imparts a desirable feel to the grain. For example, dyed leather may be treated for 45 minutes at 130°F with a fatliquor containing 2% "Antaron" FC-34 and 5% sulfated neatsfoot oil (equivalent to 40% "Antaron" FC-34 on weight of sulfated oil).

To Bend Tortoise Shell

Immerse the shell in hot linseed oil for a while; then bend to shape. On cooling, it regains hardness.

To Bend Ivory

Immerse the piece of ivory in pure phosphoric acid until it's almost transparent. Rinse in cold water, then bend. On drying it becomes hard again.

Coloring Sand and Pebbles

FORMULA NO. 1

(Gold)

Pebbles	200
"Claremont" Gold	4
"Rhoplex" B-60A	2
Water	6

Tumble until uniformly coated and dry with hot air.

NO. 2

(Green)

Silica Sand	100
Sod. Silicate K	6
Chromium Oxide	2
Water	6

Water Conditioner

Na Hexametaphosphate	350 g
Trisodium Phosphate	
H_2O soft	650 g
Tripoly Phosphate	
pH	5.5

Police Tear Gas

Chloroacetophenone	0.9–1.2
Solvent Mixture	
1,1,2-trichloro-1,2,2-trifluoroethane	70–80
1,1,2-trichoroethane, approximately	5
Kerosene, approximately	4

Pipe and Stem Sweetener

Hexynol	2.0
Propylene Glycol	5.0
Vanillin or Saccharin	0.5
Methanol	5.0
Isopropopanol	87.5

Remove stem from bowl. Pour half thimble Sweetener into bowl, drain through shank after minute. Dip pipe cleaner inside shank and bowl. Do not use on outside.

This Sweetener may be also used to clean bit and shank by dipping a bristle pipe cleaner in solution and swabing bit and shank.

Melting Salt for Road Ice

Rock Salt	3
Calcium Chloride	1

"Strike-Anywhere" Match

	Formula No. 1	No. 2
Animal Glue	11	12
Paraffin Wax	–	2
Potassium Chlorate	32	37
Phosphorus Sesquisulfide	10	3
Sulfur	–	6
Rosin	4	6
Dammar Gum	–	3
Infusorial Earth	–	3
Powdered Glass and Other Filler	33	21.5
Potassium Dichromate	–	.5
Zinc Oxide	6	1

Formula No. 1 represents the tip which ignites on any hard surface. Formula No. 2 is its base, loaded with combustibles for strong billowing flame but of low friction sensitivity.

Modelling Clay

Sulfur (powdered)	66
Kaolin	400

Lanolin	120
Glycerin	80

Imitation Milk

Water	88.2
Demineralized Whey Solids	6.5
Vegetable Fat	3.5
Sodium Caseinate	1.0
"Avicel" RC-581	0.4
"Durlac" 100	0.3
Salt	0.1
Flavor and Color	Optional

Add "Avicel" to 70°F water and agitate until dispersed. Heat dispersion to 110°F. Dry blend whey solids, sodium caseinate, and salt. Add to water dispersion. Combine vegetable fat with emulsifier, then add mixture to dispersion. Pasteurize dispersion at 160°F for 30 min. Homogenize on two-stage homogenizer at 2000 and 500 psi. Cool rapidly to 40°F before packaging.

Liquid Coffee Whitener

FORMULA	NO. 1	NO. 2
Vegetable Fat (102–104°F m.p.)	10	8.5
Corn Syrup Solids (24DE)	10	6.0
Sodium Caseinate	2.0	0.75
"Emplex"	0.2	0.2
Modified Food Starch	–	0.2
Buffer Salt (Dipotassium Phosphate)	0.15	0.15
Beta Carotene	0.0004	0.0004
Water	77.6	84.2
"Polysorbate" 60	0.1	0.1

NO. 2

	Range % (on total wt.)
Fat, Vegetable	3.0–18.0
Sodium Caseinate	1.0– 3.0

Corn Syrup Solids	1.5– 3.0
Sucrose	1.0– 3.0
Atmos® 150	0.3– 0.5
Carrageenan	0.1– 0.2
Sodium Citrate	0.1– 0.3
Flavor	q.s.
Color	q.s.
Water	q.s. to 100

The dry ingredients are blended with the fat and liquid ingredients and heated to pasteurizing temperature. (This temperature varies depending on the particular method used.) Once pasteurized the mix is pumped directly to the homogenizer and homogenized at 2,000 to 2,500 total p.s.i. (1,500 to 2,000 p.s.i. on first stage and 500 p.s.i. on second stage) on a two stage homogenizer, and 1,500 to 2,000 p.s.i. on a single stage homogenizer. Cool rapidly to 38°F. Store in refrigerator.

To Bleach Edible Fats and Oils

On the weight of the fat or oil, 3 to 5% of 130-volume hydrogen peroxide is added and the mixture is then well stirred. Increase in temperature or addition of a little alkali will assist the bleaching action. The fat or oil should be thoroughly purified before bleaching. As the only byproduct of hydrogen peroxide is water, the clarification of the oil is simple.

Appendix

Some Incompatible Chemicals

The substances in the left-hand column must be stored and handled so that they cannot come into contact with the substances in the right-hand column.

Alkaline and alkaline-earth metals, such as sodium, potassium, lithium, magnesium, calcium, aluminum	Carbon dioxide, carbon tetrachloride, and other chlorinated hydrocarbons. (Also prohibit water, foam, and dry chemical on fires involving these metals.)
Acetic acid	Chromic acid, nitric acid, hydroxyl containing compounds, ethylene glycol, perchloric acid, peroxides, and permanganates.
Acetone	Concentrated nitric and sulfuric acid mixtures.
Acetylene	Chlorine, bromine, copper, silver, fluorine, and mercury.
Ammonia (anhydr)	Mercury, chlorine, calcium hypochlorite, iodine, bromine, and hydrogen fluoride.
Ammonium nitrate	Acids, metal powders, flammable liquids, chlorates, nitrites, sulfur, finely divided organics or combustibles.
Aniline	Nitric acid, hydrogen peroxide.
Bromine	Ammonia, acetylene, butadiene, butane and other petroleum gases, sodium carbide, turpentine, and finely divided metals.
Calcium carbide	Water (See also acetylene.)
Calcium oxide	Water.
Carbon, activated	Calcium hypochlorite.
Copper	Acetylene, hydrogen peroxide.
Chlorates	Ammonium salts, acids, metal powders, sulfur, finely divided organics or combustibles.

Some Incompatible Chemicals

(continued)

Chromic acid	Acetic acid, naphthalene, camphor, glycerol, turpentine, alcohol, and other flammable liquids.
Chlorine	Ammonia, acetylene, butadiene, butane and other petroleum gases, hydrogen, sodium carbide, turpentine, benzene, and finely divided metals.
Chlorine dioxide	Ammonia, methane, phosphine, and hydrogen sulfide.
Fluorine	Isolate from everything.
Hydrocyanic acid	Nitric acid, alkalis.
Hydrogen peroxide	Copper, chromium, iron, most metals or their salts, any flammable liquid, combustible aniline, nitromethane.
Hydrofluoric acid (anhydr.) (hydrogen fluoride)	Aqueous or anhydrous ammonia.
Hydrogen sulfide	Fuming nitric acid, oxidizing gases.
Hydrocarbons (benzene, butane, propane, gasoline, turpentine, etc.)	Fluorine, chlorine, bromine, chromic acid, sodium peroxide.
Iodine	Acetylene, anhydrous or aqueous ammonia.
Mercury	Acetylene, fluminic acid, ammonia.
Nitric acid (conc.)	Acetic acid, aniline, chromic acid, hydrocyanic acid, hydrogen sulfide, flammable liquids, flammable gases, and nitritable substances.
Nitroparaffins	Inorganic bases.
Oxygen	Oils, grease, hydrogen, flammable liquids, solids or gases.
Oxalic acid	Silver, mercury.
Perchloric acid	Acetic anhydride, bismuth and its alloys, alcohol, paper, wood, grease, oils.

Some Incompatible Chemicals

(continued)

Peroxides, organic	Organic or mineral acids; avoid friction.
Phosphorus (white)	Air, oxygen.
Potassium chlorate	Acids (see also chlorate.)
Potassium perchlorates	Acids (see also perchloric acid.)
Potassium permanganate	Glycerol, ethylene glycol, benzaldehyde, sulfuric acid.
Silver	Acetylene, oxalic acid, tartaric acid, fulminic acid, ammonium compounds.
Sodium	See alkaline metals.
Sodium nitrate	Ammonium nitrate and other ammonium salts.
Sodium oxide	Water.
Sodium peroxide	Any oxidizable substance, such as ethanol, methanol, glacial acetic acid, acetic anhydride, benzaldehyde, carbon disulfide, glycerol, ethylene glycol, ethyl acetate, methyl acetate, and furfural.
Sulfuric acid	Chlorates, perchlorates, permanganates.
Zirconium	Prohibit water, carbon tetrachloride, foam, and dry chemical on zirconium fires.

COMMON HAZARDOUS CHEMICALS

Name	*Usual Shipping Container*	*Fire Hazard*	*Life Hazard*	*Storage*	*Fire-Fighting Phases*	*Remarks*
Acetic acid (glacial)	Glass carboys and barrels	Dangerous in contact with chromic acid, sodium peroxide, or nitric acid; yields moderately flammable vapors above flash point 104° F.	May cause painful burns of skin	Safeguard against mechnical injury. Isolate from oxidizing materials as noted under Fire Hazard	Extinguishing agent, water	Expands on solidification and may burst container unless kept at a temperature above 16°C. (60.8°F.)
Acetone	Carboys, steel drums, tank cars	A volatile liquid. Gives off vapors which form with air flammable and explosive mixtures. Flash point −16°C. (3° F.) (*11*). Explosive range 2.55% (*3*). The ignition to 12.8% (upward propagation) temperature is comparatively high, being within the range 538° to 566°C. 1000° to 1050° (F.), (*11*). The	Toxicity of a comparative low order (7)	Safeguard containers against mechanical injury. Only electrical equipment of the explosion-proof type, Group D classification, permitted in atmospheres containing acetone vapor in flammable proportions	Lighter than water (sp. gr. 0.792) but soluble in it in all proportions. Water, particularly in form of spray, is best extinguisher. Carbon dioxide may also be used. Automatic sprinkler systems or total-flooding carbon dioxide systems may be employed for protection in storage rooms	

Note: These columns are informative only; it is not considered necessary that the material be kept or stored only in the containers as listed, nor that each package be labeled. The requirements in the table on the storage of containers refer to chemicals in usual containers, and are not intended to apply to small bottles of chemicals such as are found in drug stores and chemical laboratories.

Name	Usual Shipping Container	Fire Hazard	Life Hazard	Storage	Fire-Fighting Phases	Remarks
		vapors are heavier than air (vapor density 2), Fire hazard slightly less than that of gasoline				
Aluminum dust	Barrels or boxes	Forms flammable and explosive mixtures with air		Keep in dry place. In case of fire do not use water; it may cause an explosion	Smother with sand, ashes, or rock dust. Do not use water, which may cause explosion	
Aluminum resinate	Wooden barrels	Combustible		Storage should be ventilated and safeguarded as for oil storage building		
Ammonium perchlorate	Wooden barrels or kegs and glass bottles	Oxidizing material. May explode in a fire. Hazard classes with potassium chlorate		Safeguard against mechanical injury. Isolate from mineral acids, also from combustibles		
Anhydrous ammonia	Steel cylinders or steel tank cars	Gas density 0.60 (air = 1). Not flammable in air except in comparatively high concentration, which is seldom encountered under practical conditions, the low limit of the flammable or explosive range being about 15 to 16% and the upper limit about 25 to 26% by volume (horizontal propagation, *8*, 14). Presence of oil will increase the	Irritant. An outstandingly serious effect produced by ammonia in concentrations of the order of 0.5% by volume for duration of exposure of the order of 0.5 hour is blindness (*12*). A concentration of 0.30% of ammonia in air for duration of exposure of the order of o.5 to 1 hour. according to Lehmann, does not cause	Safeguard against mechanical injury and excessive heating of cylinders or tanks. Fire-resistive storage recommended. In combustible buildings or if nearcombustibles, sprinklered storage recommended. Isolate from other chemicals, particularly chlorine, bromine, iodine, and mineral acids	Soluble in water. Hose streams comparatively effective in removing gas from atmosphere	

		hazard. Ammonia aqua does not burn	serious effects			
Antimony pentasulfide (golden antimony sulfide) Sb_2S_5	Fiber drums or tins	Combustible. Readily ignited by small flame. Hazardous in contact with oxidizing material. Yields flammable hydrogen sulfide on contact with mineral acid	Gaseous products of combustion contain sulfur dioxide and are irritating and corrosive	Safeguard against mechanical injury. Isolate from acids, cglorates, nitrates, and other oxidizing agents	Practically insoluble in water. Use water	Used in manufacture of matches, ammunition, fireworks, and of certain rubber compounds
Barium chlorate	Wooden boxes, barrels, or kegs	Oxidizing material. Hazard classes with potassium chlorate		Isolate	See potassium chlorate	
Barium nitarte	Wooden boxes; barrels	Oxidizing material. Hazard in class with sodium nitrate	Soluble in water. Poisonous when taken internally	Do not store with combustile materials	See sodium nitrate	
Barium peroxide	Tihgtly closed metal containers packed in wooden boxes or barrels; or in bulk in metal barrels or drums	Oxidizing material. Hazard in class with sodium peroxide		Do not store with combustile materials	Smother with sand, ashes, or rock dust. Do not use water	
Benzoyl peroxide (dry) granular or powdered (wet)	Dry granular material in individual 1-lb. containers inside wooden boxes; finely powdered material shipped wet (30% water by weight) in glass containers placed inside tightly sealed metal containers in wooden boxes or in aluminum drums	Highly flammable in the dry state. Strong supporter of combustion. Do not subject dry material to heat of friction or grinding	Dust irritating to eyes and lungs. Use goggles and dust respirator in dusty atmospheres	Store in cool ventilated place. Powder should be stored with not less than 30% water by weight. Keep away from all sources of heat and separate from all combustile materials and acids	Water, carbon dioxide, foam, sand, soda ash, or rock dust may be used as extinguishing agents	Not miscible with water

Name	Usual Spinning Container	Fire Hazard	Life Hazard	Storage	Fire-Fighting Phases	Remarks
Bleaching powder, calcium hypochlorite, chlorrinated lime, chloride of lime (incorrect name)	Air-tight tin containers, wooden barrels, and steel drums	Notcombustible but evolves chlorine and at higher temperatures oxygen. With acids or moisture evolves chlorine freely at ordinary temperatures. See chlorine	Corrosive. Irritating to skin, eyes, and lungs. See chlordine	Store in cool, dry, well-ventilated place away from combustibles. See chlorine. Rupture of drums containing bleaching powder, particularly if the chlorine contant is high, may result from exposure to heat.	Fires where compound is present may be fought with water, preferably spray. Protect eyes and skin, using gas mask of type approved by Bureau of Mines	Encountered in paper, textile, disinfectant, and alkali industries; also where water-purification processes are employed
Borneol	Barrels, kegs, boxes, and tins	Combustible. Hazard similar to camphor		Store in well-ventilated compartment or building	See camphor	
Bromine	Glass bottles: earthen jugs	Causes oxidizing effect, resulting in heating and may cause fire when in contact with organic material	Corrosive; at ordinary temperatures gives off poisonous suffocating vapor	Isolate; safeguard against mechanical injury		Bottles should be surrounded by incombustible packing
Bronze dust	Barrels or boxes	When aluminum is present forms flammable and explosive mixtures with air. Composition usually free from aluminum			Smother with sand or ashes	Bronze dust free from aluminum not considered dangerous
Butane	Steel cylinders	Flammable gas under pressure. Classes with gasoline vapor in fire hazard		Safeguard against mechanical injury		Keep cool
Calcium carbide	Iron drums and tin cans	Gives off acetylene gas on contact with water or moisture	Serious under fire conditions	Store in dry, well-ventilated place in accordance with N.F.P.A. Standards	Smother with sand or ashes. Do not use water	
Calcium oxide	Wooden barrels and bags	Heats upon contact with water or moisture and		Isolate; store in dry place away from water or		

		may cause ignition of organic material. Swells when moist and may burst container		moisture		
Camphene	Tins	When heated gives off flammable vapors. Classes with turpentine		Isolate; keep in unheated compartment away from fire or heat	Smother with sand or ashes. Avoid water	
Camphor	Tins and wooden kegs	Flammable; gives off flammable vapors when heated which may form explosive mixture with air. Flash point 180°F.		Detach from other storage. Keep in well-ventilated room remote from fire	Smother with sand or ashes. Chemical streams. Avoid water	
Carbolic acid (see phenol)						
Carbon disulfide	Small glass, earthenware, or metal containers packed in outside barrels or boxes (See I.C.C. Regulations). Steel drums, steel tank cars	Highly volatile liquid with offensive odor, giving off even at comparatively low temperatures vapors which form with air flammable and explosive mixtures. Flash point −30°C. (−22°F.) (*11*). Flammable range 1 to 50% (*1*) (upward propagation). Ignition temperature is dangerously low, about 100° to 106°C. (*11*) (212° to 223° F.). It is endothermic, and vapor may be ignited by heavy blow. Vapors are heavier than	Toxic. 3200 to 3850 parts of vapor per million (0.32 to 0.385%) by volume) may cause dangerous illness in 0.5 to 1 hour (*5*). Direct contact with skin should be avoided. Products of combustion contain sulfur dioxide, which in concentrations of 0.2% by volume in air may cause serious injury in 0.5 hour or less (*12*). Often poisonous carbon monoxide is present in products of combustion	Isolate and safeguard containers against mechanical injury and metallic blows, and keep in unheated compartment away from sunlight and any source of ignition, including electric lighting fixtures and other electrical equipment. Storage tanks should be constructed over concrete basins containing water, and carbon disulfide kept blanketed with water or inert gas at all times	Heavier than water (sp. gr. 1.29) and sparingly soluble in it. Use sand, carbon dioxide, or other inert gas as extiingisging agents. Cooling and blanketing action of water may be utilized in case of fires in metal containers or tanks. Total-flooding carbon dioxide systems may be employed for protection in storage rooms. Foam not effective. Do not use carbon tetrachloride. Use of gas masks or oxygen helmets of type approved	Carbon disulfide should never be transferred by means of air. Use inert gas, water, or pump. Use a wood measuring stick for measuring contents of storage tanks or tank cars. Tank cars when being loaded or unloaded should be well grounded. Do not dispose of carbon disulfide by pouring it on the ground. Provide a safe place for burning it.

Name	Usual Shipping Container	Fire Hazard	Life Hazard	Storage	Fire-Fiting Phases	Remarks
		air (vapor density 2.62) and may travel considerable distance to source of ignition and flash back. More hazardous than gasoline			for purpose by U.S. Bureau of Mines recommended	
Charcoal (wood)	Boxes, barrels, bags, or bulk	Spontaneously ignitable when freshly calcined and exposed to air, or when wet; hazardous when freshly ground and tightly packed	Danger from carbon monoxide poisoning during burning unless adequate ventilation is provided	Isolate; prevent dust accumulations; ventilate well; make daily inspections	Use water, completely extinguishing fire, after which storage pile should be moved	
Chinese wax	Burlap bags and wooden barrels	Combustilbe		Detach from other storage		
Chlorine	Steel cylinders and tank cars	Not combustilbe in air but reacts chemically with many common substances and may cause fire or explosion when in contact with them. See remarks under Storage	Corrosive. Irritating to eyes and mucous membrane. Toxic, 0.004 to 0.006% by volume in air causes dangerous illness in 0.5 to 1 hour (*6*)	Isolate from turpentine, ether, ammonia gas, illuminating gas. hydrocarbons, hydrogen, and finely divided metals. Safeguard against mechanical injury of containers	Use gas masks on entering atmospheres containing chlorine gas. If, however, concentration is high, or there is doubt as to degree of concentration, use oxygen helmet of type approved for such use by U.S. Bureau of Mines	Dangerous to neutralize chlorine in a room with ammonia
Chromium trioxide or chromium anhydride, CrO_3 (ofen called "chromic acid")	Iron drums and glass bottles	Oxidizing material; will ignite on contact with acetic acid and alcohol. Chars organic material such as wood, sawdust, paper, or cotton, and may cause ig-	Iriritatng to skin. Poisonous	Isolate	Use water, completely extinguishing fire, after which storage pile should be removed	Used in chromium plating, in electric batteries, and in photography

		nition. Combustible material in presence of chromium trioxide when ignited burns with great intensity. May cause explosion in fire				
Cobaltous nitrate	Wooden barrels	Oxidizing material. Classes with sodium nitrate		See sodium nitrate		
Colophony	Barrels	Combustible; gives off flammable vapors when heated		Ventilate storage; avoid dust; keep away from fire or heat		
Copper nitrate	Wooden barrels and kegs	Oxidizing material. Hazard classes with sodium nitrate	Poisonous when taken internally. Soluble in water	Safeguard against mechanical injury; isolate. See sodium nitrate		
Cyclopropane	Steel cylinders	Highly flammable gas. Forms flammable and explosive mixtures with air or oxygen. Explosive range 2.40 to 10.4% (upward propagation) (*4*)	Anesthetic	Isolate from oxygen cylinders and store in cool, well ventilated storeroom	Use water to cool cylinders not on fire. If gas is burning at valves or safety releases, usually best course to follow is not to disturb or attempt to extinguish flame. To do so will cause release of unburned gas and quickly create high dangerous explosive atmospheres. If cylinder is mounted on anesthetic machine or truck, it may be possible to move it to a safe place. Carbon	Only electrical equipment of the explosion-proof type, Group C classification, permitted in atmospheres containing cyclopropane in flammable proportions. Group C classification for cyclopropane tentative pending further tests

Name	Usual Shipping Conatiner	Fire Hazard	Life Hazard	Storage	Fire-Fighting Phases	Remarks
					dioxide or carbon tetrachloride are best extinguishing agents	
Didymium nitrate	Wooden kegs	Oxidizing material. Classes with sodium nitrate		Isolate. See sodium nitrate		
Dioxan	Glass bottle, metal cans, and metal drums	Moderately volatile flammable liquid. Flash point 54°F. Explosive range 1.97 to 22.25% (upward propagation) (*14*). Vapors heavier than air (vapor density 3.03). Capable of forming peroxides under certain conditions, and may be danger of explosion if redistilled, unless certain precautions are taken	Irritant and toci in high concentrations	Isolate and safeguard against mechanical injury	Water best extinguishing agent	Slightly heavier than and completely soluble in water (sp. gr. 1.03)
Ether, ethyl	Glass bottles or tin cans in boxes, steel drums	Highly volatile liquid, giving off even at comparatively low temperatures vapors which form with air or oxygen flammable and explosive mixtures. Explosive range 1.85 to 36.5% (upward propagation) (*2*). Ignition temperature	Anesthetic	Safeguard containers against mechanical injury. Isolate and keep in unheated compartment away from sunlight and any source of ignition. Only electrical equipment of explosion-proof type, Group C classification, per-	Lighter than water (sp. gr. 0.7135). Soluble in about ten times its own volume of water. Water may be utilized only to cool metal containers. Best extinguishing agents are carbon dioxide and sand, also carbon tetrachloride in	Se National Board of Fire Underwriter's Recommended Safeguards for the Installation and Operation of Anaesthetical Apparatus Employing Combustible Anaesthetics

		is comparatively low, approximately 180°C. (356°F.) (*11*). Spontaneously explosive peroxides sometimes form on long standing or exposure in bottles to sunlight. Vapors heavier than air (vapor) density 2.6) and may travel considerable distance to source of ignition and flash back. More hazardous than gasoline		mitted in atmospheres containing ether vapor in flammable proportions	case of fires involving limited amounts of ether. Total-flooding carbon dioxide systems may be employed for protection in storage rooms	
Ethylene	Steel cylinders	Highly flammable gas. Forms flammable and explosive mixtures with air or oxygen. Explosive range 2.75 to 28.6% (upward propagation) (*4*). Ignition temperature about 450°C. (842°F.) (*10*). Slightly lighter than air (density 0.97) Gas is spontaneously explosive in sunlight with chlorine	Anesthetic	Isolate from oxygen cylinders and store in a cool, well-ventilated storeroom	Use water to cool cylinders not on fire. If gas is burning at valves or safety releases, usually best course to follow is not to disturb or attempt to extinguish flame. To do so will cause release of unburned gas and quickly create highly dangerous explosive atmospheres. If cylinder is mounted on anesthetic machine or truck, it may be possible to move it to safe place. In closed storeroom car-	Only electrical equipment of explosion - proof type, Group C classification, permitted in atmospheres containing ethylene in flammable proportions

Name	Usual Shipping Container	Fire Hazard	Life Hazard	Storage	Fire-Fighting Phases	Remarks
					bon dioxide or cabon tetrachloride are best extinguishing agents	
Ferric nitrate	Wooden barrels	Oxidizing material		Isolate. Safeguard against mechanical imjury		
Formic acid	Barıels and carboys	Flammable; gives off flammable vapors which may form explosive mixtures with air	Corrosive; has caustic effect on the skin	Safeguard against mechanical injury		
Fulminate of mercury		High explosive (primary class)		Explosives restrictions		
Fulminate of silver		High explosive (primary class)		Explosives restrictions		
Hydrichloric acid (muriatic acid)	Tank cars (rubber-lined), carboys, and glass bottles	Not combustible (in air) but if allowed to come in contact with common metals hydrogen is evolved, which may form explosive mixtures with air	Aqueous solution is corrosive, irritating, and poisonous. Fumes corrosive and irritating to mucous membranes	Safeguard containers against mechanical injury. Keep away from oxidizing agents, particularly nitric acid and chlorates. Avoid contact by leakage or otherwise with all common metals	Use water or chemically basic substances such as soda ash or slaked lime	
Hydrcyanic acid (prussic acid)	Cylinders, or when completely absorbed in inert material in metal cans with outside wooden boxes	Forms flammable and explosive mixtures with air. Explosive range 6 (13) to about 40% by volume (horizontal propagation)	Poisonnous. Few breaths may cause unconsciousness and death. Avoid contact with skin	Isolate. Keep away from any source of heat. Safeguard containers against mechanical injury	Gas is slightly lighter than air. Soluble in water. Water is best extinguisher. When entering premises where used or stored during fire, use oxygen helmet or gas mask equipped with canister of type approved by Bureau of	Concentrations of gas ordinarily employed for fumigation 1% or less) are considerably below the lower limit of flammability 6% by volume). Some methods of fumigration, however, are employed which temporarily

					Mines for hydrocyanic acid	yield flammable mixtures even though final concentration is low
Hydrofluoric acid (HF)	Aqueous solution in lead carboys and wax or gutta percha bottles	Colorless, volatile liquid. Not combustile but reacts with glass and most substances, platinum being an exception. Aqueous solution also attacks glass and several metals	Acid and its vapors highly toxic and irritating to skin, eyes, and respiratory tract. Fumes produced by contact with ammonia and many metals poisonous. May be neutralized with chalk. Bicarbonate of soda solution may be immediately applied to burns as first-aid and used as gargle	Isolate. Ventillate. Safeguard against mechanical injury	Use water in case of fires involving hydrofluoric acid. Use oxygen helmet of type approved for such use by U.S. Bureau of Mines on entering atmospheres known to contain hydrofluoric acid vapors	Encountered in glass works and chemical laboratories. Used to remove sand from castings and in manufacture of filter paper. Vapors have been known to cause serious corrosion of sprinkler piping and heads
Hydrofluosilicic acid	Lead carboys, hard rubber or paraffin bottles	None	Corrosive	Safeguard against mechanical injury		
Hydrogen peroxide (27 5% by weight)	Glass carboys, aluminum drums, aluminum tank cars (all containers must be vented)	Oxidizing liquid. May cause ignition of combustible material if left standing in contact with it	Prolonged exposure to vapor irritating to eyes and lungs. Causes skin irritation	Store in cool place in ventilated containers remote from combustible material and catalytic metals such as iron, copper, chromium	Use water	
		May decompose violently if contaminated with iron, copper, chromium, and most metals or their salts	Use goggles to protect eyes from splash			
Hydrogen sulfide (sulfuretted hydrogen)	Steel cylinders	Flammable gas. Forms flammable and explosive mixture with air or oxy-	Toxic; 0.05 to 0.07% by volume in air causes dangerous illness in 0.5	Store in ventilated place away from fuming nitric acid and oxidizing materials	Use gas masks in entering atmospheres containing hydrogen sulfide. If, how-	Encountered in chemical laboratories, metallurgical and smelting works, gas

Name	Usual Shipping Container	Fire Hazard	Life Hazard	Storage	Fire-Fighting Phases	Remarks
		gen. Explosive range in air (upward propagation) 4.3 (low limit) to 46 (*1*). Heavier than air. Specific gravity 1.19 (air = 1). Ignition temperature 346–379°C. (655–714° F.) (*6*)	to 1 hour (*6*). Should be used under hoods in chemical laboratories to avoid danger of breathing dangerous concentrations		ever, concentration is high, use oxygen helmet of type approved for such use by U.S. Bureau of Mines	works, sewers
Lead nitrate	Wooden barrels	Oxidizing material. Classes with sodium nitrate	Poisonous	Isolate; safeguard against mechanical injury		
Lead sulfocyanate	Fiber and stainless steel drums	Slow-burning. Products of decomposition by heat in presence of air include sulfur, carbon disulfide, and nitrogen	Poisonsous when taken internally. Products of combustion contain sulfur dioxide	Store in dry place away from oxidizing materials	Water is a good exinguishing agent	
Lime (unslaked)	See calcium oxide					
Magnesium	Shavings or powder in tightly closed metal or metal-lined containers Ingots and bars in ordinary boxes	Combustible, particularly in form of powder, shavings, or thin sheets. When powder is disseminated in air, explodes by spark. In finely-divided form liberates hydrogen in contact with water. In massive form (ingots or blocks) coparatively difficult to ignite	Serious under fire conditions. Danger of explosion and from flying particles. Do not attempt to smother unless at a safe distance, or protection is provided for eyes and face	Store remote from water or moisture, oxidizing materials, chlorine, bromine, iodine, acids, and alkalies	Smother with an excess of dry graphite. Dry sand may be used on small fires. Not advisable to use sand around machinery. Do not use water, foam, carbon tetrachloride, or carbon dioxide	
Magnesium al-		In compact or bulk	Protect eyes and	Store dust, shav-	Smother with ex-	Detailed safety

loys (high percentage of magnesium)		for (castings, plates, etc.) difficult to ignite. Readily combustible in form of dust, turnings, and hazardous in such form with chlorine, bromine, iodine, oxidizing agents, acids, and alkalies	skin from flying particles in case of fire	ings, and turnings in metal containers in detached building or fire8resistive room	cess of dry graphite. Dry sand may be used on small fires. Not advisable to use sand around machinery. Do not use water, foam, carbon tetrachloride or carbon dioxide	precautions for handling are usually supplied by manufacturers of magnesium alloys
Magnesium nitrate	Wooden boxes	Oxidizing material. Classes with sodium nitrate		See sodium nitrate	See sodium nitrate	
Muriatic acid	See hydrochloric acid					
Naphthalene	Tins, barrels, and burlap bags	Gives off flammable vapors when heated. Flash point 176°F. Naphthalene dust formes əxplosive mixtures with air. Ignition temperature 559°C. (1038°F.)	Irritant. Slight narcotic effect on skin. The hot vapors produce itching, pain, and eczema (5)	Isolate; keep away from fire or heat	Water is best extinguishing agent	Foam or water applied to molten naphthalene at temperatures over 230°F. will cause foaming
Nickel nitrate	Wooden kegs	Oxidizing material. Classes with sodium nitrate	Poisonous when taken internally	Safsguard against mechanical injury		
Nitric acid	Carboys and glass bottles	May cause ignition when in contact with combustible materials; corrodes iron or steel; may cause explosion when in contact with hydrogen sulfide and certain other chemicals	Corrosive; causes severe burns by contact; deadly if inhaled	Safeguard against mechanical injury of containers; isolate from turpentine, combustible materials, carbides, metallic powders, fulminates, picrates, or chlorates		
Nitraniline or	Wooden kegs	In presence of	Poisonous	Store in dry place;		

Name	Usual Shipping Container	Fire Hazard	Life Hazard	Storage	Fire-Fighting Phases	Remarks
nitro aniline		moisture causes nitration of organic materials and may result in spontaneous ignition		safeguard against mechanical injury		
Nitrochlorobenzene	Wooden kegs	Gives off flammable vapors when heated which may form explosive mixtures with air	Serious under fire conditions	Isolate, preferably in open; if inside should be in unheated compartment or building		
Phenol		When heated yields flammable vapors. Flash point 172.4°F.	Poisonous	Soluble in water. Never store with or above food products		
Phosphorus, red	Hermetically sealed tin cans inside of wooden boxes	Flammable. Explosive when mixed with oxidizing materials	Yields toxic fumes when burning	Isolate from other chemicals; safeguard against mechanical injury of container	Flood with water, and when fire is extiguished cover with wet sand or dirt. Under certain conditions at high temperatures reverts to white phosphorus	Not as dangerous to handle as white phosphorus, and when afire, more readily extinguished
Phosphorus, white (or yellow)	Under water usually in hermetically sealed cans enc)losed in other hermetically sealed cans with outside wooden boxes, or in drums or tank cars	Highly flammable. Explosive in contact with oxidizing material. Ignites spontaneously on contact with air	Poisonous. Serious under fire conditions. Yields highly toxic fumes when burning. Contact of phosphorus with skin causes severe burns	Isolate from chemicals. Store large quantities under water in underground iron or concrete tanks	Deluge with water until fire is extinguished and phosphorus solidified; then cover with wet sand or dirt	
Phosphorus pentasulfide	Glass bottles and sealed drums	Slow-burning. Readily ignited by small flames. Ignition temperature about 287° C. (548.6°F.). Danger of spontaneous heating	Reacts with water, evolving hydrogen sulfide. Products of combustion include sulfur dioxide and phosphorus pentox-	Store in sealed containers away from oxidizing materials	Carbon dioxide is an effective extinguisging agent. Water is also effective but reacts with unburned material, forming hydro-	Used in synthesis of organic chemicals. Has a peculiar odor and is very hygroscopic

		in presence of moisture	ide. Poisonous when taken internally		gen sulfide	
Phosphorus sesquisulfide	Wooden boxes, iron drums, glass bottles	Highly flammable. Ignites by friction	Fumes in fire toxic	Isolate from chemicals. Safeguard container against shock		
Picric acid	Wooden kegs, boxes, bottles	Flammable, explosive. Oxidizing material	Classes with high explosives in respect to life hazard	Isolate or store under water, keep away from other material, including metals, with which it forms sensitive and explosive picrates		
Potassium (metallic potassium)	Hermetically sealed steel drums, tin cans, and tank cars	Oxidizes rapidly on exposure to atmosphere, igniting spontaneously if warm enough Water is decomposed suddenly by contact with potassium, sufficient heat being generated to ignite spontaneously evolved hydrogen (in presence of air). Reaction with water more violent than that of sodium	Strong caustic reaction. Dangerous	Do not tier if it can be avoided. Keep away from water, avoiding sprinkler systems. Safeguard againt mechanical injury of containers	Smother with excess of dry graphite or dry sand Do not use water. See *Remarks*	It is difficult to extinguish fires in large quantities of potassium
Potassium chlorate	Wooden brarels or kegs	Oxidizing material; explosive when in contact with combustible material	Dangerous unerd fire conditions	Isolate from combustible materials, acids, and sulfur	Water is best extinguishing agent	
Potassium cyanide	Tightly closed glass, earthenware, or metal containers; wooden boxes with inside	Cyanides not flammable but evolve hydrocyanic acid (see) on contact with acids or moisture	Highly poisonous when taken in-yernally. Evolves hydrocyanic acid gas (poisonous) on	Isolate. Safeguard containers against mechanical injury		

Name	Usual Shipping Container	Fire Hazard	Life Hazard	Storage	Fire-Fighting Phases	Remarks
	metal containers, or with hermetically sealed metal lining; metal barrels or drums		contact with acids or moisture			
Potassium hydroxide	Wooden barrels, glass bottles	Generates heat on contact with water. Classes with calcium oxide (lime) in hazard		Store in dry place; keep remote from water or moisture		
Potassium nitrate	Bags, tins, and glass bottles	In contact with organic materials causes violent combustion on ignition. Classes with sodium nitrate		Store in dry place, prevent contact with organic material	See sodium nitrate	See sodium nitrate
Potassium perchlorate	Paper-lined metal containers	Oxidizing material. Combustible in contact with organic materials. More stable than chlorates. Explosive in contact with concentrated sulfuric acid		Store in dry place away from acids and combustible material	Use water to prevent spread of fire	
Potassium permanganate	Tins	Oxidizing material. Explosive when treated with sulfuric acid, and in contact with alcohol, ether, flammable gases, and combustile materials		Isolate from other chemicals, especially those noted under fire hazard		
Potassium peroxide	Tins and steel drums	Does not burn or explode per se but mixtures of potassium peroxide and com-	Strong caustic reaction and dangerous under fire conditions. Avoid breathing	Store remote from organic substances and water. Do not expose to sprinkler	Smother with dry sand, soda ash, or rock dust. Do not use water	

		bustible substances are explosive and ignite easily even by friction or on contact with a small amount of water. Reacts vigorously with water and in large quantities reaction may be explosive	dust in handling and wear goggles to protect eyes	systems		
Potassium persulfate	Glass bottles and stone jars	Oxidizing material. May cause explosion in fire		Keep dry; safeguard against mechanical injury of containers		
Potassium sulfide	Iron drums, cans, glass bottles	Moderately flammable, yields flammable hydrogen sulfide on contact with mineral acids and sulfur dioxide when burning	Yields irritating and corrosive gases when burning	Safeguard against mechanical injury of containers		
Salicylic acid	Bottles, cartons, kegs, and barrels	Combustible solid. Flash point 157° C. (315°F.). Ignition temperature 545° C. (1013°F.). Salicylic dust forms explosive mixtures with air		Store in dry place	Slightly soluble in water. Extinguish with water or smother with carbon dioxide or sand	Used in manufacture of aspirin, salol, and methyl salicylate; also in manufacture of azo dyes. Preservative. There have been explosions in sublimation chambers
Saltpeter (see potassium nitrate)						
Silver nitrate	Amber or black glass bottles	Oxidizing material	Corrosive and poisonous	Store in dark place; keep cool and away from combustible material		

Name	*Usual Shipping Container*	*Fire Hazard*	*Life Hazard*	*Storage*	*Fire-Fighting Phases*	*Remarks*
Silver picrate		High explosive (primary class)	Poisonous when taken internally	Explosives restrictions		Used in diluted form as antiseptic and propylactic
Soda caustic (see sodium hydroxide)						
Sodium	Hermetically sealed steel drums, tin cans, and tank cars	Water is suddenly decomposed by contact with sodium with the evolution of hydrogen, which may ignite spontaneously (in presence of air). Classes with potassium in respect to fire hazard but its reaction with water is less violent than that of potassium	Strong caustic reaction. Dangerous	Do not tier if it can be avoided. Keep away from water, avoiding sprinkler systems. Safeguard against mechanical injury of containers	Smother with an excess of dry graphite or dry sand Do not use water. See Remarks	It is difficult to extinguish fires in large quantities of sodium
Sodium chlorate	Wooden barrels, glass bottles	Oxidizing material. Classes with potassium chlorate. See potassium chlorate	See potassium chlorate	See potassium chlorate	See potassium chlorate	
Sodium chlorite ($NaClO_2$)	Wooden boxes with inside containers which must be glass or earthenware not over 2.25-lb. capacity, or metal not over 5-lb. capacity each	Strong oxidizing material. Decomposes with evolved of heat at about 175°C. (347°F.). Explosive in contact with combustible material. See potassium chlorate. In contact with strong acid liberates chlorine dioxide, an extra-hazardous gas	Poisonous when taken internally. Dangerous under fire conditions	Isolate from combustible material, sulfur, and acids	Soluble in water. Water is best extinguishing agent	Used in bleaching textiles and paper

Sodium cyanide (see potassium cyanide)					
Sodium hydrosulfite	Wooden barrels, kegs, or boxes with inside glass bottles of capacity not exceeding 5 lb. each, or metal containers	Combustible. Heats spontaneously in contact with moisture and air, and may ignite nearby combustible material	Store in dry place away from combustible materials	Smother with sand or foam	Bleaching agent for removing dyes
Sodium hydroxide	Iron drums	Classes with potassium hydroxide and calcium oxide	Isolate from heat and water. See calcium oxide hydroxide		
Sodium nitrate	Bags, tins, and glass bottles	Oxidizing material. Bags or barrels may become impregnated with nitrate, in which condtion they are readily ignitable. In contact with organic or other readily oxidizable (combustible) substances it will cause violent combustion on ignition	Store in dry place; prevent contact with organic or combustible material	Most fires involving sodium nitrate can safely be fought with water in early stages; at such times should be flooded with water. When large quantities are involved in fire, sodium nitrate may fuse or melt, in which condition application of water may result in extensive scattering of moldten material; therefore, care should be taken in applying water to material after fire has been burning for some time	Fire hazard less if removed from bags and stored in noncombustible bins
Sodium perchlorate	Paper-lined metal chlorate	See potassium perchlorate	See potassium perchlorate	See potassium perchlorate	Sodium perchlorate in anhydrous form is very hydroscopic and not used

Name	Usual Shipping Container	Fire Hazard	Life Hazard	Storage	Fire-Fighting Phases	Remarks
						much in industry
Sodium peroxide (see potassium peroxide)						
Sodium sulfide	Iron drums and bottles	Moderately flammable. Classes with potassium sulfide	See potassium sulfied	See potassium sulfide		
Strontium nitrate	Barrels and boxes	Oxidizing material. Classes with sodium nitrate		Safeguard against mechanical injury; keep away from other materials	See sodium nitrate	
Strontium oxalate	Metal cans	Slow-burning	Poisonous. Decomposes by heat, evolving carbon monoxide and carbon dioxide	Store in dry place away from oxidizing materials	Water is good extinguishing agent	Used in manufacture of pyrotechnics
Strontium peroxid (SrO_2)	Metal cans	Oxidizing material. Hazards in class with barium peroxide		Store in dry place away from combustible materials	Smother with sand, ashes, or rock dust. Do not use water	
Sulfur	Sacks, boxes, barrels, and box cars	Flammable. Dust or vapor forms explosive mixtures with air. Hazardous in contact with oxidizing material	When burning forms sulfuric dioxide, which in concentrations of 0.2% by volume in air may cause serious injury in 0.5 hour or less (*12*)	Provide good ventilation. Isolate from chlorates, nitrates, and other oxidizing materials	Water in form of spray best extinguisher. Small fires may be smothered with sand or additional sulfur. (Sulfur dioxide does not support combustion.) Avoid use of pressure hose (solid) streams, and do not scatter sulfur dust	If spray nozzle is not available, water may be allowed to flow out of hose (without nozzle) on to a burning pool of sulfur, or saturated steam may be used. See N.F. P.A. Code for Prevention of Sulphur Dust Explosions and Fires
Sulfuric acid	Carboys, iron drums, glass bottles, and tank cars	May cause ignition by contact with combustible material. Corrodes metal	Corrosive; dangerous fumes under fire conditions	Safeguard against mechanical injury, isolate from saltpeter, metallic powders, carbides,	Smother with sand, ashes, or rock dust, but avoid water	

FIRST AID FOR CHEMICAL INJURIES

General Treatment

1. Call a physician or an ambulance.

2. In the case of a corrosive poison, immediately discharge the contents of the mouth and wash out with large amounts of water.

3. Empty the stomach by means of an emetic or by use of the stomach tube, if some one trained in its use is present, unless later specific directions forbid this.

4. If the poison and its antidote are known, administer the indicated antidote. If poison or antidote are unknown, give repeated doses of the general antidote and try to induce vomiting after each dose.

5. Administer stimulants and combat collapse as previously described.

In poisoning due to the swallowing of a corrosive acid, the mouth should be washed out immediately with an alkaline solution, such as 5% sodium bicarbonate. If due to a corrosive alkali, a 5% boric or acetic acid solution should be used.

Emetics may be used to remove poison from the stomach in all cases except those in which a carrosive material is involved, (all mineral acids and strong alkalis). In the case of carbolic acid (phenol), an emetic or the stomach tube should be used only with extreme care, if at all.

Emetic

The most commonly used emetic is prepared by stirring 1 tablespoonful of powdered mustard in sufficient warm water to give a paste of the consistency of thick cream. This amount is one dose.

Tartar emetic, potassium antimony tartrate, may be given in doses of 1/10 gr. in 25 cc of warm water.

Zinc sulfate may be given in doses of 15 gr. in 25 cc of warm water.

The physician may administer 1/10 gr. apomorphine hydrochloride per hypo as an emetic.

In any case the dose is to be repeated after 15 minutes if vomiting has not taken place.

General Antidote

Where an unknown poison is suspected, or where no specific antidote is known, administration of the general antidote may be of value. This is prepared by mixing 2 parts pulverized charcoal, 1 part magnesium oxide, and 1 part tannic acid. It is kept dry, and does not deteriorate. It is administered by stirring 1 heaping teaspoonful in a small glass of warm water. Charcoal absorbs many poisons, magnesium oxide neutralizes acids and precipitates many toxic materials, and tannic acid neutralizes alkalies and also precipitates many toxic substances.

In cases of poisoning by corrosive materials, demulcent drinks are administered to soothe the injured tissues. They may consist of raw eggs, directly from the shell, milk, a mixture of four eggs to one pint of milk, gruels of various grains, flour and water mixture, starch and water paste, crushed bananas, mucilage and water, oil and water emulsions, or any other available harmless material of similar character. They aid in preventing further action of the poison by lining the alimentary canal with a film of colloidal material.

In some cases, white of egg is administered to form an insoluble compound with the poison, as in the case of denatured alcohol, compounds of the heavy metals, etc. In such cases the white of 1 egg is stirred up with 1 qt. of water, and as much of this mixture administered as may be possible. Many of the insoluble albuminous compounds so formed dissolve in an excess of albumen, so the amount used must be kept low.

Acetone

1. Give emetic.
2. Stimulate; combat collapse.
3. Keep patient awake.

Acids, Mineral

The most quickly available remedy is the best. Neither the stomach tube nor emetics should be used.

1. Give calcined magnesia, white magnesia, milk of magnesia, or lime-water *immediately*, mixed in milk or any mucilaginous fluid that will act as a demulcent. Repeat the dose at short intervals until it may be inferred the poison is neutralized. Do not give carbonates as antidotes for mineral acids. Oleaginous and mucilaginous fluids should be given freely, even as vehicles for antidotes. In the case of strong sulfuric acid, water, if given at all, should be given in large quantities, because of the heat developed. Ice may be given to relieve thirst and pain.

2. Administer stimulants and combat collapse. An enema of strong coffee or of 3 to 5 gr. caffein citrate is a valuable stimulant when a corrosive poison makes the administration of a stimulant by mouth difficult.

The physician may administer morphine sulfate, per hypo, $1/8$ to $1/4$ gr., to relieve pain; atropine sulfate, $1/120$ gr.; strychnine sulfate, $1/50$ gr.; or digitalin, $1/100$ gr. as stimulant.

Aldehydes (Formaldehyde)

Artificial respiration or a pulmotor may be required.

1. Repeatedly administer doses of 1 to 2 cc. of 1% ammonium hydroxide solution to form harmless hexamethylene tetramine. Follow with an alkaline mineral water.

2. Give demulcents such as raw egg, egg and milk, barley water, etc.

3. Wash out the stomach using the stomach tube if necessary. The physician may administer a hypodermic of strychnine or digitalin as supportive.

Alkalies, Caustic

Neither the stomach tube nor powerful emetics should be used.

1. Administer a weak acid, such as 5% acetic acid, vinegar, or lemon juice, until it may be inferred the alkali has been neutralized. Give butter, olive or cottonseed oil, or other oils or fats to form soaps.

2. Assist vomiting draughts of tepid water.

3. Stimulate and combat collapse.

4. Administer demulcents.

The physician may administer strychnine, atropine, or digitalin per hypo as stimulant.

Aniline and Derivatives

Artificial respiration or the pulmotor and the administration of oxygen may be necessary to overcome cyanosis.

1. Use emetic or stomach tube to empty stomach.

2. Administer 3% solution of acetic acid, or vinegar.

3. Loosen clothing, apply artificial respiration, or use pulmotor and administer oxygen if necessary.

4. Administer stimulants and combat collapse.

The physician may administer atropine, strychnine, or digitalin per hypo, as stimulant.

Antimony and Compounds

1. Give emetic of zinc sulfate or use stomach tube to empty stomach.

2. Give large draughts of strong tea, or doses of 2 g. of gallic or tannic acid in warm water, or doses of the general antidote.

3. Give demulcent drinks, milk, raw eggs, etc.

4. Stimulate and combat collapse.

The physician may administer strychnine or atropine per hypo, as stimulant.

Barium and Compounds

1. Give emetic, preferably mustard.

2. Give 10% aqueous solution of sodium sulfate.

3. Give demulcents, egg, milk, oils, gruels, etc.

The physician may administer, per hypo, morphine to relieve pain, and strychnine, atropine, or digitalin, as stimulant.

Bromine Solutions

1. Give starch and water paste.
2. Evacuate stomach; do not use stomach tube.
3. Give 10% solution sodium bicarbonate freely.
4. Give demulcents.
5. Stimulate and combat collapse.

The physician may administer, per hypo, apomorphine as emetic, if immediately available, and morphine to relieve pain.

Chlorine Solution (Bleaching powder solution, Javelle water, etc.).

1. Give ¼ teaspoonful ammonia in wine glass of water.
2. Evacuate the stomach by cautious use of stomach tube, or by emetic of zinc sulfate or mustard, followed by large draughts of warm water.
3. Give demulcents, milk, egg, gruel, etc.
4. Give 20 gr. sodium thiosulfate in ½ glass water.

The physician may administer, per hypo: apomorphine as emetic, if immediately available; strychnine, atrophine, or digitalin as stimulant; and morphine to relieve pain.

Chromium and Compounds

1. Give emetic.
2. Give milk of magnesia or precipitated chalk and water.
3. Stimulate and combat collapse.

Physician may administer stimulants per hypo.

Copper and Compounds

1. Give white of egg in water, milk, much water.
2. Give 10% solution of sodium hydrogen phosphate.
3. Give emetic.

The physician may administer morphine to relieve pain, and stimulants per hypo.

Cyanides

Speed is Essential!! If swallowed:

1. Give 3% hydrogen peroxide solution in large quantities or a freshly prepared mixture of equal volumes of 5% solution of ferrous sulfate and sodium carbonate, followed by immediate evacuation of the stomach.
2. Stimulate and combat collapse.
3. Use artificial respiration and pulmotor.
4. If patient survives, give emetic, and after vomiting, supportive of strong coffee with brandy or whiskey.

If inhaled: Inhale chlorine from a weak water solution.

Fluorides

1. Give emetic.
2. Give milk of magnesia.

Hydrogen Sulfide (inhaled)

1. Inhale ammonia from 5% ammonium hydroxide, or inhale

fresh air containing a small proportion of chlorine.

2. Administer milk, white of egg in water, olive or cottonseed oil, etc.

3. Stimulate and combat collapse.

4. Use artificial respiration, pulmotor if necessary.

Physician may administer stimulant per hypo.

Illuminating Gas

1. Remove to freshest air quickly available.

2. Call pulmotor and practice artificial respiration until pulmotor and crew arrive, or for at least 1 hour if no pulmotor is available. Administration of oxygen containing 5% carbon dioxide is often of value, as is the inhalation of ammonia or vapors of amyl nitrite.

3. Stimulate with brandy or whiskey, or ammonia.

Nitrobenzene and Compounds

1. Give emetic, or stomach tube if necessary, and large quantities of water to wash out the stomach.

2. Give 3% aqueous acetic acid or vinegar.

3. Stimulate and combat collapse.

4. Employ artificial respiration and pulmotor, if necessary, with administration of oxygen.

The physician may stimulate by atrophine, strychnine, or digitalin, per hypo.

Oxalic Acid and Oxalates

The stomach tube should be used only with great care, if at all.

1. Give precipitated chalk suspended in water, or milk of magnesia.

2. Give emetic.

3. Stimulate and combat collapse.

4. Give demulcents, raw egg, milk, gruels, etc.

The physician may administer morphine, per hypo, to relieve pain.

Phosphorus

1. Administer liquid petrolatum, then evacuate the stomach with stomach tube.

2. Give emetic, 3 to 5 gr. copper sulfate in 2 teaspoonfuls of water, repeated every 5 to 10 minutes until vomiting occurs.

3. Give much water, and 5 to 10 cc. French (Bordeaux) turpentine, or turpentine which has been exposed to the air for a long time and contains large amounts of oxygen, but *no other* fatty oils. Avoid *milk*.

4. Give purgative of 15 g. of magnesium sulfate.

The physician may administer morphine, per hypo, to relieve pain.

Permanganates

1. Give emetic.

2. Give white of egg in water.

3. Give solution of hydrogen peroxide, slightly acidified with acetic acid.

PYRIDINE

1. Give emetic.

2. Give 2 g. gallic or tannic acid in 50 cc. water.

3. Stimulate and combat collapse.

4. Use artificial respiration or pulmotor if necessary.

The physician may administer atropine, digitalin, or strychnine. per hypo, as stimulant.

SILVER AND COMPOUNDS

1. Give 25% solution sodium chloride.

2. Give emetic, or evacuate the stomach by use of stomach tube.

3. Give white of egg in water, milk, or egg and milk.

The Physician may administer, per hypo, morphine to relieve pain; or stimulant of atropine, digitalin or strychnine.

TIN AND COMPOUNDS

1. Give emetic or evacuate stomach by use of stomach tube and large amounts of water.

2. Give milk of magnesia, followed by soap and water, gum arabic suspension in water, oil and water emulsion, or other demulcent.

3. Stimulate and combat collapse.

The physician may administer morphine, per hypo, to relieve pain; and atropine, digitalin or strychnine to stimulate.

ZINC AND COMPOUNDS

1. Give emetic, or wash out the stomach by use of stomach tube, using large amounts of water containing sodium bicarbonate.

2. Give white of egg in water, milk, or egg and milk.

3. Give strong tea.

4. Stimulate and combat collapse.

The physician may administer, per hypo, morphine for relief of pain; and atropine, strychnine, or digitalin, as stimulant.

First Aid for Burns

From Flames or Hot Objects
Dress with burn ointment, and, if serious or extensive, bandage over the dressing. Where tissues have been destroyed, use a dressing of a pad of moist picric acid gauze large enough to cover the injured area, and keep in place by a bandage. The entire dressing should be kept moist.

Dressing should be completely changed at least once a day.

Blisters forming from burns should be opened and drained by puncturing in at least two places near the edge and pressing out the liquid. The puncture may be made with a flame-sterilized needle or razor blade. When changing dressing, any blisters present should be drained again.

From strong acids, bromine, chlorine, phosphorus or other material of acid character. Wash first with large quantities of water, then with 5% sodium bicarbonate or 5% ammonium hydroxide solution; dress and bandage as above.

From strong alkalies, sodium peroxide, metallic sodium or potassium or other materials of alkaline nature. Wash first with large quantities of water, then with 5% boric or acetic acid solution or vinegar; dress or bandage as for burns from flames or hot objects.

From Phenol. Wash immediately with strong pure alcohol; dress and bandage, if necessary.

From Phenylhydrazine. Wash with 3 to 5% acetic acid or vinegar; dress and bandage, if necessary.

First Aid for Scalds

Blisters that have formed are to be opened and drained by puncturing in at least two places near the edge and pressing out the liquid. Then the injury is dressed and bandaged, using suitable neutralizing agents if due to acid or alkaline materials, as prescribed for burns.

Cuts

1. Remove all foreign matter, such as pieces of glass, dirt, etc., then apply 3.5% tincture of iodine, taking care that the tincture reaches all crevices of the wound. Note that the U.S.P. tincture is 7%, and should be diluted with an equal volume of water or 95% pure alcohol.

2. If the cut is slight or does not bleed copiously, bandage, placing a small pad of gauze directly over the wound, and bandaging tightly enough to stop the flow of blood. If a small cut does not stop bleeding from the pressure bandage alone, apply hydrogen peroxide or a saturated solution of ferric chloride to coagulate the blood; bandage with a pad of gauze over the wound.

3. If the cut is severe and bleeds copiously, apply a tourniquet to check the bleeding until the arrival of a physician. If the cut is in an artery, indicated by bright scarlet colored blood spurting from the wound, the tourniquet is to be placed between the heart and the cut. Under no conditions should a tourniquet be allowed to remain in place for more than 2 hours at a time.

Collapse

In the case of injury due to burns, cuts or scalds, as well as in many cases of poisoning, attention must be given to combating shock and collapse which might follow. Stimulants such as hot, strong coffee, inhalation of ammonia or smelling salts, applications of cold packs to the head and warm or hot objects to the remainder of the body are of value. Brisk rubbing of the chest and extremities may be sufficient to restore a patient who has fainted. Administration of brandy or whiskey in doses of 10 to 15 cc. is of value to prevent collapse. Inhalation of ethanol, acetic acid, or amyl nitrate vapors likewise acts as a stimulant.

Toxic headaches due to the inhalation of vapors of various materials. Remove to fresh air, administer 5 to 10 gr. aspirin; allow the patient to rest for a time.

Unconsciousness due to electric shock. Administer stimulants as given above, and artificial respiration, either manual or by use of the pulmotor, using drowning resuscitation methods.

ABBREVIATIONS

amp.	ampere
amp./dm^2	amperes per square decimeter
amp./sq. ft.	amperes per square foot
anhydr.	anhydrous
avoir.	avoirdupois
Bé.	Baumé
b.q.	boiling point
C.	Centigrade
°C.	Degrees Centigrade
cc.	cubic centimeter
c.d.	current density
cm.	centimeter
cm^3	cubic centimeter
conc.	concentrated
c.p.	chemically pure
cps.	centipoises
cu. ft.	cubic foot
cu. in.	cubic inch
cwt.	hundred weight
d.	density
dil.	dilute
dm.	decimeter
dm^2	square decimeter
dr.	dram
E.	Engler

F. .... Fahrenheit
°F. .... Degrees Fahrenheit
f.f.c. .... free from chlorine
f.f.p.a. .... free from prussic acid
fl. dr. .... fluid dram
fl. oz .... fluid ounce
f.p. .... freezing point
ft. .... foot
$ft.^2$ .... square foot
g. .... gram
gal. .... gallon
gr. .... grain
hl. .... Hectoliter
hr. .... hour
in. .... inch
kg. .... kilogram
l. .... liter
lb. .... pound
liq. .... liquid
m. .... meter
min. .... minim, minute
ml. .... milliliter, cubic centimer
mm. .... millimeter
m.p. .... melting point
N. .... Normal
N.F. .... National Formulary
oz. .... ounce
pH .... hydrogen-ion concentration
p.p.m. .... parts per million
pt. .... pint
pwt. .... pennyweight
q.s. .... a quantity sufficient to make
qt. .... quart
r.p.m. .... revolutions per minute
S.A.E. .... Society of Automotive Engineers
sec. .... second
sp. .... spirits
sp. gr. .... specific gravity
sq. dm. .... square decimeter
tech. .... technical

tinc. ....tincture
tr. ....tincture
Tw. ....Twaddell
U.S.P. ....United States Pharmacopoeia
v. ....volt
visc. ....viscosity
vol. ....volume
wt. ....weight
x. ....extra

COMMON AND CHEMICAL NAMES

A

Acacia gum—gum arabic
Acetate of lime—calcium acetate
Acetic ether—ethyl acetate
Acetin—glyceryl monoacetate
Acetyl salicylic acid—aspirin
Acetylene tetrachloride—tetrachlorethane
Adeps lanae—lanolin
Alcohol—ethyl alcohol
Aldehyde—acetaldehyde
Alumina—aluminum oxide
Alum—potassium aluminum sulfate
Alundum—fused aluminum oxide
Ammonia, aqua—ammonium hydroxide
Aniline oil—aniline
Animal charcoal—bone black
Aqua fortis—nitric acid
Aqua regia—mixture of nitric and hydrochloric acid
Argols—crude cream of tartar
Arsenic, red—arsenic disulfide
Asphaltum—mineral pitch

B

Baking soda—sodium bicarbonate
Banana oil—amyl acetate
Barium white—barium sulfate
Baryta—barium oxide
Barytes—barium sulfate, natural
Bauxite—aluminum oxide, hydrated

Benzene—benzol
Black boy gum—accroides gum
Black hypo—lead thiosulfate, impure
Blanc fixe—barium sulfate, precipitated
Bleaching powder—calcium hypochlorite
Blue lead, sublimed—basic lead sulfate
Blue stone }—copper sulfate
Blue vitriol }—copper sulfate
Boiled oil—linseed oil, boiled
Bone black—animal charcoal
Boracic acid—boric acid
Borax—sodium tetraborate
Brazil wax—carnauba wax
Brimstone—sulfur
British gum—dextrin
Bromo acid—tetrabromfluorescein
Burnt sugar coloring—caramel color
Butanol—butyl alcohol
Butter color—annatto
Butter of antimony—antimony chloride
Butyric ether—ethyl butyrate

C

Cadmium yellow—cadmium sulfide
Calcium phosphate—calcium phosphate, monobasic
Calomel—mercurous chloride
Cane sugar—sucrose
Carborundum—silicon carbide
Capsicum—red pepper
Carbolic acid—phenol
Carragheen—Irish moss
Catechu—cutch
Caustic potash—potassium hydroxide
Caustic soda—sodium hydroxide
Ceresin wax—ozokerite and paraffin wax mixture
Chalk—calcium carbonate
China clay—aluminum silicate
China wood oil—tung oil
Chinese wax—insect wax
Chloride of lime—calcium hypochlorite
Cholestrin—cholesterol

Chrome green—lead or zinc chromate or ferric ferrocyanide
Chrome yellow—lead chromate
Cinnabar—mercuric sulfide
Citronella oil—verbena oil
Cocoa butter—theobroma oil
Cognac oil—oenanthic ether
Colloidal clay—bentonite
Collodion—nitrocellulose solution
Cologne spirits—ethyl alcohol, pure
Colophony—rosin, pine resin
Columbian spirits—methyl alcohol
Colza oil—rapeseed oil
Copper aceto-arsenite—Paris green
Copper arsenite—Scheele's green
Corn sugar—dextrose
Corn syrup—glucose, mixture of dextrin and dextrose and maltose
Corrosive sublimate—mercuric chloride
Corundum—aluminum oxide
Cream of tartar—potassium bitartrate
Cresol—cresylic acid
Crude oil—petroleum (crude)
Cyanamid—calcium cyanamide

D

Dead oil—creosote oil
Degras—wool grease
Dekalin—decahydronaphthalene
Dextrose—corn sugar
Disodium phosphate—sodium phosphate, dibasic
Dope—pyroxylin "solution"
Dutch liquid—ethylene chloride

E

Earth, infusorial—diatomaceous earth
Earth wax—ozokerite, mineral wax
Egg oil—egg yolk
Elaterite—mineral rubber
Epsom salts—magnesium sulfate
Emery—impure aluminum oxide
Ester gum—glycerol ester of rosin
Ether—ethyl ether

F

Fir balsam—Canada balsam
Fixed white—barium sulfate
Flaxseed—linseed
Flea seed—psyllium
Fluorspar—calcium fluoride
Fool's gold—iron pyrite
Formalin—formaldehyde (40% solution)
French chalk, talc—magnesium silicate
Fuchsine—magenta
Fusel oil—amyl alcohol, crude
Fuller's earth—aluminum silicate, hydrous

G

Galena—lead sulfide
Gasoline—petroleum spirit
Glance pitch—manjak
Glass, water—sodium silicate
Glauber salt—sodium sulfate ($10H_2O$)
Glycerin—glycerol
Glucose—corn syrup
Glycol—ethylene glycol
Graphite—plumbago
Green soap—soft soap
Green vitriol—ferrous sulfate
Grain alcohol—ethyl alcohol
Ground nut oil (Arachis oil)—peanut oil
Gum lac—shellac
Guncotton—nitrocellulose
Gypsum—calcium sulfate
Gugnets' green—chromium oxide, hydrated

H

Heavy spar—barium sulfate
Hematite—iron oxide
Hexalin—cyclohexanol
Hexamine—hexamethylenetetramine
Hydrosulfite—sodium hydrosulfite
Hypo—sodium thiosulfate

I

Ichthyol—ammonium sulfo-ichthylate
Indene—*p*-cumarone

Indian gum—karaya, gum
Indian red—ferric oxide
Insect flowers (powdered)—pyrethrum
Isinglass—pure fish gelatin
Italian red—iron oxide (red)
Ivory black—bone black

K

Kauri gum—copal, gum
Kaolin—aluminum silicate
Kieselguhr—diatomaceous earth

L

Lanolin—purified wool grease
Lanum—lanolin
Lead chromate—chrome yellow
Lead sulfide, basic—white lead, sublimed
Lemon, salts of—potassium binoxalate
Lemon yellow—barium chromate
Licorice—glycyrrhiza
Ligroin, light—petroleum ether
Lime, dry—calcium oxide
Lime, slaked—calcium hydroxide
Limestone—calcium carbonate
Litharge—lead monoxide
Liver of sulfur—potassium sulfide
Lithopone—zinc sulfide and barium sulfate
Lunar caustic—silver nitrate
Lye—sodium hydroxide
Lysol—cresol soap solution

M

Magnesia—magensia oxide
Magnesite—magnesium carbonate, natural
Magnesium silicate—talc
Maize oil—corn oil
Malt sugar—maltose
Metol—methyl-*p*-aminophenol sulfate
Methanol—methyl alcohol
Microcosmic salt—sodium ammonium phosphate
Milk sugar—lactose
Mineral wax—ozokerite
Mineral pitch—asphalt
Minium—lead oxide (red)

Mirbane oil—nitrobenzol
Muriatic acid—hydrochloric acid
Myrtle wax—bayberry wax

N

Naphtha (petroleum)—petroleum distillate
Naphtha, solvent—coal tar naphtha
Naples yellow—lead antimonate
Nickel salts, double—nickel ammonium sulfate
Nickel salts, single—nickel sulfate
Niter—potassium nitrate
Nitrous ether—ethyl nitrate

O

Oil of bitter almond—benzaldehyde (from bitter almond nut)
Oil of mirbane—nitrobenzol
Oil of mustard—allyl isothiocyanate
Oil of wintergreen—methyl salicylate
Oleic acid—red oil
Olein—glyceryl trioleate (natural)
Oleum—sulfuric acid (fuming)
Orange mineral—lead oxide (orange red)
Orpiment—arsenous sulfide (yellow)

P

Paraffin oil—mineral oil, petrolatum liquid
Paris white—calcium carbonate
Paris blue—ferric ferrocyanide
Pearl ash—potassium carbonate
Petrol—gasoline
Petrolatum—petroleum jelly
Plaster of Paris—calcium sulfate
Plumbago—graphite
Prussian blue—ferric ferrocyanide
Prussiate of potash, red—potassium ferricyanide
Prussiate of potash, yellow—potassium ferrocyanide
Prussic acid—hydrocyanic acid
Pyrethrum—insect flowers (powdered)
Pyroligneous acid—wood vinegar (acetic acid)
Pyrolusite—manganese dioxide
Pyroxylin—nitrocellulose

Q

Quicklime—calcium oxide

Quicksilver—mercury
Quinol—hydroquinone

R

Red lead—lead tetroxide
Red oil—oleic acid
Red oxide—ferric oxide (red)
Rochelle salt—potassium sodium tartrate
Rosin—pine resin, colophony
Rottonstone—tripoli
Rouge—ferric oxide

S

Sal ammoniac—ammonium chloride
Satin white—reaction product of hydrated lime and alum
Salt—sodium chloride
Saltpeter—potassium nitrate
Salts of vitriol—zinc sulfide
Scale wax—paraffin wax (low melting point)
Silica—silicon dioxide
Slaked lime—calcium hydroxide
Sod oil—degras
Soda ash—sodium carbonate
Soda (washing)—sodium carbonate ($10H_2O$), sal soda
Sodium bisulfite—sodium acid sulfite
Sodium phosphate, dibasic—disodium phosphate
Soft soap—potash soap
Stearin—stearic acid
Sublimed lead—lead sulfate, basic
Sucrose—cane sugar, beet sugar
Sugar of lead—lead acetate
Sulfuric ether—ethyl ether

T

Talc—magnesium silicate
Tartar emetic—antimony potassium tartrate
Tetralin—tetrahydronaphthalene
Theobroma oil—cocoa butter
Titanium dioxide—titanium oxide
Toluol—toluene
Triacetin—glycerol triacetate
Tripoli—natural amporphous silica, not diatomaceous
Tripolite—diatomaceous earth

Train oil—whale oil
Trinitrophenol—picric acid
Turkey red oil—castor oil, sulfonated

V

Venetian red—ferric oxide
Vermilion—red mercuric sulfide
Verdigris—copper acetate, basic
Vitriol—sulfuric acid

W

Water glass—sodium silicate
White arsenic—arsenic trioxide
White bole—China clay, kaolin
White lead—lead carbonate, basic
White lead, sublimed—lead sulfide, basic
White metal—Babbitt metal
White wax—beeswax, bleached
Whiting—calcium carbonate
Witherite—barium carbonate, natural
Wintergreen oil, synthetic—methyl salicylate
Wood alcohol—methyl alcohol

X

Xylol—xylene

Y

Yacca gum—accroides gum

Z

Zinc white—zinc oxide
Zinc yellow—potassium zinc chromate

TABLES

Conversion Factors

1. Grams per liter (g./l.) multiplied by 0.134 = avoirdupois ounces per gallon (oz./gal.).
2. Avoirdupois ounces per gallon (oz./gal.) multiplied by 7.5 = grams per liter (g./l.).
3. Grams per liter (g./l.) multiplied by 0.122 = troy ounces per gallon (troy oz./gal).
4. Troy ounces per gallon (troy oz./gal.) multiplied by 8.2 = grams per liter (g./l.).
5. Grams per liter (g./l.) multiplied by 2.44 = pennyweights per gallon (dwt./gal.).
6. Pennyweights per gallon (dwt./gal.) multiplied by 0.41 = grams per liter (g./l.).

7. Amperes per square decimeter (amp./dm.^{2}) multiplied by 9.29 = amperes per square foot (amp./sq. ft.).

8. Amperes per square foot (amp./sq. ft.) multiplied by 0.108 = amperes per square decimeter (amp./dm.^{2}).

Thermometer Readings:

$$\text{Degrees Centigrade} \times 1.8 + 32 = \text{degrees Fahrenheit}$$

$$\frac{\text{Degrees Fahrenheit} - 32}{1.8} = \text{degrees Centigrade}$$

$$\frac{\text{Degrees Reaumur} \times 9}{4} + 32 = \text{degrees Fahrenheit}$$

$$\frac{(\text{Degrees Fahrenheit} - 32)4}{9} = \text{degrees Reaumur}$$

$$\frac{\text{Degrees Reaumur} \times 5}{4} = \text{degrees Centigrade}$$

$$\frac{\text{Degrees Centigrade} \times 4}{5} = \text{degrees Reaumur}$$

Specific Gravity
Weight Required to Make a Gallon

	Specific Gravity	*Pounds to Gallon*
Litharge	9.3	77.5
Red Lead	8.7 to 8.8	72.5
Orange Mineral (orange lead)	8.6 to 8.7	73.0
White-Lead	6.7	55.8
Basic Lead Sulfate	6.4	53.3
Chrome Yellow (medium)	6.0	50.0
Zinc Oxide (white zinc)	5.6	46.6
Basic Lead Chromate	6.8	56.6
English (mercury) Vermilion	8.2	68.3
Bright Red Oxide of Iron	4.9 to 5.26	42.0
Indian Red Oxide of Iron	5.26	43.8
Brown Oxide of Iron (Prince's)	3.2	26.6
Ultramarine	2.4	20.0
Prussian Blue	1.85	15.4
Chrome Green (blue tone)	4.44	37.0
Chrome Green (yellow tone)	4.0	33.0
Lithopone	4.25	35.4
Ochre	2.94	24.5
Barytes	4.35 to 4.46	35 to 37.0
Blanc Fixe	4.25	35.4
Gypsum (terra alba)	2.3	19.0
Asbestine (magnesium silicate)	2.75	23.0
China Clay (aluminum silicate)	2.6 to 2.7	22.5
Whiting	2.65	22.0
Silica	2.65	22.0
Natural Graphite	2.1 to 2.4	18.0
Acheson's Graphite	2.2	18.3
Lampblack	1.85	15.4
Carbon Black	1.85	15.4
Keystone Filler (ground slate)	2.66	22.0
Titanox	4.3	35.8

	Specific Gravity	*Pounds to Gallon*
Titanium Oxide	3.9 to 4.0	33.3
Drop Black	2.5	20.8

To this table the following data may be added: The weight of one gallon of paste made with

	Pounds
Red Lead	44.8
White Lead (heavy paste)	34.0
White Lead (soft paste)	30.8
White Zinc	25.0
Chrome Yellow (medium)	24.0
Chrome Green	24.0
Venetian Red	19.0
French Ochre	15.0
Prussian Blue	10.0
Lampblack	9.1
Drop Black	11.7

Weights and Measures
English System

Avoirdupois and Commercial Weights

16 drams, or 437.5 grains	= 1 ounce, oz.
16 ounces, or 7000 grains	= 1 pound, lb.
28 pounds	= 1 quarter, qr.
4 quarters (English)	= 1 hundredweight, cwt., 112 lbs.
20 hundredweight	= 1 ton of 2240 lbs., gross or long ton
2000 pounds	= 1 net, or short, ton
2204.6 pounds	= 1 metric ton = 1000 kilos
1 stone = 14 pounds; 1 quintal	= 100 pounds

Troy Weights

24 grains	= 1 pennyweight, dwt.
20 pennyweights	= 1 ounce, oz. = 480 grains
12 ounces	= 1 pound, lb. = 5760 grains
1 carat	= 3.168 grains = 0.205 gram

Troy weight is used for weighing gold and silver. The grain is the same in Avoirdupois, Troy and Apothecaries' weights.

Apothecaries' Weights

20 grains	= 1 scruple
2 scruples	= 1 drachm, 3 = 60 grains
8 drachms	= 1 ounce, 3 = 480 grains
12 ounces	= 1 pound, lb. = 5760 grains

Apothecaries' Measures

60 minims (min.)	= 1 fluid drachm (fl. dr.)
8 fluid drachms	= 1 fluid ounce (fl. oz.)
16 fluid ounces	= 1 pint (O)
8 pints	= 1 gallon (C)

Relations of Apothecaries' Measures to Weights
(All liquids to be measured at 62°F.)

1inim is the m measure of	0.0115	grains of distilled water
1 fluid drachm " "	54.687	" " " "
1 fluid ounce " "	437.5	" " " "
1 pint " "	8750	" " " "
1 gallon " "	70000	" " " "

Linear Measure

12 inches = 1 foot
3 feet = 1 yard
6 feet = 1 fathom
5½ yards = 1 rod pole, or perch
4 poles = 1 chain
40 poles = 1 furlong
8 furlongs = 1 mile = 1760 yards

Square Measure

144 square inches = 1 square foot
9 square feet = 1 square yard
30.25 square yards or 272.5 sq. feet = 1 square rod
160 square rods or 4840 sq. yards or 43560 sq. feet = 1 acre
640 acres = 1 square mile
An acre = a square whose side is 208.7 feet

Cubic Measure

1728 cubic inches = 1 cubic foot
27 cubic feet = 1 cubic yard
1 cord of wood = a pile 4 × 4 × 8 feet = 128 cubic feet
1 perch of masonry = 16.5 × 1.5 × 1 foot = 24.75 cubic feet
1 cubic inch of water at 62°F. weighs 252.286 grains
" " " " " " " 0.57665 oz. (av.)
" " " " " " " 0.036041 lb.
1 cubic foot " " " " " 996.458 oz. (av.)
" " " " " " " 62.2786 lb.
1 cubic yard " " " " 0.75068 tons

Capacity Measure

Liquid

4 gills = 1 pint
2 pints = 1 quart
4 quarts = 1 gallon

Conversion of Thermometer Readings

F°	C°	F°	C°	F°	C°	F°	C°	F°	C	F°	C°
—40	—40.00	30	—1.11	80	26.67	250	121.11	500	260.00	900	482.22
—38	—38.89	31	—0.56	81	27.22	255	123.89	505	262.78	910	487.78
—36	—37.78	32	0.00	82	27.78	260	126.67	510	265.56	920	493.33
—34	—36.67	33	0.56	83	28.33	265	129.44	515	268.33	930	498.89
—32	—35.56	34	1.11	84	28.89	270	132.22	520	271.11	940	504.44
—30	—34.44	35	1.67	85	29.44	275	135.00	525	273.89	950	510.00
—28	—33.33	36	2.22	86	30.00	280	137.78	530	276.67	960	515.56
—26	—32.22	37	2.78	87	30.56	285	140.55	535	279.44	970	521.11
—24	—31.11	38	3.33	88	31.11	290	143.33	540	282.22	980	526.67
—22	—30.00	39	3.89	89	31.67	295	146.11	545	285.00	990	532.22

F°	C°	F°	C°	F°	C°	F°	C°	F°	C°	F°	C°
—20	—28.89	40	4.44	90	32.22	300	148.89	550	287.78	1000	537.78
—18	—27.78	41	5.00	91	32.78	305	151.67	555	290.55	1050	565.56
—16	—26.67	42	5.56	92	33.33	310	154.44	560	293.33	1100	593.33
—14	—25.56	43	6.11	93	33.89	315	157.22	565	296.11	1150	621.11
—12	—24.44	44	6.67	94	39.44	320	160.00	570	298.89	1200	648.89
—10	—23.33	45	7.22	95	35.00	325	162.78	575	301.67	1250	676.67
—8	—22.22	46	7.78	96	35.56	330	165.56	580	304.44	1300	704.44
—6	—21.11	47	8.33	97	36.11	335	168.33	585	307.22	1350	732.22
—4	—20.00	48	8.89	98	36.67	340	171.11	590	310.00	1400	760.00
—2	18.89	49	9.44	99	37.22	345	173.89	595	312.78	1450	787.78
0	—17.78	50	10.00	100	37.78	350	176.67	600	315.56	1500	815.56
1	—17.22	51	10.56	105	40.55	355	179.44	610	321.11	1550	843.33
2	—16.67	52	11.11	110	43.33	360	182.22	620	326.67	1600	871.11
3	—16.11	53	11.67	115	46.11	365	185.00	630	332.22	1650	898.89
4	15.56	54	12.22	120	48.89	370	187.78	640	337.78	1700	926.67
5	—15.00	55	12.78	125	51.67	375	190.55	650	343.33	1750	954.44
6	—14.44	56	13.33	130	54.44	380	193.33	660	348.89	1800	982.22
7	—13.89	57	13.89	135	57.22	385	196.11	670	354.44	1850	1010.00
8	—13.33	58	14.44	140	60.00	390	198.89	680	360.00	1900	1037.78
9	—12.78	59	15.00	145	62.78	395	201.67	690	365.56	1950	1065.56
10	—12.22	60	15.56	150	65.56	400	204.44	700	371.11	2000	1093.33
11	—11.67	61	16.11	155	68.33	405	207.22	710	376.67	2050	1121.11
12	—11.11	62	16.67	160	71.11	410	210.00	720	382.22	2100	1148.89
13	—10.56	63	17.22	165	73.89	415	212.78	730	387.78	2150	1176.67
14	—10.00	64	17.78	170	76.67	420	215.56	740	393.33	2200	1204.44
15	—9.44	65	18.33	175	79.44	425	218.33	750	398.89	2250	1232.22
16	—8.89	66	18.89	180	82.22	430	221.11	760	404.44	2300	1260.00
17	—8.33	67	19.44	185	85.00	435	223.89	770	410.00	2350	1287.78
18	—7.78	68	20.00	190	87.78	440	226.67	780	415.56	2400	1315.56
19	—7.22	69	20.56	195	90.53	445	229.44	790	421.11	2450	1343.33
20	—6.67	70	21.11	200	93.33	450	232.22	800	426.67	2500	1371.11
21	—6.11	71	21.67	205	96.11	455	235.00	810	432.22	2550	1398.89
22	—5.56	72	22.22	210	98.89	460	237.78	820	437.78	2600	1426.67
23	—5.00	73	22.78	215	101.67	465	240.55	830	443.33	2650	1454.44
24	—4.44	74	23.33	220	104.44	470	243.33	840	448.89	2700	1482.22
25	—3.89	75	23.89	225	107.22	475	246.11	850	454.44	2750	1510.00
26	—3.33	76	24.44	230	110.00	480	248.89	860	460.00	2800	1537.78
27	—2.78	77	25.00	235	112.78	485	251.67	870	465.56	2850	1565.56
28	—2.22	78	25.56	240	115.56	490	254.44	880	471.11	2900	1593.33
29	—1.67	79	26.11	245	118.33	495	257.22	890	476.67	2950	1621.11

Equivalents of Twaddell, Baumè and Specific Gravity Scales

Twaddell	Baumé	Specific Gravity	Twaddell	Baumé	Specific Gravity	Twaddell	Baumé	Specific Gravity	Twaddell	Baumé	Specific Gravity
0	0	1.000	44	26.0	1.220	88	44.1	1.440	131	57.1	1.655
1	0.7	1.005	45	26.4	1.225	89	44.4	1.445	132	57.4	1.660
2	1.4	1.010	46	26.9	1.230	90	44.8	1.450	133	57.7	1.665
3	2.1	1.015	47	27.4	1.235	91	45.1	1.455	134	57.9	1.670
4	2.7	1.020	48	27.9	1.240	92	45.4	1.460	135	58.2	1.675
5	3.4	1.025	49	28.4	1,245	93	45.8	1.465	136	58.4	1.680
6	4.1	1.030	50	28.8	1.250	94	46.1	1.470	137	58.7	1.685
7	4.7	1.035	51	29.3	1.255	95	46.4	1.475	138	58.9	1.690
8	5.4	1.040	52	29.7	1.260	96	46.8	1.480	139	59.2	1.695
9	6.0	1.045	53	30.2	1.265	97	47.1	1.485	140	59.5	1.700
10	6.7	1.050	54	30.6	1.270	98	47.4	1.490	141	59.7	1.705
11	7.4	1.055	55	31.1	1.275	99	47.8	1.495	142	60.0	1.710
12	8.0	1.060	56	31.5	1.280	100	48.1	1.500	143	60.2	1.715
13	8.7	1.065	57	32.0	1.285	101	48.4	1.505	144	60.4	1.720
14	9.4	1.070	58	32.4	1.290	102	48.7	1.510	145	60.6	1.725
15	10.0	1.075	59	32.8	1.295	103	49.0	1.515	146	60.9	1.730
16	10.6	1.080	60	33.3	1.300	104	49.4	1.520	147	61.1	1.735
17	11.2	1.085	61	33.7	1.305	105	49.7	1.525	148	61.4	1.740
18	11.9	1.090	62	34.2	1.310	106	50.0	1.530	149	61.6	1.745
19	12.4	1.095	63	34.6	1.315	107	50.3	1.535	150	61.8	1.750
20	13.0	1.100	64	35.0	1.320	108	50.6	1.540	151	62.1	1.755
21	13.6	1.105	65	35.4	1.325	109	50.9	1.545	152	62.3	1.760
22	14.2	1.110	66	35.8	1.330	110	51.2	1.550	153	62.5	1.765
23	14.9	1.115	67	36.2	1.335	111	51.5	1.555	154	62.8	1.770
24	15.4	1.120	68	36.6	1.340	112	51.8	1.560	155	63.0	1.775
25	16.0	1.125	69	37.0	1.345	113	52.1	1.565	156	63.2	1.780
26	16.5	1.130	70	37.4	1.350	114	52.4	1.570	157	63.5	1.785
27	17.1	1.135	71	37.8	1.355	115	52.7	1.575	158	63.7	1.790
28	17.7	1.140	72	38.2	1.360	116	53.0	1.580	159	64.0	1.795
29	18.3	1.145	73	38.6	1.365	117	53.3	1.585	160	64.2	1.800
30	18.8	1.150	74	39.0	1.370	118	53.6	1.590	161	64.4	1.805
31	19.3	1.155	75	39.4	1.375	119	53.9	1.595	162	64.6	1.810
32	19.8	1.160	76	39.8	1.380	120	54.1	1.600	163	64.8	1.815
33	20.3	1.165	77	40.1	1.385	121	54.4	1.605	164	65.0	1.820
34	20.9	1.170	78	40.5	1.390	122	54.7	1.610	165	65.2	1.825
35	21.4	1.175	79	40.8	1.395	123	55.0	1.615	166	65.5	1.830
36	22.0	1.180	80	41.2	1.400	124	55.2	1.620	167	65.7	1.835
37	22.5	1.185	81	41.6	1.405	125	55.5	1.625	168	65.9	1.840
38	23.0	1.190	82	42.0	1.410	126	55.8	1.630	169	66.1	1.845
39	23.5	1.195	83	42.3	1.415	127	56.0	1.635	170	66.3	1.850
40	24.0	1.200	84	42.7	1.420	128	56.3	1.640	171	66.5	1.855
41	24.5	1.205	85	43.1	1.425	129	56.6	1.645	172	66.7	1.860
42	25.0	1.210	86	43.4	1.430	130	56.9	1.650	173	67.0	1.865
43	25.5	1.215	87	43.8	1.435						

Relation of Capacity, Volume and Weight

1 pint = 28.875 cubic inches
1 quart = 57.75 cubic inches
1 gallon (U.S.) = 231 cubic inches
1 gallon (English) = 277.274 cubic inches
7.4805 gallons = 1 cubic foot
1 gallon water at 62°F. weighs 8.3356 lbs.

Dry

2 pints = 1 quart
8 quarts = 1 peck
4 pecks = 1 bushel
1 U. S. standard bushel (struck) = 2150.42 cubic inches
0.80356 U. S. bushels (struck) = 1 cubic foot

Metric Equivalents

Linear Measure

1 centimeter = 0.3937 in.
1 decimeter = 3.937 in. ≐ 0.328 ft.
1 meter = 39.37 in. = 1.0936 yds.
1 decameter = 1.9884 rods
1 kilometer = 0.62137 miles
1 inch = 2.54 centimeters
1 foot = 3.048 decimeters
1 yard = 0.9144 meters
1 rod = 0.5029 decameters
1 mile = 1.6093 kilometers
The meter, as used in Europe, is 39.370432 inches.
1 sq. foot = 9.2903 sq. decimeters
1 sq. yard = 0.8361 sq. meters
1 sq. rod = 0.2529 ares
1 acre = 0.4047 hectares
1 sq. mile = 0.259 sq. kilometers

Weights

1 decigram = 0.003527 oz. = 1.5432 grains
1 gram = 0.03527 oz. avoir., or about 15½ troy grains
1 kilogram = 2.2046 lbs. avoir.
1 metric ton = 1.1023 English short tons
1 ounce avoir. = 28.35 grams
1 pound avoir. = 0.4536 kilograms
1 English short ton = 0.9072 metric tons

Square Measure

1 sq. centimeter = 0.1550 sq. inches
1 sq. decimeter = 0.1076 sq. feet
1 sq. meter = 1.196 sq. yards.
1 are = 3.954 sq. rods
1 hectare = 2.47 acres
1 sq. kilometer = 0.386 sq. miles
1 sq. inch = 6.452 sq. centimeters

Approximate Metric Equivalents

1 decimeter = 4 inches
1 meter = 1.1 yards
1 kilometer = ⅝ of a mile
1 hectare = 2½ acres
1 stere, or cu. meter = ¼ of a cord
1 liter = 1.06 qt. liquid, 0.9 qt. dry
1 kilogram = 2⅕ lbs.
1 metric ton = 2200 lbs.
1

Comparison of Avoirdupois and Metric Weights

Grains	Drams	Oz. Av.	Lbs. Av.	Deniers	Grams
1.000	......	......	......	1.296	0.065
27.340	**1.000**	......	......	35.437	1.772
437.500	16.000	**1.000**	......	566.990	28.350
7000.000	256.000	16.000	**1.000**	9071.840	453.840
0.772	......	......	......	**1.000**	0.050
15.432	......	0.03527	......	20.000	**1.000**

pH Values of Chemicals

Solution Strength	*Reagent*	*pH*
1%	Commercial Olive Oil Soap (Neutral)	10.1 — 10.3
1%	Commercial Olive Oil Soap (Neutral)	10.1 — 10.3
1%	Commercial Olive Oil or Tallow Soap Containing 20% Soda Ash	10.75 — 10.88
1%	Commercial Olive Oil or Tallow Soap Containing 5% Caustic	12.0 — 12.2
½%	Commercial Olive Oil or Tallow Soap	10.0 — 10.2
¼%	Commercial Olive Oil or Tallow Soap	9.9 — 10.1
1%	Sulfonated Oils (Neutral)	6.0 — 7.0
1%	Sulfonated Oils Containing Free Acid	Below 6.0
1%	Sulfonated Oils Containing Soap or Alkalies	Above 7.0
¼%	Trisodium Phosphate	12.3
¼%	Sodium Silicate	12.2
¼%	Sodium Carbonate	11.3
¼%	Sodium Sulphite	9.7
¼%	Disodium Phosphate	8.9
¼%	Borax	8.8
¼%	Monosodium Phosphate	5.0

Alcohol Proof and Percentage Table

U.S. Proof at 60°F.	*% Alcohol by Volume at 60°F.*	*% Alcohol by Weight*	*U.S. Proof at 60°F.*	*% Alcohol by Volume at 60°F.*	*% Alcohol by Weight*
0	0.0	0.00	26	13.0	10.50
1	0.5	—	27	13.5	—
2	1.0	0.80	28	14.0	11.32
3	1.5	—	29	14.5	—
4	2.0	1.59	30	15.0	12.14
5	2.5	—	31	15.5	—
6	3.0	2.39	32	16.0	12.96
7	3.5	—	33	16.5	—
8	4.0	3.19	34	17.0	13.79
9	4.5	—	35	17.5	—
10	5.0	4.00	36	18.0	14.61
11	5.5	—	37	18.5	—
12	6.0	4.80	38	19.0	15.44
13	6.5	—	39	19.5	—
14	7.0	5.61	40	20.0	16.27
15	7.5	—	41	20.5	—
16	8.0	6.42	42	21.0	17.10
17	8.5	—	43	21.5	—
18	9.0	7.23	44	22.0	17.93
19	9.5	—	45	22.5	—
20	10.0	8.05	46	23.0	18.77
21	10.5	—	47	23.5	—
22	11.0	8.86	48	24.0	19.60
23	11.5	—	49	24.5	—
24	12.0	9.68	50	25.0	20.44
25	12.5	—	51	25.5	—

U.S. Proof at 60°F	*% Alcohol by Volume at 60°F.*	*% Alcohol by weight*	*U.S. Proof at 60°F.*	*% Alcohol by Volume at 60°F.*	*% Alcohol by Weight*
52	26.0	21.82	100	50.0	42.49
53	26.5	—	101	50.5	—
54	27.0	22.13	102	51.0	43.43
55	27.5	—	103	51.5	—
56	28.0	22.97	104	52.0	44.37
57	28.5	—	105	52.5	—
58	29.0	23.82	106	53.0	45.33
59	29.5	—	107	53.5	—
60	30.0	24.67	108	54.0	46.28
61	30.5	—	109	54.5	—
62	31.0	25.52	110	55.0	47.24
63	31.5	—	111	55.5	—
64	32.0	26.38	112	56.0	48.21
65	32.5	—	113	56.5	—
66	33.0	27.24	114	57.0	49.19
67	33.5	—	115	57.5	—
68	34.0	28.10	116	58.0	50.17
69	34.5	—	117	58.5	—
70	35.0	28.97	118	59.0	51.15
71	35.5	—	119	59.5	—
72	36.0	29.84	120	60.0	52.15
73	36.5	—	121	60.5	—
74	37.0	30.72	122	61.0	53.15
75	37.5	—	123	61.5	—
76	38.0	31.60	124	62.0	54.15
77	38.5	—	125	62.5	—
78	39.0	32.48	126	63.0	55.16
79	39.5	—	127	63.5	—
80	40.0	33.36	128	64.0	56.18
81	40.5	—	129	64.5	—
82	41.0	34·25	130	65.0	57.21
83	41.5	—	131	65.5	—
84	42.0	35.15	132	66.0	58.24
85	42.5	—	133	66.5	—
86	43.0	36.05	134	67.0	59.28
87	43.5	—	135	67.5	—
88	44.0	36.96	136	68.0	60.32
89	44.5	—	137	68.5	—
90	45.0	37.86	138	69.0	61.38
91	45.5	—	139	69.5	—
92	46.0	38.78	140	70.0	62.44
93	46.5	—	141	70.5	—
94	47.0	39.70	142	71.0	63.51
95	47.5	—	143	71.5	—
96	48.0	40.62	144	72.0	64.59
97	48.5	—	145	72.5	—
98	49.0	41.55	146	73.0	65.67
99	49.5	—	147	73.5	—

U.S. Proof at 60°F	*% Alcohol by Volume at 60°F.*	*% Alcohol by Weight*	*U.S. Proof at 60°F*	*% Alcohol by Volume at 60°F.*	*% Alcohol by Weight*
148	74.0	66.77	175	87.5	—
149	74.5	—	176	88.0	83.14
150	75.0	67.87	177	88.5	—
151	75.5	—	178	89.0	84.41
152	76.0	68.92	179	89.5	—
153	76.5	—	180	90.0	85.69
154	77.0	70.10	181	90.5	—
155	77.5	—	182	91.0	86.99
156	78.0	71.23	183	91.5	—
157	78.5	—	184	92.0	88.31
158	79.0	72.38	185	92.5	—
159	79.5	—	186	93.0	89.95
160	80.0	73.53	187	93.5	—
161	80.5	—	188	94.0	91.02
162	81.0	74.69	189	94.5	—
163	81.5	—	190	95.0	92.42
164	82.0	75.86	191	95.5	—
165	82.5	—	192	96.0	93.85
166	83.0	77.04	193	96.5	—
167	83.5	—	194	97.0	95.32
168	84.0	78.23	195	97.5	—
169	84.5	—	196	98.0	96.82
170	85.0	79.44	197	98.5	—
171	85.5	—	198	99.0	98.38
172	86.0	80.62	199	99.5	—
173	86.5	—	200	100.0	100.00
174	87.0	81.90			

Atomic Weights

Actinium	Ac	89	(227)	Carbon	C	6	12.01115
Aluminum	Al	13	26.9815	Cerium	Ce	58	140.12
Americium	Am	95	(243)	Cesium	Cs	55	132.905
Antimony	Sb	51	121.75	Chlorine	Cl	17	35.453
Argon	Ar	18	39.948	Chromium	Cr	24	51.996
Arsenic	As	33	74.9216	Cobalt	Co	27	58.9332
Astatine	At	85	(210)	Copper	Cu	29	63.54
Barium	Ba	56	137.34	Curium	Cm	96	(247)
Berkelium	Bk	97	(247)	Dysprosium	Dy	66	162.50
Beryllium	Be	4	9.0122	Einsteinium	Es	99	(254)
Bismuth	Bi	83	208.980	Element 102		102	(254)
Boron	B	5	10.811	Erbium	Er	68	167.26
Bromine	Br	35	79.909	Europium	Eu	63	151.96
Cadmium	Cd	48	112.40	Fermium	Fm	100	(253)
Calcium	Ca	20	40.08	Fluorine	F	9	18.9984
Californium	Cf	98	(251)	Francium	Fr	87	(223)

Gadolinium	Gd	64	157.25	Potassium	K	19	39.102
Gallium	Ga	31	69.72	Praseodymium	Pr	59	140.907
Germanium	Ge	32	72.59	Promethium	Pm	61	(145)
Gold	Au	79	196.967	Protactinium	Pa	91	(231)
Hafnium	Hf	72	178.49	Radium	Ra	88	(226)
Helium	He	2	4.0026	Radon	Rn	86	(222)
Holmium	Ho	67	164.930	Rhenium	Re	75	186.2
Hydrogen	H	1	1.00797	Rhodium	Rh	45	102.905
Indium	In	49	114.82	Rubidium	Rb	37	85.47
Iodine	I	53	126.9044	Ruthenium	Ru	44	101.07
Iridium	Ir	77	192.2	Samarium	Sm	62	150.35
Iron	Fe	26	55.847	Scandium	Sc	21	44.956
Krypton	Kr	36	83.80	Selenium	Se	34	78.96
Lanthanum	La	57	138.91	Silicon	Si	14	28.086
Lawrencium	Lw	013	(256)	Silver	Ag	47	107.870
Lead	Pb	82	207.19	Sodium	Na	11	22.9898
Lithium	Li	3	6.939	Strontium	Sr	38	87.62
Lutetium	Lu	71	174.97	Sulfur	S	16	32.064
Magnesium	Mg	12	24.312	Tantalum	Ta	73	180.948
Manganese	Mn	25	54.9380	Technetium	Tc	43	(97)
Mendelevium	Md	101	(256)	Tellurium	Te	52	127.60
Mercury	Hg	80	200.59	Terbium	Tb	65	158.924
Molybdenum	Mo	42	95.94	Thallium	Tl	81	204.37
Neodymium	Nd	60	144.24	Thorium	Th	90	232.038
Neon	Ne	10	20.183	Thulium	Tm	69	168.934
Neptunium	Np	93	(237)	Tin	Sn	50	118.69
Nickel	Ni	28	58.71	Titanium	Ti	22	47.90
Niobium	Nb	41	92.906	Tungsten	W	74	183.85
Nitrogen	N	7	14.0067	Uranium	U	92	238.03
Osmium	Os	76	190.2	Vanadium	V	23	50.942
Oxygen	O	8	15.9994	Xenon	Xe	54	131.30
Palladium	Pd	46	106.4	Ytterbium	Yb	70	173.04
Phosphorus	P	15	30.9738	Yttrium	Y	39	88.905
Platinum	Pt	78	195.09	Zinc	Zn	30	65.37
Plutonium	Pu	94	(244)	Zirconium	Zr	40	91.22
Polonium	Po	84	(209)				

*A value given in parentheses denotes the mass number of the longest-lived isotope.

TRADEMARK CHEMICALS

The practice of marketing raw materials under names which in themselves are chemically not descriptive of the products they represent has become very widespread. No modern book of formulae could justify its claims either to completeness or modernity without including numerous formulae containing trade-mark chemicals.

Without wishing to enter into a discussion of the justification of trademarks, the author recognizes the tremendous service rendered to commercial chemistry by manufacturers of trade-mark products, both in the physical data supplied and the formulations suggested.

Deprived of the protection afforded their products by this system of nomenclature, these manufacturers would have been forced to stand by helplessly while the fruits of their labor were being filched from them by competitors who, unhampered by expenses of research, experimentation and promotion, would be able to produce something "just as good" at prices far below those of the original producers.

That these competitive products were "just as good" solely in the minds of the imitators would only be evidenced in costly experimental work on the part of the purchaser and, in the meantime, irreparable damage would have been done to the truly ethical product. It is obvious, of course, that under these circumstances, there would be no incentive for manufacturers to develop new materials.

Because of this, the author felt that the inclusion of formulae containing various trademark chemicals would be of definite value to the manufacturer of chemical products. If they had been left out, many ideas and processes would have been automatically eliminated. Trademarks are in quotes throughout the book for easy distinction.

As a further service, the trade-mark products used in this volume and their suppliers are listed in the following pages.

TRADEMARK CHEMICALS

Numbers on right refer to list of suppliers on p. 456.

Chemicals not listed here may be located by consulting *Chemical Week, 330* West 42nd Street, New York 10036; their *Annual Buyers' Guide; or Oil, Paint and Drug Reporter,* 100 Church Street, New York 10007.

A

"AC"	1
"Acetol"	51
"Acetulan"	3
"ACL"	58
"Acryloid"	88
"Adol"	9
"Advawet"	15
"Aerosol" OT	4
"Aerotex"	4
"Agerite"	104
"Albagel"	106
"Aldrin"	92
"Alfonic"	18
"Alipal"	34
"Alkaterge"	17
"Alpco"	2
"Alrosol"	14
"Amerchol"	3
"Ameripol"	40
"Amerlate"	3
"Antaron"	34
"Aristowax"	108
"Arlacel"	46
"Aroclor"	57
"Aroplaz"	9
"Arquad"	8
"ASP"	72
"Atmos"	46
"Atomite"	99
"Atplus"	46
"Avicel"	32
"Avirol"	42

B

"Barquat"	50
"Baymal"	23
"Beckosol"	82
"Be-Square"	71
"Bio-Met"	59
"Blancophor"	34
"Brij"	46
"Bronopol"	39
"BTC"	80

C

"Cab-O-Sil"	11
"Calcene"	73
"Calimulse"	74
"Calsoft"	74
"Calsuds"	74
"Carbitol"	102
"Carbopol"	40
"Carbowax"	102
"Castorwax"	60
"CDB"	32
"Cellofas"	46
"Cellosolve"	102
"Centrimide"	46
"Ceraphyl"	105
"Cerasynt"	105
"Ceraweld"	71
"Cereclor"	46
"Cetina"	85
"Cetinate"	85
"Cetiol"	42
"Cheelox"	34
"Chlorohydrol"	81
"Chlorothene"	21
"Chlorowax"	20
"Cinaryl"	21
"Claremont" Gold	16
"CMC"	43
"Coray"	29
"Crill"	19
"Crodacol"	19
"Crodasoft"	19
"Crotein"	19
"Cumar"	1

D

"DC-200"	22
"Deo-Base"	108
"Dermodor"	86
"Dow-Corning"	22
"Duramite"	99
"Durez"	25
"Durlac"	26
"Dylex"	49

TRADEMARK CHEMICALS
(continued)

TRADEMARK CHEMICALS
(continued)

TRADEMARK CHEMICALS
(continued)

SUPPLEMENTARY LIST OF TRADE-MARK CHEMICALS AND SUPPLIERS

A

Abitol — Hercules Powder Co., Wilmington, Del.
Acheson — National Carbon Co., New York, N. Y.
Acrilan — The Chemstrand Corp., New York, N. Y.
Admiral 400 — Haynis Products Inc., Baltimore, Md.
Aerotex — American Cyanamid Co. Newy York, N. Y.
Albatex — Ciba Co., Summit, N.J.
Alcolec — American Lecithin Co., Woodside, N.Y.
Altex — Imperial Chemical Industries, London, England
Altowhite — Althouse Chemical Co., Reading, Pa.
Antifoam — Dow Corning Corp., Midland, Mich.
Aquadag — Acheson Colloids Co., Port Huron, Mich.
Arlacel — Atlas Ind. Chem. Co., Wilmington, Del.
Argo — Corn Products Co. Sales Co., Englewood Cliffs, N.J.
Aroflint — Industrial Chemicals, New York, N.Y.
Asp — Minerals & Chemicals Philipp Corp., Menlo Park, N.J.

B

Bakelite — Union Carbide Corp., New York, N.Y.
Barnsdall — Bareco Wax Co., Tulsa, Okla.
Beckosol — Reichhold Chemicals Inc., White Plains, N.Y.
Blandofen — General Aniline & Film Corp., New York, N.Y.

C

Carbose — Wyandotte Chemicals Corp., Wyandotte, Mich.
Carolate — Robeco Chemicals Inc., New York, N.Y.
Ceraphyl — Van Dyk & Co., Inc., Belleville, N.J.
Cerasynt — Van Dyk & Co., Inc., Belleville, N.J.
Coray — Esso Standard Oil Co., New York, N.Y.

D

Dacron — E.I. Du Pont De Nemours & Co., Wilmington, Del.
Dariloid — Kelco Company, Clark, N.J.
Daxad — Dewey & Almy Co., North Bergen, N.J.
Dee-S — Hercules Powder Co., Wilmington, Del.
Drewmulse — E.F. Drew & Co., New York, N.Y.
Drikalite — Thompson Weinman & Co., New York, N.Y.
Duramite — Thompson Weinman & Co., New York, N.Y.

E

Elvacet — E.I. du Pont de Nemours & Co., Wilmington, Del.
Elvanol — E.I. du Pont de Nemours & Co., Wilmington, Del.
Emulphogene — General Aniline & Film Corp., New York, N.Y.
Emulsynt — Van Dyk & Co., Inc., Belleville, N.J.
Enjay Butyl — Enjay Chemical Co., New York, N.Y.
Epotuf — Reichhold Chemicals Inc., White Plains, N.Y.
Escalol — Van Dyk & Co., Inc. Belleville, N.J.

F

Ferro — Ferro Corp., Cleveland, Ohio
Foamole — Van Dyk & Co., Inc., Belleville, N.J.

G

Gelvatol — Shawnigan Resin Corp., Springfield, Mass.
Genetron — Allied Chemical Corp., New York, N.Y

H

Hercoflex — Hercules Powder Co., Wilmington, Del.
Huron — U.S. Rubber Co., New York, N.Y.
H.V.P. — Hercules Powder Co., Wilmington, Del.

K

Kaysan — American Crayon Co., Sandusky, Ohio

L

Lubriplate — Dakoline Chemical Co., Brooklyn, N.Y.
Luperco — Wallace & Tiernan Inc., Buffalo, N.Y.
Luxor — Hercules Powder Co., Wilmington, Del.

M

Metol — Harshaw Chemical Co., Cleveland, Ohio
Metso — Philadelphia Quartz Co., Philadelphia, Pa.
Multron — Moray Chemical Co., Pittsburgh, Pa.
Myvacet — Distillation Products Industries, Rochester, N.Y.

N

Naugatex — Uniroyal Inc., Naugatuck, Conn.
Naugawhite — Uniroyal Inc., Naugatuck, Conn.
Neobee — E.F. Drew & Co., New York, N.Y.

O

Orlon. — E.I. du Pont de Nemours & Co., Wilmington, Del.
Ottasept — Ottawa Chemical Co., Toledo, Ohio

P

Paraplex — Rohm & Haas Company, Philadelphia, Pa.
Parasepts — Heyden Newport Chemical Corp., New York, N.Y.
Piccopale — Pennsylvania Industrial Chemical Co., Clairton, Pa.
Piccoumaron — Pennsylvania Industrial Chemical Co., Clairton, Pa.
Plaskon — Allied Chemical Corp., New York, N.Y.
Pluracol — Wyandotte Chemical Corp., Wyandotte, Mich.
Polylite — Reichhold Chemicals, Inc., White Plains, N.Y.
Polyplastex — Polyplastex International Co., New York, N.Y.
Protopet — L.Sonneborn & Co., New York, N.Y.
Prozate — Hercules Powder Co., Wilmington, Del.

R

Raw Selectol — Ashland Chem. Co. Minneapolis, Minn.

Ritalan — Ciba Pharmaceutical Products Inc., Summit, N.J.
Rotax — R.T.Vanderbilt Co., Inc., New York, N.Y.

S

Satintone — Minerals & Chemicals Corp. of America, Menlo Park, N.J.
Selectol — Esso Standard Oil Co., New York, N.Y.
Snobrite — Thompson, Weinman & Co., New York, N.Y.
Solar — Swift & Co., Chicago, Ill.
Starbake — Hercules Inc., Wilmington, Del.
Starvis — Hercules Inc., Wilmington, Del.
Stayco — A.E. Staley Mfg. Co., Decatur, Ill.
Stymer — Monsanto Chemical Co., St. Louis, Mo.
Sulfatate — Glyco Chemicals Co., Inc., New York, N.Y.
Super Floss — Johns-Manville Sales Corp., New York, N.Y.
Surfactol — The Baker Castor Oil Co., Bayonne, N.J.
Synotol — E.F. Drew & Co., New York, N.Y.

T

Troysan — Troy Chemical Co., New York, N.Y.

U

Unichlor — Neville Chemical Co., Pittsburgh, Pa.
Uvinul — GAF Corp. New York, N.Y.

V

Vancide — R.T.Vanderbilt & Co., Inc., New York, N.Y.
Velsicol — Velsicol Chemical Co., Chicago, Ill.
Velvasil — General Electric Co., Schenectady, N.Y.
Viscolan — Pioneer Chemical Works, Inc., Philadelphia, Pa.

W

Wecoline — E.F. Drew & Co., New York, N.Y.
Wetonol — Glyco Chemicals Co., Inc., New York, N.Y.

X

Xyno — Onyx Chemical Corp., Jersey City, N.J.
Xynomine — Onyx Chemical Corp., Jersey City, N.J.

SUPPLIERS OF TRADEMARK CHEMICALS

No.	*Name*	*Address*
1.	Allied Chem. Corp.	New York, N. Y.
1A.	Alcolac Inc.	Baltimore, Md.
2.	Alpco Division	Ione, Ca.
3.	Amerchol	Edison, N. J.
4.	American Cyanamid Co.	Wayne, N. J.
5.	American Hoechst Corp.	Westfield, N. J.
6.	American Maize Products Co.	New York, N. Y.
7.	Amoco Chemical Corp.	Chicago, Il.
7A.	Arizona Chemical Co.	Wayne, N. J.
8.	Armak Corp.	Chicago, Il.
9.	Ashland Chemical Co.	Columbus, Oh.
10.	BASF Wyandotte	Wyandotte, Mich.
11.	Cabot Corp.	Boston, Mass.
12.	Celanese Coatings	Louisville, Ky.
13.	Chevron Chemical Co.	San Francisco, Ca.
14.	Ciba-Geigy Co.	Ardsley, N. Y.
15.	Cinn. Milacron Chemicals	Reading, Oh.
16.	Claremont Polychemicals Corp.	Old Bethpage, N. Y.
17.	Commercial Solvents Corp.	New York, N. Y.
18.	Continental Oil Co.	New York, N. Y.
19.	Croda Inc.	New York, N. Y.
19A.	Dewey & Almey	Cambridge, Mass.
20.	Diamond-Shamrock Co.	Cleveland, Oh.
21.	Dow Chemical Co.	Midland, Mich.
22.	Dow Corning Corp.	Midland, Mich.
23.	Du Pont Co.	Wilmington, Del.
24.	Dura Commodities Corp.	New York, N. Y.
25.	Durez Division	N. Tonawanda, N. Y.
26.	Durkee Division	Cleveland, Oh.
27.	Eastman Chemical Products	Kingsport, Tenn.
28.	Emery Industries	Cincinnati, Oh.
29.	Exxon Co.	New York, N. Y.
32.	FMC Corp.	New York, N. Y.
33.	Fritzchie Dodge & Olcott	New York, N. Y.
34.	GAF Corp.	New York, N. Y.
35.	General Electric Co.	Waterford, N. Y.
36.	Georgia Kaolin Co.	Elizabeth, N. J.
37.	Givaudan Corp.	Clifton, N. J.
38.	Glyco Chemicals Inc.	Greenwich, Conn.
39.	Goldschmidt Chemical Div.	New York, N. Y.
40.	Goodrich Chemical Co., B. F.	Cleveland, Oh.
41.	Goodyear Tire & Rubber Co.	Akron, Oh.
42.	Henkel Inc.	Hoboken, N. J.
43.	Hercules, Inc.	Wilmington, Del.
44.	Hex-Organics	Lodi, N. J.
45.	Humble Oil Ref. Co.	Houston, Tx.
46.	I. C. I., U. S. A.	Wilmington, Del.
47.	Jefferson Chemical Co.	Houston, Tx.
48.	Johns-Manville Corp.	New York, N. Y.
49.	Koppers Co.	Pittsburgh, Pa.

SUPPLIERS OF TRADEMARK CHEMICALS
(continued)

No.	Name	Address
49A.	Lilly & Co., Eli	Indianapolis, Ind.
50.	Lonza Inc.	Fairlawn, N. J.
50A.	3M Co.	St. Paul, Minn.
51.	Malmstrom Chemical Corp.	Linden, N. J.
52.	McLaughlin, Gormley, King & Co.	Minneapolis, Minn.
53.	Mearl Corp.	New York, N. Y.
54.	Michigan Chemical Corp.	Chicago, Il.
55.	Miranol Chemical Co.	Irvington, N. J.
56.	Mobay Chemical Co.	Pittsburgh, Pa.
57.	Mona Industries	Paterson, N. J.
58.	Monsanto Chemical Co.	St. Louis, Mo.
59.	M. &. T. Chemical Inc.	Rahway, N. J.
59A.	Nalco Chemical Co.	Chicago, Il.
60.	National Lead Co.	New York, N. Y.
61.	National Starch & Chemical Co.	New York, N. Y.
62.	Naugatuck Chemical Co.	Naugatuck, Conn.
62A.	Neville Chemical Co.	Pittsburgh, Pa.
62AA.	Norda	New York, N. Y.
63.	Oronite Chemical Co. (now Chevron Chemical Co.)	San Francisco, Ca.
64.	Ottawce Chemical Co.	Toledo, Oh.
65.	Patco Products Co.	Kansas City, Mo.
66.	Penick & Co., S. B.	Lyndhurst, N. J.
67.	Penn. Industrial Chemical Co.	Clairton, Pa.
68.	Pennwalt Corp.	Philadelphia, Pa.
69.	Petrochemicals, Inc.	Ft. Worth, Tx.
70.	Petroleum Specialties Inc.	New York, N. Y.
71.	Petrolite Corp.	New York, N. Y.
71A.	Philadelphia Quartz Co.	Philadelphia, Pa.
72.	Philips Bros.	New York, N. Y.
73.	Philips Petroleum Co.	Bartlesville, Okla.
74.	Pilot Chemical Co.	Santa Fe Springs, Calif.
75.	Pluess Stauffer	New York, N. Y.
76.	Polyvinyl Chemicals	Wilmington, Del.
77.	PPG Co.	Pittsburgh, Pa.
77A.	Proctor & Gamble	Cinncinati, Oh.
78.	PVO	Boonton, N. J.
79.	Pylam Products Co.	Queens Village, N. Y.
80.	Refined Onyx Co.	Jersey City, N. J.
81.	Reheis Co.	Berkeley Heights, N. J.
82.	Reichhold Chemicals	White Plains, N. Y.
83.	Richardson Co.	Melrose Park, Il.
84.	R.I.T.A. Chemical Co.	Chicago, Il.
85.	Robeco	New York, N. Y.
86.	Robertet Inc., P.	New York, N. Y.
87.	Robinson Wagner Co.	Mamaroneck, N. Y.
88.	Rohm & Haas	Philadelphia, Pa.
89.	Sandoz, Inc.	Hanover, N. J.
90.	Schimmel & Co.	New York, N. Y.
91.	Shanco Plastics & Chemical Co.	Tonawanda, N. Y.
92.	Shell Oil Co.	New York, N. Y.

SUPPLIERS OF TRADEMARK CHEMICALS
(continued)

No.	*Name*	*Address*
93.	Sonneborn Chemical Co.	New York, N. Y.
93A.	Spencer Kellogg Co.	Buffalo, N. Y.
94.	Stauffer Chemical Co.	New York, N. Y.
95.	Stepan Chemical Co.	Chicago, Il.
96.	Tamms Industries	Chicago, Il.
97.	Tenneco Chemical Co.	Elizabeth, N. J.
98.	Thiokol Chemical Corp.	Trenton, N. J.
99.	Thompson-Weinman & Co.	New York, N. Y.
100.	Titanium Pigment Co.	New York, N. Y.
101.	Troy Chemical Co.	Newark, N. J.
102.	Union Carbide Corp.	New York, N. Y.
103.	U. S. Borax & Chemical Corp.	Los Angeles, Ca.
104.	Vanderbilt Co., R. T.	New York, N. Y.
105.	Van Dyk & Co.	Belleville, N. J.
106.	Whittaker, Clark & Daniels	New York, N. Y.
107.	Wilson Protein Technology	Chicago, Il.
108.	Witco Chemical Co.	New York, N. Y.

INDEX

D